An Environmental History of the World

This second edition of *An Environmental History of the World* continues to present a concise history, from ancient to modern times, of the interactions between human societies and the natural environment, including the other forms of life that inhabit our planet. Throughout their evolutionary history, humans have affected the natural environment, sometimes with a promise of sustainable balance, but also in a destructive manner. This book investigates the ways in which environmental changes, often the result of human actions, have caused historical trends in human societies. This process has happened in every historical period and in every part of the inhabited earth.

The book is organized into ten chapters. The main chapters follow a chronological path through the history of humankind, in relationship to ecosystems around the world. The first explains what environmental history is, and argues for its importance in understanding the present state of the world's ecological problems. Chapters 2 through 8 form the core of the historical analysis, each concentrating on a major period of human history (pre-civilized, early civilizations, classical, medieval, early modern, early and later twentieth century, and contemporary) that has been characterized by large-scale changes in the relationship between human societies and the biosphere, and each gives several case studies that illustrate significant patterns occurring at that time. The chapters covering contemporary times discuss the physical impacts of the huge growth in population and technology, and the human responses to these problems. Our moral obligations to nature and how we can achieve a sustainable balance between technology and the environment are also considered. This revised second edition takes account of new research and contains new sections on global warming, the response of New Orleans to the hurricanes Katrina and Rita, and the experience of the Dutch people in protecting their low-lying lands against the encroachments of rivers, lakes, and the North Sea. New material is also offered on the Pacific Islands, including the famous case of Easter Island.

This is an original work that reaches further than other environmental histories. Rather than looking at humans and the environment as separate entities, this book places humans within the community of life. The relationship between environmental thought and actions, and their evolution, is discussed throughout. Little environmental or historical knowledge is assumed from the reader in this introduction to environmental history. We cannot reach a useful understanding of modern environmental problems without the aid of perspective provided by environmental history, with its illustrations of the ways in which past decisions helped or hindered the interaction between nature and culture. This book will be influential and timely to all interested in or researching the world in which we live.

J. Donald Hughes is John Evans Distinguished Professor of History, University of Denver, USA. Author of *What is Environmental History?* (Polity, 2006), *The Mediterranean: An Environmental History* (ABC-CLIO, 2005), and *Pan's Travail: Environmental Problems of the Ancient Greeks and Romans* (Johns Hopkins, 1994), he is a founding member of the American and European Societies for Environmental History and past editor of *Environmental History*.

An Environmental History of the World

Humankind's changing role in the community of life

Second edition

J. Donald Hughes

LONDON AND NEW YORK

First edition published 2001
Second edition published 2009 by Routledge
2 Park Square, Milton Park, Abingdon, Oxon, OX14 4RN

Simultaneously published in the USA and Canada
by Routledge
270 Madison Avenue, New York, NY 10016

Routledge is an imprint of the Taylor & Francis Group, an informa business

© 2009 J. Donald Hughes

Typeset in ITC Galliard by Saxon Graphics Ltd, Derby
Printed and bound in Great Britain by TJ International Ltd, Padstow, Cornwall

British Library Cataloguing in Publication Data
A catalogue record for this book is available from the British Library

Library of Congress Cataloguing in Publication Data
Hughes, J. Donald (Johnson Donald), 1932-
   An environmental history of the world : humankind's changing role in the community of life / J. Donald
Hughes.
        p. cm. -- (Routledge studies in physical geography and environment)
   "Simultaneously published in the USA and Canada"--T.p. verso.
   Includes bibliographical references and index.
 1. Human ecology--History. 2. Nature--Effect of human beings on--History. 3. Biotic communities--
History. I. Title.
   GF13.H83 2009
   304.2'8--dc22
                                        2008053780

ISBN 13: 978-0-415-48149-6 (hbk)
ISBN 13: 978-0-415-48150-2 (pbk)
ISBN 13: 978-0-203-88575-8 (ebk)

ISBN 10: 0-415-48149-X (hbk)
ISBN 10: 0-415-48150-3 (pbk)
ISBN 10: 0-203-88575-9 (ebk)

Contents

Figures

All photographs are originals taken by J.Donald Hughes, except where noted in the full
figure captions.

Acknowledgments

Many communities and individuals gave me guidance and support in writing both the first and second editions of this book. Most important among these are the American Society for Environmental History, my professional association for more than thirty years, and its sister organizations, the European Society for Environmental History and the Forest History Society. Two among my generous colleagues, Donald Worster and John McNeill, read an earlier version of the manuscript of this book and gave helpful criticisms that greatly improved it. Others who read all or part of the manuscript and provided valuable comments include Ari Kelman, James O'Connor, Mikko Saikku, Petra J.E.M. van Dam, and B.A. Haggart.

The first edition had its genesis in a grant from the Charles A. and Anne Morrow Lindbergh Foundation in 1987 that made possible my initial research trips to the Soviet Union and China. Our association did not end with the year of the grant; Reeve Lindbergh, Clare Hallward, and others provided continuing encouragement in periodic reunions of the grantees.

My university and department have supported me in many ways. The chairs of the Department of History over the years – Allen D. Breck, Robert Roeder, Charles P. Carlson, John Livingston, Michael Gibbs, and Ingrid Tague – have been untiring and understanding advocates on behalf of my efforts in research. The Faculty Research Committee, the PROF Grant Committee, the Office of Internationalization and Vice Provost Ved Nanda, Dean Roscoe Hill, Provosts William Zaranka and Gregg Kvistad, and Chancellors Dan Ritchie and Robert Coombe made time and resources available for this effort. My distinguished faculty colleagues elected me to the Evans Professorship, which includes a modest budget for research that helped to make this book possible. Phyllis Corchary served as my research assistant during the preparation of the first edition, and found a wealth of sources for it.

Special thanks are due to James O'Connor and Barbara Laurence and to their journal, *Capitalism, Nature, Socialism*, where I was able to try out the ideas for several sections of this book in the form of a series of columns.

Thanks also go to the students in my World Environmental History course, who have read and commented on drafts of this book and discussed them in class. They too are my colleagues; I have learned from them.

I am grateful to many members of the worldwide community of scholars who have willingly been my hosts, guides, and counselors during travel and research. There are more than I can list, and for the sake of brevity I must reluctantly leave out my numerous friends in the US and Canada, but some around the world have given me so much that they must be mentioned here: in Australia, John Dargavel and Denise Gaughwin; in Austria, Verena Winiwarter; in China, Mei Xueqin and Bao Maohong, in Egypt, Shafik Farid; in Finland,

Yrjö Vasari, Mikko Saikku, and Simo Laakkonen; in France, Patrick Petitjean, and (at UNESCO, Paris), Marie Roué and Thomas Schaaf; in Germany, Achim Köddermann and Heinrich Rubner; in Greece, Niki Goulandris, Giorgios Nakos, and John Rendall; in India, M.D. Subash Chandran, Madhav Gadgil, Deepak Kumar, Ajay Rawat, Rana Singh, M.L.K. Murty, and P.S. Ramakrishnan; in Kenya, J. Otike; in Italy, Mauro Agnoletti and Marcus Hall; in the Netherlands, Petra J.E.M. van Dam; in Panama, Stanley Heckadon-Moreno; in New Zealand, Helen Leach, Judith Bennett, Tom Brooking, Eric Pawson, and James Belich; in Pohnpei, Francis X. Hezel and Willy Kostka; in Polynesia, Patrick Nunn, Chris Stevenson, Jim Delle, and Andrew Lockwood; in Portugal (Madeira), Alberto Vieira and Henrique Costa Neves; in Russia, Ludmila Zhirina, Galina Krivosheina, Igor Shpilenok, and Galina Putyatina; in Sweden, Anders Öckerman, Sverker Sörlin, and Alf Hornborg; in Turkey, Carolyn Aslan and Murat Aydin; in the United Kingdom, I.G. Simmons, Fiona Watson, and Paul Warde; and in several places around the world, including India, England, Australia, Madeira, and Sweden, my dear friend Richard Grove. While some of the strengths of this work may be attributed to the people mentioned above, they are not responsible for any of its weaknesses.

Above all I thank my wife, Pamela L. Hughes, who has accompanied me to various parts of the planet, doing her research on world music history while I did mine on world environmental history, and who has given me every encouragement and consideration during the long process of writing.

Permissions

The author and publishers would like to thank the following publications for permission to use copyright material:

"The Sacred Groves of South India: Ecology, Traditional Communities and Religious Change," by M.D. Subash Chandran and J. Donald Hughes, *Social Compass* (Brussels) 44, 3, September 1997, 413–27.

J. Donald Hughes as sole author of:

"Nature and Culture in the Pacific Islands," *Leidschrift, Historisch Tijdschrift* (Leiden) 21, 1, April 2006, 129–43.
"Scenery versus Habitat at the Grand Canyon," in Michael F. Anderson, ed., *A Gathering of Grand Canyon Historians: Ideas, Arguments, and First-Person Accounts*, Grand Canyon, AZ, Grand Canyon Association, 2005, 105–10.
"Global Dimensions of Environmental History," *Pacific Historical Review* 70, 1, February 2001, 91–101.
"Francis of Assisi and the Diversity of Creation," *Environmental Ethics* 18, 3, Fall 1996, 311–20.
"Ecology and Development as Narrative Themes of World History," *Environmental History Review* 19, 1, Spring 1995, 1–16.
"Sustainable Agriculture in Ancient Egypt," *Agricultural History* (University of California Press) 66, 2, Spring 1992, 12–22.

The Guilford Press, 72 Spring Street, New York, NY 10012 for:

"The Dams at Aswan: Does Environmental History Inform Decisions?" *Capitalism, Nature, Socialism* 11, 4, December 2000, 73–81.
"The European Biotic Invasion of Aztec Mexico," *Capitalism, Nature, Socialism* 11, 1, March 2000, 105–12.
"Conservation in the Inca Empire," *Capitalism, Nature, Socialism* 10, 4, December 1999, 69–76.
"The Serengeti: Reflections on Human Membership in the Community of Life," *Capitalism, Nature, Socialism* 10, 3, September 1999, 161–7.
"Darwin in the Galápagos," *Capitalism, Nature, Socialism* 10, 2, June 1999, 107–14.
"The Classic Maya Collapse," *Capitalism, Nature, Socialism* 10, 1, March 1999, 81–9.
"Bryansk: The Aftermath of Chernobyl," *Capitalism, Nature, Socialism* 9, 4, December 1998, 95–101.

"Medieval Florence and the Barriers to Growth Revisited," *Capitalism, Nature, Socialism* 9, 3, September 1998, 133–40.

"A Sense of Place," *Capitalism, Nature, Socialism* 9, 2, June 1998, 91–6.

"The Preindustrial City as Ecosystem," *Capitalism, Nature, Socialism* 9, 1, March 1998, 105–10.

"Sacred Groves and Community Power," *Capitalism, Nature, Socialism* 8, 4, December 1997, 99–105.

"Mencius, Ecologist," *Capitalism, Nature, Socialism* 8, 3, September 1997, 117–21.

"Rome's Decline and Fall: Ecological Mistakes?" *Capitalism, Nature, Socialism* 8, 2, June 1997, 121–5.

"Ancient Egypt and the Question of Appropriate Technology," *Capitalism, Nature, Socialism* 8, 1, March 1997, 125–30.

"Now That the Big Trees Are Down," *Capitalism, Nature, Socialism* 7, 4, December 1996, 99–104.

"Classical Athens and Ecosystemic Collapse," *Capitalism, Nature, Socialism* 7, 3, September 1996, 97–102.

"Bali and the Green Witch of the West," *Capitalism, Nature, Socialism* 7, 2, June 1996, 139–45.

"Medieval Florence and the Barriers to Growth," *Capitalism, Nature, Socialism* 7, 1, March 1996, 63–8.

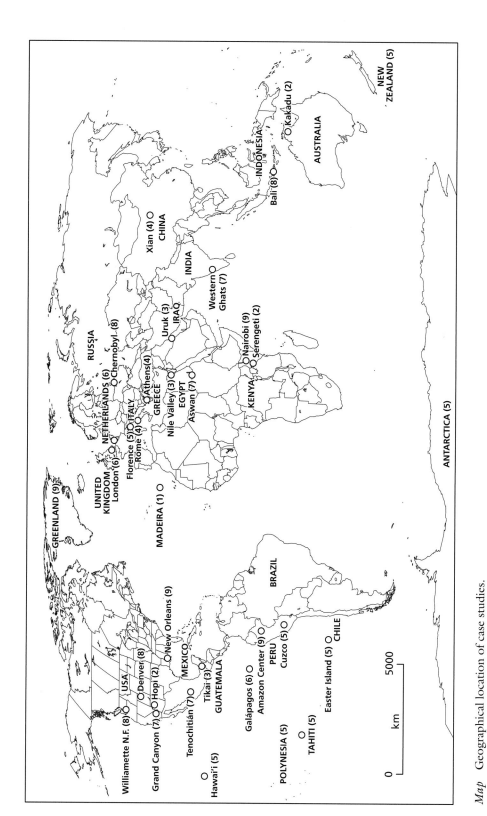

Map   Geographical location of case studies.

1 Introduction
History and ecology

History is a saga of change. If people and the world around them remained the same from year to year and generation to generation, or merely repeated a cycle of growth and decay that offered no escape, there would be no history worth being written – or read. But fortunately or unfortunately, change is an inescapable phenomenon in human societies and the world of nature, and in the relations linking them. Challenges appear sometimes in the form of natural catastrophes that threaten the survival of communities, and at other times in the form of cultural and economic choices that threaten the ability of natural systems to endure and to provide necessary support for those communities. The past offers many instances of antagonism between humankind and nature, and other cases of restoration and hope. Ecological process has helped to shape the course of human history. Humans have made major changes in their environments. They have had to adapt to the changes they made, by altering the patterns of their societies, or to decline or even to disappear. This has happened in every historical period and in every part of the inhabited Earth. Dealing with the threats of the present and making informed choices for the future both depend on understanding the environmental experiences of the past.

I saw a many-faceted example of this in the volcanic island of Madeira, which rises in the Atlantic Ocean 1,200 km (750 mi) southwest of Lisbon and 765 km (475 mi) off the African coast. It is a spectacular island; its highest peak reaches 1,861 m (6,106 ft), and on the northern coast, swells from the open ocean often produce thundering surf. Madeira was uninhabited until 1425, when João Gonçalves Zarco founded a Portuguese settlement under Prince Henry the Navigator.[1] At the time, most of the island was covered by the *laurissilva*, a thick forest of native laurel trees.[2] It was this forest that gave occasion to the island's name: Madeira, the Wooded Isle. There were no mammals except bats and the colonies of monk seals on the coast.[3] Birds, especially marine species, were plentiful; there were also a few species of land birds that occurred only in Madeira. The numerous species of insects fascinated Charles Darwin when he read about them; he pointed out that a surprising proportion of them, in the isolated safety of the island environment, were flightless or unusually large, or both.[4] The settlers began an attack on the forest, hewing down the trees for export and starting fires to clear land for agriculture: sugar cane[5] at first and then grapevines that yielded the famous Madeira wine.[6] A folk story says that the forests burned continually for seven years. An unknown number of native species perished from the fires and forest removal. Many non-native species were introduced, some intentionally and others by accident. Fifteen years after settlement, colonists found that cattle had escaped, gone wild, and become so numerous that they could kill them with ease.[7] Along with goats, they decimated the vegetation, further reducing the habitat for wildlife. Once introduced to the nearby island, Porto Santo, rabbits swarmed everywhere, eating everything and driving the

Figure 1.1 A landscape transformed by human actions. A native forest of laurel and other trees covered these mountain slopes on the north coast of Madeira Island before the fifteenth century. Then Portuguese settlers arrived, constructed terraces, and planted vineyards whose grapes were used to produce the well-known Madeira wines. Photograph taken in 1999.

human residents off the island for a time. Cats, mice, and rats destroyed birds that were not used to mammalian predators. The Madeira wood pigeon and possibly three flightless rails became extinct.[8] Plants alien to Madeira, from showy garden flowers to aggressive weeds (sometimes the same plant is both), were introduced by the hundreds. I visited some of the few remaining stands of the laurel forest, which are now protected, and was told by Henrique Costa Neves, the director of the National Park, that a major project of eradication has to be waged against invading species, particularly the *bananilha* (Malayan ginger), a plant that escaped from gardens, forms thickets that choke out other plants, and in a similar invasion has virtually taken over the Azores Islands in recent years.[9] On Madeira's neighboring islet, Deserta Grande, a campaign in 1996 eradicated rabbits, and probably mice and goats as well, and both vegetation and native birds are making a remarkable recovery.[10] The outlying Selvagens, the least disturbed islands in the North Atlantic, are now protected by the Madeiran Park Service and are home to thriving sea bird colonies.[11] But the native ecosystem of Madeira itself has been irreparably disrupted.

As an example offered at the beginning of this ecological history of the world, Madeira presents a question of scale. Madeira is a small island, only 57 km (35 mi) long and 22 km (14 mi) wide. The changes that take place there are local in scale, although they reflect events worldwide in scope, such as the colonial expansion of Europeans and the introduction of non-native species to formerly isolated lands. To talk only about planetary processes in a history like this one would be too abstract, too generalized. To use only local examples would be to lose the major themes in a mass of detail. Therefore chapters in this book contain both general narratives on a global scale and case studies on local and intermediate scales that illustrate the larger picture.

Egypt provides an example of ecological processes on a regional scale, that of an immense river valley. For thousands of years the Nile rose annually in a flood that watered and renewed the soil, depositing a layer of rich sediment. Then in the mid-twentieth century, a high dam constructed at Aswan ended the flooding. The structure itself, which I have seen from the river and from the air, is intimidating, vaster than the pyramids, but its effects on the land and people were even more enormous. Nubians who lived in the area of the new reservoir had to move elsewhere, and Egyptian farmers modified their systems of cropping and fertilization. A rising water table, salt accumulation, and other environmental problems appeared. As a result of these changes and the governmental policies that helped to produce them, and population increase, Egypt ceased being a net exporter of food and began to depend on imports to feed its people.

An example on the continental scale may be found in the British seizure of Australia. When they established penal colonies in the eighteenth century, they brought not only prisoners but also domestic animals and plants, along with exotic organisms such as rats and other mammals (later including, disastrously, rabbits), foreign trees, weeds, and diseases, all new to the ecosystems and formerly unknown to the aboriginal inhabitants. Within a few decades, the indigenous population fell to a fraction of its former number, and the landscape was transformed by deforestation,[12] overgrazing, and soil compaction. The changes are not finished; when I was in Kakadu in the Northern Territory, a tribal elder told me of the damage done by water buffaloes in the wetlands, and the fear that large cane toads, introduced into Queensland to control insects, but which have devoured native wildlife there, may spread into his homeland. The ecological changes in Australia were as great as the societal alterations, and intensified them.

To give an example on the global scale, the explosion caused by human error and negligence at the Chernobyl nuclear power station in the Soviet Union in 1986 produced heavy fallout over hundreds of square kilometers and made a portion of it uninhabitable. Radiation may be invisible, but its effects are often visible. Trees have died, plants have been observed to grow in strange sizes and shapes, animals have been born with mutations, and abandoned houses stand with children's toys still on the window sills. Those whom circumstances forced to stay in contaminated areas suffered radiation-induced illnesses. Radioactive particles fell over much of Europe, making crops and milk too dangerous to use for a time, and lesser increases in radiation were detected throughout the Northern Hemisphere. The event and its aftermath caused concern around the world and contributed to a sharp drop in the number of new nuclear facilities approved in many nations in the years following.

These are examples of humans producing environmental changes that had major effects, intended or unintended. There have also been many cases in which natural causes have seriously affected human history. These include climatic changes, such as the Little Ice Age that forced the Norse abandonment of Greenland in the fifteenth century; volcanic eruptions like the explosion of Krakatoa in 1883 that destroyed the island, killed more than 36,000 people, and produced worldwide changes in the atmosphere; earthquakes as severe as the one (with an attendant tsunami or tidal wave) in 1755 that reduced Lisbon to ruins; cycles of population in various species, as for example the periodic outbreaks of locusts that have destroyed the crops in east Africa and other continental areas; and outbreaks of epidemics, of which the most famous is the Black Death that killed at least a quarter of Europe's population between 1347 and 1351 and altered the economic and political structure of late medieval times. The study of past events in which people have altered the environment, and in which environmental influences have changed human society, is the aspect of environmental history that is the subject of this book.

Environmental history

The task of environmental history is the study of human relationships through time with the natural communities of which they are part, in order to explain the processes of change that affect that relationship. As a method, environmental history is the use of ecological analysis as a means of understanding human history. It studies the mutual effects that other species, natural forces, and cycles have on humans, and the actions of humans that affect the web of connections with non-human organisms and entities. Environmental historians recognize the ways in which the living and non-living systems of the Earth have influenced the course of human affairs.[13] They also evaluate the impacts of changes caused by human agency in the natural environment. These processes occur at the same time and are mutually conditional.

William Green, in *History, Historians, and the Dynamics of Change*,[14] observed that no approach to history is more perceptive of human interconnections in the world community, or of the interdependence of humans and other living beings on the planet, than environmental history, which supplements and often challenges traditional economic, social, and political forms of historical analysis.

An environmental historical narrative should be an account of changes in human societies as they relate to changes in the natural environment. In this way, its approach is close to those of the other social sciences.[15] One good example of this would be Alfred Crosby's *The Columbian Exchange*,[16] which showed how the European conquest of the Americas was more than a military, political, and religious process, since it involved invasion by a European "portmanteau biota" including domestic species and opportunistic animals. Eurasian plants, whether cultivated ones or weeds, he noted, replaced native species, and the impact of introduced microorganisms on the indigenous human population was even more devastating than warfare.

Like history itself, environmental history is also a humanistic inquiry. Environmental historians are interested in what people think about nature, and how they have expressed those ideas in folk religions, popular culture, literature, and art. That is, at least in one of its aspects, environmental history can be a history of culture and ideas. It asks how attitudes affect human actions in regard to natural phenomena, and in search of an answer, describes what the significant views were on the part of individuals and societies.

Environmental history is derived in part from a recognition of the implications of ecological science for the understanding of the history of the human species. This was appropriate, because ecology, in the sense that it studies sequential changes in natural communities, is a historical science. Paul Sears called ecology a "subversive" science,[17] and when taken seriously by historians, it has certainly subverted the accepted view of world history as it was up to the mid-twentieth century. The older history made little contact with nature; it was concerned mainly with the political activities of human beings. When it dealt with nature at all, it portrayed the advance of culture and technology as releasing humans from dependence on the natural world and providing them with the means to manage it. It positively celebrated human mastery over other forms of life and the natural environment, and expected technological improvement and economic growth to continue to accelerate. Environmental history, however, recognizes the biological fact that humans are dependent on natural factors and subject to ecological principles. For example, it is an ecological principle that the ability of any organism to increase in number and total biomass, and spread geographically, will eventually encounter one of several environmental factors that prevent further increase. Growth is limited by the least available factor, and no resource is infinite. An ecologist viewing any other species increasing at the present human rate, and using a comparable proportion of the energy in an ecosystem, would predict imminent collapse. Also, ecology points out the value of biological diversity, which helps to maintain the balance and productivity of an ecosystem in reaction to moderate stress. The older history saw human replacement of natural diversity with monoculture, in terms of agriculture and civilization, as desirable. Environmental history looks at the land, with its human and nonhuman inhabitants, as a varied and changing mosaic in space and time.

Most importantly, a significant group of ecologists took the natural community as the subject of their science. The older history saw no important relationships beyond those within human society, but environmental history emphasizes in its narratives the importance of the interrelationships of the human species to other species and the conditions that make life possible. The older history, when it recognized that nature and the environment were present, treated them as a backdrop, but environmental history treats them as active forces.[18]

The community of life

This book endeavors to give an account of environmental history that portrays major ecological processes that were at work in each period from human origins to the present. The narrative is not an attempt to give a neutral account of past events. It is interpretive, and I owe to the reader an explanation of the point of view that guides my interpretation. It is my view that historical explanations must take account of the fact that the human species is part of ecological systems.[19] What has happened to human societies, and continues to happen to them, is in many important ways an ecological process. The distinction, first made by the ancient Greeks, between "nature" (*physis*, what exists and grows of itself) and "culture" (*nomos*, what human societies create) is not an absolute one; in an important sense, culture is part of nature because culture is the product of a species of animal, the human species.

Nature consists of dynamic systems with many parts and functions. Among these systems are ecosystems. Humans are part of ecosystems, and participate in the processes that change them through time. History must take account of the importance and complexity of these processes.

The human species evolved within the community of life by competing against, cooperating with, imitating, using, and being used by other species. Thus our species is an offspring of the interacting forms of life on Earth. This means not only that human bodies achieved their forms through evolution, but that the ecosystems of the Earth provided our ancestors with sustenance, set problems for them, sharpened their wits, and to a large extent showed them the way they must go.[20] Humans, to more impressive degree than any other species, have made ecosystems what they are. That is, humans and the rest of the community of life have been engaged in a process of coevolution. That process continues to the present day. History's job includes examining the record of the changing roles the human species has enacted within the biotic community, some of them more successful than others, and some more destructive than others.

The idea of environment as something separate from the human, and offering merely a setting for human history, is misleading. Whatever humans have done to the rest of the community has inevitably affected themselves. The living connections of humans to the communities of which they are part must be integral components of the historical account. In this book, I will use "environment" in an inclusive sense, with no intent to imply that humans are exempt from the ecological principles that govern all species. They operate *within* the principles of ecology, and must continue to do so as long as the species is to survive.

That all human societies, everywhere and throughout history, have existed within and depended upon biotic communities is true of huge cities as well as small farming villages and hunter clans. The connectedness of life is a fact. Humans never existed in isolation from the rest of life, and could not exist alone, because they depend on the complex and intimate associations that make life possible. To a very large extent, ecosystems have influenced the patterns of human events. Consequently, the narratives of history must place human events within the context of local and regional ecosystems, and world history must in addition place them within the ecosphere, the worldwide ecosystem.

This is not merely a novel way of looking at history, but a recognition of how things have happened. As Aldo Leopold wrote,

> One of the anomalies of modern ecolog[ical thought] is that it is the creation of two groups, each of which seems barely aware of the existence of the other. The one studies the *human community*, almost as if it were a separate entity, and calls its findings sociology, economics, and history. The other studies the *plant and animal community* and comfortably relegates the hodgepodge of politics to 'the liberal arts.' The inevitable *fusion* of these two lines of thought will, perhaps, constitute the outstanding advance of the present century.[21]

More pointedly, John McNeill said,

> The enormity of ecological change [today] strongly suggests that history and ecology, at least in modern times, must take one another properly into account. Modern history

written as if life-support systems of the planet were stable, present only in the background of human affairs, is not only incomplete but is misleading. Ecology that neglects the complexity of social forces and dynamics of historical change is equally limited. Both history and ecology are, as fields of knowledge go, supremely integrative. They merely need to integrate with one another.[22]

I believe environmental history offers the opportunity for that integration.

Community ecology and history

This approach to history is to some extent the result of the interest of historians in the discoveries made by ecologists, particularly those who study community ecology. Environmental history is not a branch of ecology; as Stephen Dovers remarked, "To contribute to the reconciliation of environment and humans, environmental history needs to be more than merely a subset of either history or ecology."[23] It discovers new perspectives by combining the insights of both. Before the twentieth century, it would have been very difficult to give an account of history like the one in this book because ecological science had not advanced enough to demonstrate its possibility and the need for it. Ecology describes nature as consisting of complex systems with many parts and reciprocal functions. Among these are biological communities, which are interacting groups of organisms, and ecosystems, which are biological communities together with their nonliving environments.[24]

Ernst Haeckel, a German biologist, coined the term *ökologie* (ecology) in 1866.[25] He subsequently defined the new science:

> By ecology we mean the body of knowledge concerning the economy of nature – the investigation of the total relations of the animal both to its organic and inorganic environment; including above all, its friendly and inimical relations with those animals and plants with which it comes directly or indirectly into contact – in a word, ecology is the study of all those complex interrelations referred to by Darwin as the conditions of the struggle for existence.[26]

This was less a definition of an already existing science than an agenda for future investigation. Other scientists pursued the agenda.

In describing the totality of all organisms growing in an oyster reef, the zoologist Karl Möbius in 1877 coined the word *biocœnosis*, sometimes rendered *biocenosis* or *biocœnose*.[27] It is derived from two Greek words, meaning "life" and "community." It came to mean any interacting assemblage of animals and/or plants, whether defined geographically or according to habitat type. It is a beautiful word in Greek, and is used by ecologists in some European languages, but it looks and sounds formidable in English. Some possible alternatives are "biome," "biotic community," or "biocommunity." Biotic communities can be of various sizes, from the life in a small pond to all life on Earth. The largest community, including all life on Earth, is often called the "biosphere."

Victor Shelford, a leading ecologist in the early twentieth century, asserted, "Ecology is a science of communities."[28] A similar assertion can be made about environmental history; that is, that human relationships to the environment must be understood in the context of ecological communities. "Ecosystem" was first used in 1935 by the ecologist Arthur G. Tansley, who defined it as "the whole *system* (in the sense of physics) including not only the organism-complex, but also the whole complex of physical factors forming what we call the environment of the biome – the habitat factors in the widest sense."[29] Thus "ecosystem" includes the biotic community but is broader, including also nonliving factors such as water, gases, the mineral substrate, and energy in its various forms. Like biocommunities, ecosystems can be of many different sizes, from a small marsh to the "ecosphere," the entire surface and atmosphere of the Earth with all its inhabitants.

Ecological process

The narrative of world history must have ecological process as a major theme. The story of world history, if it is balanced and accurate, will consider the natural environment and the myriad ways in which it has both affected and been affected by human activities. The theme of the interaction of human events and nature has been operative during every chronological period. It modifies or determines other organizing principles. Political and economic histories ignore geography, geology, and biology to their peril, since the latter reveal aspects of the order of things within which the former operate, and upon which they depend. Economics, trade, and world politics are regulated, whether humans wish it or not, and whether or not they are conscious of it, by the availability, location, and limits of what, in language derived from economics, are called "natural resources."

Ecological process is a dynamic concept. It implies that the interrelationship of humans and the natural environment undergoes continual changes. These changes make environmental history just as necessary as ecological science in explaining the predicament of humankind and nature. Past changes help to explain the present, and lead us to expect further changes.

The idea of "balance" is too often taken to connote an unchanging status quo. It is to counter this misunderstanding that the word "process" is used here. Balance is a useful idea in environmental theory, as long as what is intended is not stagnation. The concept of ecological process implies that balance can accommodate change. Conversely, "sustainability," in spite of its misuse in present developmental rhetoric, should not imply an ever-growing economy, but one in which use of resources varies within the capacity of the ecosystem to supply them without permanent damage. Environmental history describes an ecological process that has sometimes moved toward balance and sustainability, and has often moved away from it. But move it has, and always does. Human actions can divert, but not retrieve, time's arrow. Still, there is some hope of diverting it in a better direction, and that may prove to be one of the uses of environmental history.

What is attempted here is a world history that adopts ecological process as its organizing principle, keeping the ecological context and the operation of environmental changes constantly in the forefront. What are the principles of ecological process, as they apply to human history? Some of them can be suggested briefly. The human species is part of nature, and nature consists of systems with many parts and functions. Among these are ecosystems, which also include the elements of the environment with which life interacts. These systems undergo changes through time. Human actions produce many of these changes. Changes are always complex, so that different changes are results of the same actions. Some changes are within the capacity of an ecosystem to absorb and compensate for, and remain healthy. Others may go beyond that capacity, and erode or transform the ecosystem, even so completely as to destroy it. Changes may go so far as to interfere with the functioning of local ecosystems, and even of the planetary system.

Human population growth tends to multiply effects on ecosystems. In some cases it may carry changes beyond the point of sustainability. At that point, a difference of degree becomes a difference of kind. Technology also may accentuate human impacts, making possible much more rapid changes and producing effects at great distance from the particular humans who cause them.

Sometimes humans have taken steps in accord with their knowledge and ethics to produce desirable changes and to ameliorate or arrest undesirable changes. Unfortunately, this is all too seldom the case. Often humans continue actions that threaten sustainability. Well-intentioned actions may be undertaken with insufficient knowledge. Many humans, particularly those who wield power, decide that other values, such as short-term survival or profit, are more important to them than long-term survival and sustainability. Philosophical and religious ideas often affect practices of people that make changes in ecosystems. They have sometimes had positive effects, especially in isolated areas with homogeneous populations and traditional ways of life. They do not generally aid sustainability, however, due to formulaic inflexibility that does not appropriately adapt to change and to a variety of ecosystems, and due to ways in which unscrupulous people evade them, or exploitative forces defeat them.

As Douglas Weiner asserted, "Every environmental struggle is, at its foundations, a struggle among interests about power."[30] In most societies, a minority that exploits resources has usurped power from a majority whose genuine, if not always conscious, interest is to maintain the sustainability of resources. Legislation and international agreements are effective only to the extent that they are observed and enforced. At least since the early twentieth century, governments have tended to regard economic growth as the highest good. As Richard Grove observed, "States will act to prevent environmental degradation only when their economic interests are shown to be directly threatened. Philosophical ideas, science, indigenous knowledge and threats to people and species are, unfortunately, not enough to precipitate such decisions."[31] Multinational corporations, which almost by definition want economic growth, are at times more powerful than governments. Garrett Hardin pointed out that when a resource is generally available, each person, corporation, or nation that has access to it tends to maximize its use as long as there is a marginal return, regardless of sustainability or its cost to a larger social unit. He called this "the tragedy of the commons."[32] Similarly, when unrestrained the global marketplace assures that the cost of a scarce resource will rise to a level making possible its continued exploitation until extreme depletion or exhaustion. To avoid that result, Hardin advises "mutual coercion, mutually agreed upon." But in practice, coercion by the instruments of the world market economy is rarely exercised to conserve resources; rather, it is used to reduce barriers to the increased production and exchange of commodities.

Most world histories in the second half of the twentieth century adopted "development" as their narrative theme. The word has an interesting history.[33] Aristotle and other ancient philosophers used the verb *phyein*, which means "to grow" as a plant or other organism grows, according to a purpose or pattern which is inherent within the organism. "Nature," *physis*, is the process of growth, or what manifests development. When this biological image was adopted to explain the course of history, however, a change occurred in its meaning. In a natural organism, birth is followed by the vigorous growth of youth, then by maturity, decay, and death. But in the view of history most widely accepted today, unending economic growth is essential to development, and "development" itself is the goal. Development in this sense does not mean primarily improvement in the arts and sciences, a healthier and more abundant environment, or social relations that are more authentic and fulfilling. Indeed, these things are considered subsidiary to, and may be eroded by the overriding "need" for ever-increasing production and trade. If ecological process is adopted as the major theme of world history, development will not disappear from the story; it will, however, need careful redefinition. Studies of economic growth in a world of limited resources have recognized the need for such revision.[34] Development as economic growth that refuses to recognize limits, and therefore is destructive of the community of life, has been a questionable blessing. Mere growth in quantity, driven by exponential expansion of human population and increasingly powerful technologies, has come close to collision with the limits of the Earth. But development as improvement in quality, development as advancement in the arts of living, development as the discovery of ways to use resources more efficiently, more creatively, and less wastefully, and development in moral inclusiveness has a positive role. Such development could be directed in ways that conserve and are not destructive of Earth's biosphere.

In these pages, I follow a path through the history of humankind in relationship to ecosystems around the whole Earth. The subject is vaster than a rainforest, so I cannot offer a comprehensive superhighway, but only a series of footprints. After this introductory chapter, the itinerary is roughly chronological. Each of the eight chapters that follow concentrates on a general period in human history. These are periods characterized by large-scale changes in the relationship of human societies to the biosphere. To provide greater depth and to give concrete historical examples, each chapter contains case studies that deal with issues or environmental problems as enacted on local or regional scales. The case studies are not intended to cover everything important that happened in a given time frame, but to illustrate significant patterns that were occurring then. Each of these sections is centered in, but not entirely limited to, a particular place that serves as an example of a process affecting the role of humans in living communities. I began with the determination to use only places that I have seen and studied, and have been able to keep that resolution with few exceptions.[35] Each example can be regarded as in some ways typical of many others in various parts of the world.

I hope that this journey through environmental history can take us beyond the distinction between human-centered and environmentally-centered discourse to a broader view that recognizes and embraces the community of life, surrounding, including, and in relationship to human beings.

Notes

1   T. Bentley Duncan, *Atlantic Islands: Madeira, the Azores and the Cape Verdes in Seventeenth-Century Commerce and Navigation*, Chicago, University of Chicago Press, 1972, 7.

2   Henrique Costa Neves, *Laurissilva da Madeira: Caracterização Quantitativa e Qualitativa*, Madeira, Governo Regional, Parque Natural da Madeira, 1996.

3   William M. Johnson and David M. Lavigne, *Monk Seals in Antiquity*, Mededelingen No. 35, Leiden, Netherlands Commission for International Nature Protection, 1999.

4   Sherwin Carlquist, *Island Biology*, New York, Columbia University Press, 1974, 501.

5   Alberto Vieira, ed., *Slaves With or Without Sugar*, Madeira, Atlantic History Study Centre, 1996.

6   Alberto Vieira, Constantino Palma, and Homem Cardoso, *Madeira Wine*, Lisbon, Chaves Ferreira, 1998.

7   Charles Verlinden, *The Beginnings of Modern Colonization: Eleven Essays with an Introduction*, Ithaca, NY, Cornell University Press, 1970, 210.

8   Harald Pieper, "The Fossil Land Birds of Madeira and Porto Santo," *Bocagiana* 88, October 1985, 2; noted in Mikko Saikku, "Extinction in the Areas of European Expansion: Imperiled Bird Life of the North Atlantic Region," given at the International Seminar on Environmental History and European Expansion, Funchal, Madeira, April 8, 1999.

9   *Medidas de gestão e de conservação da floresta laurissilva da Madeira*, código 45.62, Madeira, Project Life, Contrato B4-3200/97/281, August 1998.

10 Brian D. Bell and Elizabeth A. Bell, *Habitat Restoration: Deserta Grande, Madeira*, Report prepared for Parque Natural da Madeira, Funded by the Life Project Grant, Wellington, New Zealand, 1997.

11 *Selvagens Islands Nature Reserve, Portugal*, European Diploma Series No. 36, Strasbourg, Council of Europe Press, 1993.

12 Michael Williams, *Deforesting the Earth: From Prehistory to Global Crisis*, Chicago, University of Chicago Press, 2003.

13 For accounts of the work of environmental historians up to the early twenty-first century, see J.R. McNeill, "Observations on the Nature and Culture of Environmental History," *History and Theory* 42, December 2003, 5–43; and J. Donald Hughes, *What is Environmental History?*, Cambridge, UK, Polity, 2006.

14 William A. Green, "Environmental History," in *History, Historians, and the Dynamics of Change*, Westport, CT, Praeger, 1993, 167–90.

15 Mei Xueqin, "From the History of the Environment to Environmental History: A Personal Understanding of Environmental History Studies," *Frontiers of History in China* 2, 2, 2007, 121–4.

16 Alfred W. Crosby, Jr., *The Columbian Exchange: Biological and Cultural Consequences of 1492*, Westport, CT, Greenwood Press, 1972 (new edn, Praeger, 2003).

17 Paul B. Sears, "Ecology: A Subversive Subject," *BioScience* 14 , 7, 1964, 11; Paul Shepard and Daniel McKinley, eds, *The Subversive Science: Essays toward an Ecology of Man*, Boston, MA, Houghton Mifflin Co., 1969.

18 For the formulation in this sentence, I am indebted to James O'Connor. See his essay, "What is Environmental History?", in *Natural Causes: Essays in Ecological Marxism*, New York, Guilford Press, 1998, 48–70.

19 William Cronon, "A Place for Stories: Nature, History, and Narrative," *Journal of American History* 78, 4, March 1992, 1347–76, 1372.

20 Edmund Russell, "Evolutionary History: Prospectus for a New Field," *Environmental History* 8, 2, April 2003, 204–28.

21 Aldo Leopold, "Wilderness" (undated fragment), Leopold Papers 10–6, 16, 1935. Quoted in Curt Meine, *Aldo Leopold: His Life and Work*, Madison, University of Wisconsin Press, 1988, 359–60.

22 John R. McNeill, *Something New Under the Sun: An Environmental History of the Twentieth-Century World*, New York, W.W. Norton, 2000, 362.

23 Stephen Dovers, "Australian Environmental History: Introduction, Review and Principles," in Stephen Dovers, ed., *Australian Environmental History*, Melbourne, Oxford University Press, 1994, 2–20, 6.

24 Frank Benjamin Golley, *A History of the Ecosystem Concept in Ecology: More Than the Sum of the Parts*, New Haven, CT, Yale University Press, 1993.

25 Ernst Haeckel, *Generelle Morphologie der Organismen*, Berlin, Reimer, 1866.

26 W.C. Allee, A.E. Emerson, O. Park, T. Park, and K. Schmidt, *Principles of Animal Ecology*, Philadelphia, PA, Saunders, 1949, frontis., quoted in Robert McIntosh, *The Background of Ecology: Concept and Theory*, Cambridge, Cambridge University Press, 1985, 7–8.

27 Karl Möbius, *Die Auster und die Austernwirtschaft*, Berlin, Wiegandt, Hempel, and Pary, 1877, 41, cited in Robert McIntosh, *The Background of Ecology: Concept and Theory*, Cambridge, Cambridge University Press, 1985, 52.

28 Victor E. Shelford, *Laboratory and Field Ecology*, Baltimore, MD, Williams & Wilkins, 1929, 608.

29 Arthur George Tansley, "The Use and Abuse of Vegetational Concepts and Terms," *Ecology* 16, 1935, 284–307, quoted in McIntosh, *Background of Ecology*, 193.

30 Douglas R. Weiner, "A Death-Defying Attempt to Articulate a Coherent Definition of Environmental History," *Environmental History* 10, 3, July 2005, 404–20, at 408.

31 Richard H. Grove, "Origins of Western Environmentalism," *Scientific American* 267, 1, July 1992, 42–7, 47.

32 Garrett Hardin, "The Tragedy of the Commons," *Science* 162, 1968, 1243.

33 Gilbert Rist, *The History of Development: From Western Origins to Global Faith*, London, Zed Books, 1997, 25–46.

34 Mark Sagoff, *The Economy of the Earth: Philosophy, Law, and the Environment*, Cambridge, Cambridge University Press, 1988.

35 Political conditions prevented my travel to Iraq, but the importance of Mesopotamia to the subject of Chapter 3 demanded that it be included, so I "set" that section in the great collection of archeological treasures from Uruk and other sites in ancient Iraq that is in the Pergamon Museum in Berlin.

2 Primal harmony

There are still a few places in the world where one can sense what Earth was like before the advent of human beings. In the aisles of a tropical rainforest such as the one that flourishes by the Río Napo in Peru, there are so many species of trees that often one has to walk some distance before finding the same one twice, and the variety of iridescent butterflies, mantises, and other insects is incredible. In a cave under the coastal cliffs of Oregon, open to the breakers of the seemingly changeless ocean, the great sea lions bark clouds of steam above pools where mussels and anemones cling amid a constantly moving throng of crustaceans. At evening in springtime around a desert water hole in Organ Pipe Cactus National Monument, Arizona, bright flowers spice the air as bighorn sheep lower their heads and bats dive to the surface, drinking on the wing. Earth before humankind appeared was a place of abundant biodiversity and of dynamic balance among species and elements.

An environmental history has to begin with the environment. There is a history of the environment before the human species evolved into its present form. Indeed, the appearance of *Homo sapiens* came only recently in the long story of the Earth's geology and biology, a story to which many scientists apply the term "environmental history" in a broader sense than it is used in this book.

What is the natural state of Earth? This is a question that must be answered before it is possible to understand how the human species relates to, and changes, the natural world. Some early ecologists argued that selected areas ought to be preserved as much as possible from human disturbance to show how natural living systems operate when compared with areas that have suffered from various kinds of interference.[1] While it is important today to preserve habitats for animal and plant species, it is also increasingly apparent that no place on Earth is really unaffected by human activity; none has escaped such widespread effects as air pollution, intensification in the acidity of precipitation, radioactive fallout, and the penetration of ultraviolet radiation due to the depletion of the ozone layer in the high atmosphere. This means that historians must look to evidence of the deep past to find out how nature operated without humankind, and use that as a baseline or control against which to judge the changes brought about since the beginning of human history.

Contemplating the immense age that Earth had reached before humans appeared can provide perspective. The planet condensed into its nearly spherical shape, seas and continents formed, the phyla of the animal and vegetable kingdoms appeared, and living species evolved ways of interacting with the physical matrix and with each other over hundreds of millions of years. The result was an ecological balance that sustained the conditions for life. Natural laws may, according to the new views of cosmological physics, change as the universe unfolds, but they do not apparently make exceptions for individuals or species. Humans, whose written history has spanned only the last few thousand years, must live within the conditions of the physical and biological universe and take careful account of the balance that is the natural state of Earth.

Ecological balance is dynamic, not static. It operates through change. It is not rest, but harmony in movement. It is not the stable condition of block resting on block in a pyramid, nor the unstable equilibrium of scales where a weight added on one side will bring one arm down and the other up, but the poise of an eagle flying, adjusting her wings to carry her body evenly through shifting currents of air. A living creature allows for changes that come from inner states and outer forces, adapting to sustain its life.

The idea that Earth is a living organism is very ancient. Such was the intuitive understanding of the earliest people whose thought can be fathomed, the tribal hunters and farmers. They regarded Earth as mother, and worshipped her as a goddess. Plato and other Greek philosophers maintained that the cosmos is alive, as we, who are among its constituent parts, are alive.[2] In this century, the atmospheric chemist James Lovelock has enunciated a theory that all life on Earth acts together like a great living organism to influence temperature, atmospheric composition, and other physical factors so as to maintain optimum conditions for itself. As the name for this organism, Lovelock selected "Gaia," the Greek name for the goddess Earth.[3] This idea, called the Gaia hypothesis, is a seminal concept, but should be used critically and carefully. In what sense is Earth alive?

Earth, viewed as an entire planet, does seem to be alive. Time-lapse films taken from artificial satellites show the great cycles of weather systems streaming like the currents of cytoplasm in a cell. The seas also circulate. Geologists have detected a much slower recycling called plate tectonics, in which the renewal of the sea beds, welling up from under the crust and being swallowed millions of years later by subduction back to underworld places of melting heat, moves the continental masses, splitting and joining, in ever-changing patterns that look like living processes. It can be maintained with good reason that the entire planet is alive, that just as a body includes seemingly nonliving parts like bones and blood serum, so a living planet includes air, sea, and rocks. Ecological science shows us how animals and plants interact with each other and their environments, forming larger units called ecosystems. Through reproduction, the food chain, and the cycles of elements and energy, in an immensely complex set of relationships, species increase and decrease in number, but the ecosystem as a whole continues. In this sense, ecosystems are organisms, and Gaia, or the biosphere of Earth, is the largest ecosystem. This does not mean, however, that Gaia is an organism in just the same way that the human body is an organism. To explain this, one can look at the relationship between a single cell and the body. Both are alive, but the body is not just a large cell. The body is an immense community of living cells, related to one another in myriads of ways. The whole is greater than the sum of parts. Similarly, Gaia is a community that includes billions of living bodies, but the structure of that living community is as much more complex than that of the body as the structure of the body is more complex than that of the cell. The body is a somatic organism, but Gaia is an ecological organism. Thus defined, Gaia is more than a metaphor. The physiological processes of Gaia are the interrelationships defined and studied by ecology.

It is possible to examine the natural state of Earth in realistic ecological terms. Though it was undeniably less polluted and more profuse in living things than today, Earth before humans was not a boring Eden. There were sudden and immense changes: volcanic eruptions, earthquakes, tsunamis, tornadoes, forest fires, floods, and droughts. The wounds left by these traumas healed, as life, often finding the devastated areas enriched by mineral and organic deposits, reclaimed them in the stages of natural succession. Great changes were wrought by the working of the ecological processes themselves. The populations of some species built up cyclically to insupportable numbers, depleting their food supplies, and then suddenly declined. Some species became extinct while others evolved. There were those that, like beavers and termites, greatly altered their environments over large areas. But changes prepared the way for new forms of life. Life was sustainable and, above all, abundant. The air thronged with billions of birds, compared to which the present avian population is a sad remnant. The plains themselves must have seemed to move with herds of herbivores, followed by their predators. Schools of fish silvered the sea, while the great whales rejoiced in numbers unseen in more modern times. Even in the late twentieth century, one could still gain an idea of what the primal state of Earth was like by visiting the savannas of East Africa or bird colonies on isolated islands. Although these regions are still impressive, they have suffered diminution, and some degree of imagination is necessary to appreciate the abundance and diversity of life as it existed before human beings evolved.

The Serengeti: kinship of humans with other forms of life

Floating above the Maasai Mara in a hot air balloon, as I did early one morning, affords a wide prospect of the mosaic of the Serengeti–Mara ecosystem. The number and variety of large animals visible from the air amazed me. Someone who had seen this Edenic remnant of the Pleistocene fauna in earlier decades might have noticed some diminution since then, but I could not complain of any lack of abundance. The Serengeti–Mara ecosystem embraces an area of 26,000 sq km (10,000 sq mi) in Tanzania and Kenya. A large part of it is protected: the Maasai Mara Game Reserve in Kenya includes 1,500 sq km (600 sq mi) of the ecosystem's northern extension, and Tanzania's Serengeti National Park, along with the Ngorongoro Conservation Area, comprises a biosphere reserve of 23,000 sq km (8,900 sq mi).[4] But with agricultural developments moving ever closer, and poachers at their destructive work, the future of the wildlife is in question.

Seen from the air, hundreds of wildebeest grazed in vast, irregular circular formations on the open grassland. Zebra inside thorny thickets had their hiding places revealed. Giraffe, hyena, ostrich, many kinds of antelope, and one enormous lion came into view, but none seemed to pay attention to the hundred-foot-high cloud-like balloon drifting overhead, nor to the thunder-like roar of its generator.

Then I began to notice the patterns of vegetation. Four major ecological types interpenetrate one another there: grassland, savanna, thorn woodland, and the gallery forests along the watercourses. Most extensive is the grassland, which supports the greatest concentration of large mammals in the world.[5] So many herbivores can utilize the grasses because each species has a different preference in food, with mouth shape and teeth suited to its diet. Zebras, for example, can digest the coarse stems of tall grass, while Thompson's gazelles prefer tender herbs and new shoots. The annual migration of hundreds of thousands of grazing animals between the southern and northern sections of the Serengeti–Mara ecosystem follows the seasonal green wave of new vegetation nourished by the rains.[6] Herbivores attract the interest of carnivores: wild dogs, hyenas, leopards, cheetahs, and lions among them.[7] Scavengers are not far behind; jackals, vultures, and other eaters of carrion assure that no flesh of fallen animals will remain for long, and hyenas have jaws strong enough to shatter the bones. The most spectacular wildlife phenomenon at Maasai Mara, and indeed in the world, is this annual migration of hundreds of thousands of wildebeest, gazelle, zebra, and impala into the area at the beginning of the dry season.

The savanna is grassland dotted with trees such as acacias and the desert date (*Balanites*). Typically, these trees rise on bare trunks and branches up to 5–6 m (15–20 ft) above the

Figure 2.1 An example of one of many species adapted to the East African dry forest ecosystem. This is part of a group of twenty-eight Maasai giraffes in Amboseli National Park, Kenya. Photograph taken in 1989.

ground, where the horizontal crown appears just above the level that giraffes can easily reach. Elephants, when moved to do so, break off branches or knock over trees to get at the leaves.[8]

The thorn woodland is dominated by shrubs and trees that tolerate dry conditions. These include *Croton*, acacias, and succulent *Euphorbia* (the "candelabra" tree). Buffalo and black rhinoceros find cover here, but along with the elephant, make paths that encourage the growth of grass, so fire, too, can follow. Browsers such as impala, bushbuck, kudu, gerenuk, and dikdik wander in the bushland. In recent years, this plant community has been shrinking due to repeated fires set by humans, and to the invasion of elephants that have been pushed out of the spreading agricultural lands bordering the reserves.[9]

Gallery forests line the banks of watercourses. These riverine communities are composed of water-dependent trees, of which *Euclea*, *Diospyros*, and greenheart, along with African olive, figs, and palm, predominate. I stayed at a lodge that provided a raised wooden walk through the trees of the gallery forest, where one could go to watch birds, monkeys, and tree hyrax. In the evenings, giraffe and waterbuck would saunter by. Other animals that frequent these narrow strips are river dwellers such as hippopotamus and crocodile. Bird life is especially abundant along the rivers, including weavers, whose numerous hanging nests are prominent in the trees, hornbills, egrets, fish eagles, and Egyptian geese. The great annual migration of grazing animals must cross the Mara River. There, hundreds of wildebeest drown, their bodies floating downstream to be devoured by vultures and other scavengers. When I was there, huge numbers of animals had already arrived. Views across the plains revealed vast herds of them, and the river banks were covered with corpses. The hippos and crocodiles, however, did not seem to suffer.

It was in a place like this, with different habitats interspersed, a landscape that ecologists call "ecotonal," that the first families of *Homo sapiens* evolved.[10] There is general agreement among anthropologists, archaeologists, and geneticists that the ancestral home was Africa, its eastern and southern parts. The environment of the first "modern" humans, a number of hundred thousand years ago, was like the Serengeti–Mara ecosystem. Of course it was not exactly the same; there were more woodlands and the animals were of somewhat different forms, some larger.[11] The climate has passed through periods of change, and this changing climate had important influences on the evolution of hominid and human groups.[12] But the similarities are important enough to make the comparison apt. "East Africa ... contains an intricate network of equatorial habitats that preserve many of the features of the environmental tapestry through which the strands of early human evolution were woven."[13] It was, and is, varied country with accessible borders between forests of different kinds and open country. There was a high degree of biodiversity, which means the number, variability, and variety of forms of life. In the time of their emergence, humans interacted with many different kinds of plants and animals. Since they could pass quickly from grassland, say, to forest, early humans encountered hundreds of species of plants. Even more impressive was the richness of animal life. Early humans hunted, gathered, dwelt, and served as prey in that nexus of constant interaction between species.

Humans have never been alone on the Earth. Their lives – culture, technology, and art – have been immeasurably enriched because they learned to watch, listen to, and imitate the other animals that shared the land and sea with them. So the ancient Greek philosopher Democritus thought. He speculated that people learned to weave from spiders, and how to sing from songbirds, swans, and nightingales. They got the inspiration to build houses of clay from watching swallows at work on their nests. "In the most important concerns," he wrote, "we are pupils of the animals."[14] A more recent author, Steven Lonsdale, argued in a book filled with examples from every part of the world that dance owes its origin and elaboration to human imitation of the varied movements of mammals, reptiles, amphibians, fish, birds, and even invertebrates.[15] The idea of the impacts of other forms of life on humans can be followed even further. The human species, from earliest times down through history, gained more from the others than a few crafts. Interaction with countless kinds of animals and plants largely created the shapes of human bodies and minds, gave direction to cultures, and in an important sense made us what we are. The diminishment or loss of that interaction has affected and will affect us more deeply than we commonly think.

The human species and human culture evolved through natural selection that took place because humans during their history as a species were part of biotic communities where their interactions with other kinds of animals and plants decided whether or not they survived and reproduced. One of the greatest mistakes made by humans today is to think about themselves as existing and acting without reference to other forms of life. No species exists in complete isolation; every one relates to others in a living system.[16] This is common knowledge in biology, but must also be recognized as a basic fact of history.

A large and subtle brain seems to have given some humans a survival advantage in dealing with complex vegetation, devising tools, and outwitting herd animals and predators, so it seems that human intelligence is a response to the challenges offered by living among many other species. To quote Edward O. Wilson,

> How could it be otherwise? The brain evolved into its present form over a period of about two million years, from the time of *Homo habilis* to the late stone age of *Homo sapiens*, during which people existed in hunter-gatherer bands in intimate contact with the natural environment. Snakes mattered. The smell of water, the hum of a bee, the directional bend of a plant stalk mattered. ... The glimpse of one small animal hidden

in the grass could make the difference between eating and going hungry in the evening ... The brain appears to have kept its old capacities, its channeled quickness. We stay alert and alive in the vanished forests of the world.[17]

During the course of history, human development has been so deeply affected by interrelationships with other forms of life that it can be described as a process of complex coevolution, genetically and culturally. Ever since human beings began to reflect on what they are, they have pondered their relationship to other parts of nature. Ancient and modern philosophers have attempted definitions to divide the human species from the others. Many of them held that the distinctive human characteristic was abstract: the soul, or reason, or a self-concept. Sometimes this seems to have been an attempt to exempt us from the laws of nature, as if humans had a special relationship to the incorporeal and could somehow avoid the inexorable operation of the principles of ecology. Aristotle said that the rational soul is the unique possession of humankind, but he granted souls of lesser character to animals and plants. He believed human beings also had "sensitive" (animal) and "nutritive" (vegetable) souls, however, thus maintaining a connection with non-human species.[18] Others have proposed social structure, language, tool-using or tool-making, [19] and erect posture as unique. The studies of behavioral scientists and others have shown that none of these, at least the ones that can be observed, is an exclusively human trait.

The ostrich and kangaroo stand erect. Birds such as Darwin's finches use thorns and stones as tools. Jane Goodall has observed and photographed chimpanzees stripping twigs in order to draw termites from their nests, and making sponges of leaves to get water out of difficult places.[20] Thus they are tool-makers as well as tool-users. Whales, dolphins, and some birds communicate by making sophisticated sounds. Bees, ants, and termites have social structures and create architecture. As for the ability to reason, wild animals have been observed to solve complex problems, and anyone who has lived with cats and dogs must have noticed that they can figure things out to an amazing degree. The Greek writer Plutarch turned the question around in his essay, "Beasts are Rational," and asked if the actions of people indicate that they are more or rather less rational than most intelligent animals.[21] Consciousness and the ability to use language belong to the great apes, as indicated by studies in which experimenters taught sign language to gorillas and chimpanzees. These animals not only repeat the individual signs and sentences they have learned, using them appropriately, but also invent their own sentences, words, and signs, and use them to talk with each other as well as their trainers. They apparently form abstractions, remember the past, and plan for the future. They tell jokes, and even lie and swear, other characteristics once thought to be exclusively human. The student of animal behavior Francine Patterson reported a conversation with a gorilla named Koko in which Koko indicated that she had a self-concept. When asked by Patterson, "Are you an animal or a person?" she replied, "Fine animal gorilla."[22] Gorillas, in studies cited by Patterson, have been observed to communicate by signs of their own in the wild.

In coming to understand the place of the human species in the history of Earth, it may be more instructive to examine resemblances to the rest of creation than differences. All things considered, we are more alike than unlike the other animals. In the evolution of humans from primate ancestors, at no point did a transformation occur that justifies an attitude of dismissal to the non-human natural world. On the most basic level, human beings are part of the material cosmos. Historians must never forget that the human body is composed of physical elements, made of the same stuff the stars, and therefore Earth, are made of. It takes up space, has weight, and shares tangible existence with rocks and rivers. The human species has mass and dimensions; that there are so many human bodies in the world is an inescapable fact of economic geography. Next, humans are alive in the same way that plants are alive: growing, exerting force, incorporating material from outside, respiring, and reproducing. One can recognize this when one sees a tree and watches it through the seasons and years. We are living things as are lichens and lianas. On yet another level, humans share movement, consciousness, and the experience of the senses with other animals. We know what it is to be animals, for in fact we are animals.

There are those who think that all this is somehow demeaning, a lowering of human beings to the level of animals, but they could not be more wrong. It need not lower humanity; it can raise nature. Indeed it exalts humanity, and it may indicate a true uniqueness, that humans are able to recognize kinship with other beings and to admire their freedom, beauty, and autonomous roles in the balance of nature. The human species does not need to use its preeminent intelligence to be nature's tyrant, but may decide to be a skilled, unique partner with other living beings. Just as many men in the late twentieth century have found that the increased freedom and expanded roles of women have not subtracted from their masculinity, but have added new perspective to what it means to be a man, so many men and women may find in the ecological vision of a structured but non-hierarchical living world not a less important role for humanity, but a dimension added to what it means to be human.

The human species evolved in the context of the ecological systems of Earth. Humans are adapted to live within many of them, though not all. Human existence is linked to those ecosystems; any traveler outside them must carry parts of them along: oxygen, water, food manufactured by plants and perhaps animals, and ways of maintaining temperatures within a certain range. Anyone who wants to live in space, or like Jacques Cousteau and his colleagues in stations on the sea bottom,[23] must reproduce at least the minimum conditions that sustain life in the parts of Earth's surface to which humans are adapted. For any extended stay, this requires resupply. Until another inhabitable, and reachable, planet is found, humankind is totally dependent upon Earth. If Earth's biosphere were to become uninhabitable, all species would become extinct. For us to live, other animals and plants must continue to live, too. Humans need other living things as much as we need other humans.

Fossil evidence at the present time indicates that ancestral humans evolved in East Africa in an environment where rich tropical forests intergraded with savannas, and where there were numerous lakes and streams. Other animals, including large mammals, were abundant. The ecological niche filled by the human species was that of a large, mobile omnivore. Human teeth and stomach indicate neither carnivorous nor herbivorous specialization of diet. From the beginning, then, humans gathered edible plants, caught fish and crustaceans, and hunted mammals. Humans are slower than many of the animals they killed and ate, but humans are also very persevering; in more recent times a skilled but unarmed Native American Indian hunter was able to track and follow a deer until it fell exhausted, then smother it to death by holding its mouth and nostrils closed.[24] Humans also cooperated with one another in the hunt, setting ambushes, digging pits, and driving animals over cliffs.

It was not necessary for humans to use only their own arms and legs in hunting, of course, due to a way of adaptation that, although not absolutely unique to humans, was capable of producing an unusual number of changes: culture in general, and technology in particular. Humans formed patterns of behavior they had not been born with, and passed these to others, especially to the new generations, by means of demonstration and language. They invented new tools. The australopithecines in East Africa began the technology of making hand axes from stones. Later, *Homo erectus* ("Man the Upright"), a species that lived across much of the Old World, knew the use of fire in cooking and keeping warm, and possibly in hunting as well. The human species presently surviving is called *Homo sapiens* ("Man the Wise"), and may or may not actually have been wise, but was certainly technologically cunning from the start. Our clever forebears made grinding stones for wild seeds. They invented the spear-thrower (atlatl), and later the bow and arrow. A cultural tradition appeared and was handed down, including stories, customary views of the world, and teachings about family relationships, the gods, creation, birth, and death. About many of these matters it is necessary to speculate, since writing had not yet appeared, although some idea of what early cultures were like can be gained from works of art and from studies of similar groups that survived down to modern times. Archaeology clearly indicates that technological knowledge was passed from parent to child, and that slowly at first and then more rapidly it became increasingly complex and sophisticated. Culture and technology represent adaptations to the natural environment. Although they have changed in form and degree, often becoming more powerful, they have not freed human beings from reliance on the natural world. The human species has been able, through them, to use an ever larger proportion of the matter and energy in Earth's ecosystems, but it has not been able to declare independence from the working of ecological principles.

Kakadu, Australia: the primal tradition

Following a white-haired tribal elder, I walked through the acacia–eucalyptus forest at Manyallaluk ("Frog Dreaming Place"), in northern Australia. He explained the uses of many of the plants and insects we came across; it was evident that he had an encyclopedic knowledge of the local ecosystem. Another man, an artist, showed me how to grind red and white ocher on a stone to use as paint. Such grindstones have been found in archaeological sites not far away, dated at least eighteen thousand years before the present, the oldest artists' palettes known in Australia.[25] Later a young hunter demonstrated how to throw a spear with the wooden spear-thrower, an implement that has been in use for at least seven to nine thousand years in that Arnhem Land district.[26] The spear soared an incredible distance and hit its target. Weapons like it were terribly effective in early times against large marsupials. Later we joined in a feast where the main dish was kangaroo meat cooked over a wood fire.

The primal[27] tradition of the human race is the culture of hunting, fishing, and gathering. This was the only way of life for more than nine-tenths of the time that *Homo sapiens* has existed. It spread everywhere with the peopling of the Earth, was typical of early groups in Africa, Eurasia, America, and Australia, and persisted in places such as Manyallaluk up to the twentieth century. This is a substratum for all later stages, and still lies importantly below the civilized veneer of modern societies. Specific expressions of this ancestral tradition vary greatly among tribes and other groups, but certain broad characteristics mark it wherever it appears, and it can be well illustrated by the traditional ways of life of the aboriginal northern Australians. People for whom it is a living culture see the world as filled with spiritual power and populated by spirit beings. These are the beings who enacted the stories of origins in the ancient Australian Dreamtime, forming the landscape and placing their names upon its features. The universe is a sacred place. As Silas Roberts, an Aboriginal leader, expressed it, "Our connection to all things natural is spiritual."[28] All beings are alive and sentient, including the Earth and sky. Hunters must approach animals and plants with reverence, kill them only when necessary, and treat them with honor even after killing them.[29] A human being is primarily a member (in the old sense of "an integral part") of a tribe, not a separate individual, so that community is a given in primal experience. Members of the community must defend it against human enemies from outside. Individuals may go out

Figure 2.2 An elder in the forest near Manyallaluk, which means "Frog Dreaming Place" in Northern Territory, Australia. He possessed knowledge of the characteristics and human uses of hundreds of species of plants and animals. Photograph taken in 1996.

into the world of nature, or into the world of dreams, to gain spiritual power, and this power is to be used for the benefit of the tribal group. Elders are respected and protected, since they embody the wisdom and memory of the community. The community, through oral tradition, traces its origin to ancestral living beings, animals, and plants who are the Dreamtime founders of the way things are in the world.[30] "These totems are used to distinguish social groupings and can be influenced by ceremonies conducted by their human 'kinsmen,' such as ceremonies to maintain the natural species."[31] Ancestral history is embodied in features of the landscape, sacred sites which must be treated with the utmost care and never violated.[32] But Dreamtime is not only in the past; Aboriginal people maintain their identification with the land and other living things. As Big Bill Neidjie, elder of a Gagudju clan, put it, "That tree same as me. This piece of ground he grow you."[33]

Each of these cultural attitudes helps to adapt humans to the environment and to keep them in balance with it. For example, hunters had an extraordinary reverence for the animals they hunted. The magnificent paintings on cliffs and overhangs in Australia display this feeling, and it is explicit in descriptions provided by primal hunters of the modern period when questioned by explorers and anthropologists. It may seem incredible to those who have never encountered it, because urban people tend to think that a hunter would regard his prey as an enemy, an inferior victim, or a prize of sport, as many "civilized" hunters do. The primal hunters approached animals with deep respect. They prepared themselves for the hunt with purification, fasting, and taboos against such things as casually mentioning the animal's name. The hunter implored the animal to give itself freely, with the plea that the hunter killed only out of great need, and would honor the animal's gift. The hunter tried to make the kill as painlessly as possible, and treated the dead animal with deference, addressing it kindly, cutting up the carcass carefully in a ritual manner, and putting the remains, especially the skull, in a tree or other place of honor. Strictures like these surrounded the hunt: do not kill more than you need; do not kill the first one of the species that you see; do not take a mother together with her offspring; do not kill all of a given herd; use everything you take. The expressed hope was that the spirit of the animal, happy with its treatment, would tell the others of its species about it, and consent to be reborn, returning to be killed again. Further, they believed in spiritual protectors of the animals, like Masters or Mistresses of Beasts, who watched carefully, rewarded respectful behavior, and punished the careless or irreverent hunter.[34] Some groups also followed the practice of hunting in only a portion of their territories each year, thus allowing the animal species in any given area to remain undisturbed for one or more years.

An attitude of reverence encouraged practices of conservation that would tend to sustain the wild animals upon which the hunters depended. Similar attitudes and practices also are typical of gatherers and their relationship to plants and the smaller animals they collected. The primal hunters and gatherers were not aimless wanderers; generally they lived within home territories which they knew well. Their own food supply and therefore their numbers and health depended on the condition of the ecosystem. A subsistence hunting–gathering economy provides a relatively immediate feedback; if hunters kill too many of a critical prey species, they will suffer. Doubtless such things did happen in the time tens of thousands of years ago, after the ancestral people entered Australia. Many important large animal species disappeared, so that over the generations, myths and taboos against careless killing may have come about as a result of experience with depletion and extinction. "The individual acquires this knowledge progressively and cumulatively during a lifetime punctuated by periods of intense learning now described in many parts of Australia as 'going through The Law.'"[35] Primal peoples' treatment of the natural environment showed care, and was guided by attitudes that might today be called religious, but which from the standpoint of their own cultures were simply an integral part of their whole pattern of life.

This does not mean that peoples who lived according to the primal tradition left nature undisturbed. They had a visible effect on their environments. The kangaroos in aboriginal Australia, for example, were doubtless swifter and more wary animals because skilled hunters killed the slower and less alert ones. Aboriginal people introduced the dingo, which became feral and added to the pressure of predation on the native species, while competing with native predators such as the thylacine (or marsupial "tiger") and the Tasmanian devil.[36] Gatherers removed some species, but also may have scattered the seeds of desirable plants to encourage their growth. More frequent fires altered the habitats; as the forest historian John Dargavel observed, fire "at times could escape even the most careful controllers."[37] Some of the largest animals became extinct within the context both of hunting and a changing climate. These extinctions are part of the experience that led to the establishment of the primal hunters' code. Ancient people were wise enough to understand the role of fire in improving forage for animals. They knew the places and the times of year to set fires so that they would be helpful, not destructive. Why would they want to do something that would harm their own hunting territories, upon which they depended so completely? Aboriginal Australians systematically burned the countryside in a mosaic pattern based on knowledge of seasonal weather patterns and the varying fire-tolerance of plants including the way in which various species regrew afterwards and served as food for herbivores.[38] Fire is a natural phenomenon, and in areas that are not burned periodically, wildfire can become a holocaust. If the primal hunters and gatherers made a major impact on natural systems, however, they usually intended and managed to maintain a balance with them. Of course, they had little choice, because if they did upset the balance within their own territory, the ecology might restore a balance by failing to provide enough food to support human numbers. The group could try to move, but the available land would probably already be occupied by similar groups and conflict would occur. A group might drive out its rivals, but would find its original task intact: how to survive within the ecological requirements of the environment it inhabited.

All of the attitudes and practices just described are the results of long cultural experience in the environments where primal groups are located. They came from trial and error over millennia. It seems that the first hunters to populate a new land, such as the pioneer Australian aborigines who arrived on the continent perhaps 60,000 years ago, had not yet evolved such a careful lifestyle. Within a relatively short time after their advent, they had decimated some of the native species and extirpated others, especially the large marsupials.[39] Destroying one's own food supply is, however, a self-defeating practice. Eric Rolls notes, "By the time smaller animals took over, … Aborigines had learnt to husband game. Apart from killing only what was needed, they devised systems of taboo, forbidding certain foods to certain people as a method of control."[40] Cultural traditions that tended to keep the resource sustainable would have increased the chances of survival of the groups that adopted them, and therefore they would have persisted.

Hopi, Arizona: agriculture in the spirit of the land

My first visit to a Hopi town was in 1960, to attend the Niman kachina ceremony at the sandstone village of Mishongnovi in northern Arizona. I had slept overnight on the desert in order to be there at dawn. As the sun rose, a long line of masked kachina dancers in white kilts came up over the mesa edge into view, and filed into the plaza bearing gifts, prominent among which were stalks of early corn. As they danced, they chanted to the rhythm of deer scapulas scraped along notched sticks with hollow gourds as sounding boxes. The priests anointed the dancers with corn meal. Most impressive were the masks of the kachinas, crowned with transverse crests or tablas that were decorated with fertility symbols – the clouds that bring the rain necessary to vegetative growth in this high, arid region – and topped by plant shoots standing upright.

The iconography of the kachinas made two major purposes of the ceremony quite clear: to assure the productiveness of the fields and to call for rain. The translation of a kachina song expresses this:

> The green prayer-stick brings the water,
> For the earth and its vegetation are combined in it.
> From the four corners come the clouds –
> Come together, gather over us.

> The green prayer-sticks bring the water,
> From the four directions in which we planted them.

The Spirit of the Rain passes over the prayer-sticks,
And their feathers are stirred.

We have found the water,
It has entered to the roots,
All things are beautiful,
All things are glad.[41]

The Hopi, before cultural changes that came from contact with Europeans and European-Americans in the nineteenth and twentieth centuries, may serve as an example of a people who lived by subsistence agriculture, an invention of the Neolithic Age.

The invention of agriculture marks a major change in human lifestyle, and although it took thousands of years to occur, it is even so one of the most revolutionary happenings in human history. As part of this transformation, plants and animals were, in a word, domesticated. The ancient Greek philosopher Theophrastus commented on this: "It is mankind, alone among all living things, to which the term 'domesticated' is perhaps strictly appropriate."[42] After all, it was humans who built houses (*domus*, the Latin root of "domestication," means "house"), and in this process they became more sedentary, locating in more permanent places. The newly cultivated plants were planted and harvested by humans, and their forms changed due to seed selection until some, like maize, became entirely dependent on human agriculture for their survival. Animals altered their patterns of movement under domestication, and also their shapes; the varied breeds of dogs and pigeons, for example, are the result of human preference. Stephen Budiansky has suggested that domestication of

Figure 2.3 The Hopi town of Moenkopi in Arizona. In the surrounding fields, traditional agriculture is practiced including the staples maize, beans, and squash, along with plants introduced in the Spanish period such as peaches and chilis. Photograph taken in 1961.

animals represents as much an adaptation of the domesticated species both in behavior and in evolution as it does human choice.[43] Ways of life changed radically for some humans as well. When farmers planted crops, they had to live near them during the growing season to guard them against birds, animals, and human opportunists. Their technology changed; for example, they began to make and decorate pottery, an art that is difficult for people who move often because its products are heavy and breakable.

Domestication of animals probably began first, and the first animal to be domesticated was the dog, long before the agricultural revolution. It often happened that hunters captured wild puppies and kept them to raise, and the pack instincts of the young animals became imprinted on the human group. There were also hunters who followed herds of animals and gradually began to control them. In the Near East, this happened with goats and sheep. Subsequently humans tamed cattle and pigs, and still later, donkeys and horses.[44] Similarly, northern peoples controlled reindeer and Andean folk domesticated llamas and alpacas. The herders originated a pastoral life style, living in shelters such as tents or yurts that they could move as the needs of the animals for grazing or browsing dictated. They moved regularly to the highlands in summer and the lowlands in winter; to call them "nomads," implying aimless wandering, is misleading. Sheep's wool and goats' hair provided ample fiber for weaving, another technological achievement of this age. With animal domestication, the human ability to change the natural environment increased. Herders became a force that could destroy vegetation, setting fires in order to open forested areas for their animals and overgrazing some hillsides. Indeed, when numbers of sheep increased the danger of overgrazing appeared, since they eat grasses and herbs, roots and all, and their sharp hooves tear up the sod. Goats eagerly consume brush and tree seedlings, preventing forest regeneration, and can climb to the tops of some kinds of trees to eat the foliage. Cattle munch all the palatable green things they can reach, including leaves on the lower branches of trees. With the denuding of the soil came erosion. On the credit side, grazing animals fertilized the soil with nutrient-rich manure, and the movement of herds to different pastures at certain times of the year made the damage less intensive.

Farming began independently in several parts of the world. Nikolai Vavilov identified seven centers of origin of domesticated plants.[45] There is archaeological evidence of very early experimentation with planting and harvesting from Egypt around 12,500 BC and from southeast Asia around 10,000 BC, before the appearance of numerous farming villages in many parts of southwest Asia (the Fertile Crescent) between 7000 and 5000 BC, where the major crops were wheat, barley, and legumes.[46] Another invention of agriculture occurred in the New World between 5000 and 3000 BC, based on maize, beans, squashes, and (in South America about 2000 BC) potatoes. At about the same time, or perhaps earlier, farmers domesticated rice in tropical south Asia. In addition to those centers, Vavilov listed east Asia, the wider Mediterranean (he included Egypt as part of this center), and Abyssinia (Ethiopia and the southern Arabian peninsula).

Early farming used the simple hoe and digging stick, which disturbed the soil but not to an excessive degree. In a more destructive practice, some planters learned to clear land for agriculture with fire, and observed that ashes seemed to encourage the crops. In some forested zones, shifting cultivators used a pattern of clearing, planting for a few years until harvests declined, and then moving to another tract, allowing the first one to regenerate. This worked as long as overall populations were small and plenty of unoccupied land was available. In contrast, subsistence farmers who lived in one area for generations usually cared for the soil, guarding and restoring its fertility. One such method is terracing on hilly ground, which reduces erosion. A second is fallowing, or letting the land rest for one or more years between crops. A third is the use of manure and other fertilizers, including the planting of crops that enrich the soil, such as beans. Native American Indians noticed that corn and beans grow well together, the cornstalks holding up the bean vines and the beans helping the corn be more vigorous. The Hopi juxtaposed the two plants in art and talked of their spiritual unity. All told, Neolithic farmers learned by experience. Some made errors that resulted in starvation. Others managed to remain in balance with their slowly changing environment and, like the Hopi, endured.

Agriculture enabled an increase and concentration in human populations; farming villages were generally larger than hunting settlements. But the conditions of life for human individuals did not ordinarily improve. Systematic studies of skeletal materials from burials show that among Neolithic farmers, both men and women were not as tall as the Palaeolithic hunters, had less healthy teeth and bones, and lived shorter lives.[47] Being more crowded, they were more subject to communicable diseases. Without necessarily choosing to do so, the farmers had sacrificed a degree of their health, physique, and life expectancy for numbers and greater security of the group. Still, they made relatively few negative impacts on the environment. It was not agriculture in itself that destroyed the land, but some of the more intensive practices that were still to come, such as plowing and irrigation, combined with population pressures that made continuous agricultural use of the land necessary. Some scholars of human–environmental relations have suggested that "the agricultural revolution may prove to be the greatest mistake that ever occurred in the biosphere – a mistake not just for *Homo sapiens*, but for the integrity of all ecosystems."[48] In other words, they consider it to have been the environmental "original sin." That questionable metaphor might be more appropriate if applied to the Urban Revolution, the next major change in human lifestyle, which is discussed in the following chapter.

The subsistence farmers of the Neolithic Age extended the respect felt by the hunter–gatherers for wild animals and plants to the domestic species that increasingly supported their lives. They continued to hunt as well, but now to supplement the food they raised. They honored grain plants such as "Mother Corn," a goddess like the Greek Demeter later on, who was regarded not just as the "spirit" of grain, but also as identical with the plants themselves and their seed. Planting and harvest became the great festivals of the year. Domestic animals, especially such a powerful creature as the bull, attracted esteem as possessors of power and fertility. To kill them for food was a great act of religious sacrifice. As long as their population density remained low, it was still possible for people living close to natural cycles, closely dependent on the annual crops and the increase of the herds, to maintain a balance with Earth.

For centuries before the twentieth, the Hopi had been, and to some degree still are, a people who provide their own food by farming with digging sticks and hoes in small plots located on slopes, sand dunes, and alluvial fans near their villages.[49] Large and accessible springs were used to irrigate some fields.[50] It was an "intensive agricultural system that required substantial labor to construct and maintain."[51] Farming was done by Hopi men, but land tenure followed a clan system that descended in the female line, and land was considered to belong to women. Clan lands were assigned by clan mothers.[52] Their agriculture was based on maize (corn), beans, and squashes. They distinguished at least six different kinds of maize, of almost as many different colors. Maize was brought from its earlier area of domestication in Mexico[53] to the area where the Hopi now live, the mesas of northern Arizona, sometime before AD 100.

The Hopi country is at the outer edge of the area where maize can be grown; the growing season of 130 days is barely long enough, and rainfall – averaging 1330 mm (3 in) annually

– and other sources of water are highly variable, scarcely adequate in many years.[54] The other traditional crops, including beans (kidney, tepary, and lima), melons, and squash, are also difficult to grow in the desert climate. In addition to food crops, the Hopi grew a species of cotton native to the new world which bears the scientific name *Gossypium hopi*.

The Hopi adapted such cultivated species as maize and cotton to local conditions by centuries of selection. Maize seeds must be planted a foot or so deep under a sand layer in order to have moisture for growth during the dry spring weeks. Hopi corn sends up a long shoot to the surface before leafing out, and sends down a long single root to find moist soil and anchor the plant.[55] The aboveground part of the plants is relatively short and bushy, a necessity in the windy conditions to which they are exposed because they are planted far enough apart so as not to compete too much for water. Hopi cotton will grow in aridity that would kill other varieties.

Hopi agriculture was guided by a thorough knowledge of the high desert ecosystem. Fields were small and had to be located carefully, taking into account the water supply, danger of frost, and soil. Often farmers identified the best places to establish fields by the wild vegetation that grew there. Rabbitbrush, which often grows on the alluvial fans of tributary watercourses, was a good indicator. Indeed, the Hopi can name and identify uses for two or three hundred species of native plants. They collect seed from a number of wild plants, including mint, bee plant, wild potatoes, and devil's claw, that are valuable as foods or are used in basket making, and plant them in their gardens.

Hopi cultural attitudes held that humans are part of a community of living things, and must strive to cooperate with the other members of that community in order to thrive. "The whole universe is enhanced with the same breath, rocks, trees, grass, earth, all animals, and human beings,"[56] said Intiwa. This idea was especially strong for the major plants in their agriculture; a man from Oraibi asked, "Do we not live on corn, just as the child draws life from the mother?"[57] Every child received a special ear of corn that symbolized its "corn mother." The Hopi feeling of comradeship for animals, and of respect and awe of their power, helps to explain their use of animals in ceremonies, such as the live snakes that are carried in the Snake Dance and released to take the prayers of the people for rain to the powers of nature. Like most of the early agricultural people, the Hopi also maintained hunting as a subsistence activity, and they maintained attitudes and practices similar to those of the hunters and gatherers. For example, when Hopi hunters caught a herd of mountain sheep, they always released two, a male and female, "to make more."[58] Similarly, when gathering, they had rituals to honor the other beings in the environmental community. When they quarried grinding stones (metates) or grill slabs for frying the thin bread called *piki*, they left offerings of cornmeal and feathers and said prayers to the rock that had given of itself. When they needed cottonwood roots for carving kachina dolls, they preferred to gather them as driftwood along stream banks rather than to harm living trees.

The highly developed Hopi ceremonial system embodied the structure and cycles of the local ecosystem. It might be expected that such a system of practice and belief would accord with a traditional agriculture that was sustainable, and the evidence of its long history supports that expectation. Hopi farmers suited their methods to living within the high desert ecosystem, not trying to conquer or transform it. The complex Hopi structure of ritual and agricultural practice was derived from long experience, and displayed certain general principles. These included preserving varieties of seeds and genetic lines of plants adapted to local conditions through many generations of selection, knowledge of and respect for native species, close observation of the growing season and agricultural calendar, locating fields so as to take advantage of moisture reserves and flooding, and careful use and conservation of water. Most importantly, their sustainable agriculture was based on comprehensive knowledge of the ecosystem of the immediate environment and how to use it for sustenance without destroying it. Although they were visited by Spanish explorers accompanying Coronado's expedition in 1540, and received some new plants and animals, such as peaches and donkeys, from the Spanish settlers in New Mexico after 1600, they stoutly maintained their independence and their ways of life.[59] European cultural influences touched them relatively lightly until recently.[60]

Conclusion

This chapter, short as it is, covers a period of time longer than the next seven chapters combined. It describes some of the last places on Earth where nature and human cultures survive in ways reminiscent of earlier times. These places, and the humans that continue to inhabit them, demonstrate that our heritage includes not only our immediate predecessors, but the hominids and other animals that lived within the ancient environments, which for that very reason deserve study and preservation. Contemplation of the many ways in which interaction with them made us what we are can show us again that we are members of the great community of life, and are not exempt from the principles that govern it. The boundaries of human history, like those of many species, are permeable. At the beginning of their story, humans needed the animals, plants, and natural elements that sustained life and challenged their creativity. From the times of gathering and scavenging, through hunting and dawning agriculture, humans adapted to the environment even as they changed it, and maintained a dynamic balance with it.

Notes

1   The idea that a natural area could be preserved as a standard, or *etalon,* where changes in the ecosystem undisturbed by human intervention could be observed, was advanced as early as 1908 by Russian ecologist Grigorii Aleksandrovitch Kozhevnikov. See Douglas R. Weiner, "The Historical Origins of Soviet Environmentalism," *Environmental Review* 6, Fall 1982, 45.
2   Plato *Timaeus* 30D.
3   James E. Lovelock and Sidney Epton, "The Quest for Gaia," *New Scientist* 65, 1975, 304.
4   A.R.E. Sinclair and Peter Arcese, "Serengeti in the Context of Worldwide Conservation Efforts," in A.R.E.Sinclair and Peter Arcese, eds, *Serengeti II: Dynamics, Management, and Conservation of an Ecosystem,* Chicago, University of Chicago Press, 1995, 31–46.
5   John Reader and Harvey Croze, *Pyramids of Life: An Investigation of Nature's Fearful Symmetry,* London, Collins, 1977, 19.
6   S.J. McNaughton, "Grassland–Herbivore Dynamics," in A.R.E. Sinclair and M. Norton-Griffiths, eds, *Serengeti: Dynamics of an Ecosystem,* Chicago, University of Chicago Press, 1979, 46–81, especially 53–7.
7   Brian Jackman, *The Marsh Lions: The Story of an African Pride,* Boston, MA, David R. Godine, 1983.
8   Cynthia Moss, *Elephant Memories: Thirteen Years in the Life of an Elephant Family,* New York, W. Morrow, 1988.
9   A.R.E. Sinclair, "The Serengeti Environment," in A.R.E. Sinclair and M. Norton-Griffiths, eds, *Serengeti: Dynamics of an Ecosystem,* Chicago, University of Chicago Press, 1979, 31–45, 40.
10   Richard E. Leakey and Roger Lewin, *Origins,* New York, E.P. Dutton, 1977, 117; Donald C. Johansen and James Shreeve, *Lucy's Child: The Search for Our Beginnings,* New York, Morrow, 1989, 209.
11   D.A. Burney, "Paleoecology of Humans and their Ancestors," in T.R. McClanahan and T.P. Young, eds, *East African Ecosystems and Their Conservation,* New York, Oxford University Press, 1996, 19–36.
12   D.A. Livingstone, "Historical Ecology," in T.R. McClanahan and T.P. Young, eds, *East African Ecosystems and their Conservation,* New York, Oxford University Press, 1996, 3–18.

13   Glynn Ll. Isaac and Elizabeth R. McCown, eds, *Human Origins: Louis Leakey and the East African Evidence*, Menlo Park, CA, Staples Press, 1976, xiii.

14   Democritus fr. 154. Philip Wheelwright, *The Presocratics*, New York, Odyssey Press, 1966, 184.

15   Steven Lonsdale, *Animals and the Origin of Dance*, London, Thames & Hudson, 1981.

16   Frank Benjamin Golley, *A History of the Ecosystem Concept in Ecology: More Than the Sum of the Parts*, New Haven, CT, Yale University Press, 1993.

17   Edward O. Wilson, *Biophilia*, Cambridge, MA, Harvard University Press, 1984, 101.

18   Aristotle *De Anima*, 414–15.

19   Kenneth Page Oakley, *Man the Tool-Maker*, 6th edn, London, British Museum (Natural History), 1972, 1–3.

20   Jane Goodall, *Through a Window: My Thirty Years with the Chimpanzees of Gombe*, Boston, MA, Houghton Mifflin, 1990, 5, 18–23, 58–9.

21   Plutarch *Moralia*, 985D–992E, in Plutarch's *Moralia*, Vol. 12, tr. by Harold Cherniss and William C. Helmbold, Cambridge, MA, Harvard University Press, Loeb Classical Library, 1957, 492–533.

22   Francine Patterson and Ronald H. Cohn, "Conversations with a Gorilla," *National Geographic* 154, October 1978, 465; see also Francine Patterson and Wendy Gordon, "The Case for the Personhood of Gorillas," in Paolo Cavalieri and Peter Singer, eds, *The Great Ape Project: Equality beyond Humanity*, New York, St. Martin's Press, 1993, 58–77, at 76.

23   Jacques-Yves Cousteau, "At Home in the Sea (Underwater Lodge)," *National Geographic* 125, April 1964, 465–507.

24   Ernest Beaglehole, *Hopi Hunting and Hunting Ritual*, London, Oxford University Press, 1936, 4–7.

25   Josephine Flood, *Archaeology of the Dreamtime*, Honolulu, University of Hawaii Press, 1983, 86–7.

26   Ibid., 134, 223.

27   The word "primal" is used here to refer to the modes of thought and expression characteristic of people with hunter–gatherer–fisher lifestyles, or subsistence farmers and herders unaffected by urban ways, in order to avoid the pejorative connotations of words such as "primitive" or "precivilized."

28   Keith Cole, *The Aborigines of Arnhem Land*, Adelaide, Rigby, 1979, 162.

29   See text

30   Olga Gostin and Alwin Chong, "Living Wisdom: Aborigines and the Environment," in Colin Bourke, Eleanor Bourke, and Bill Edwards, eds, *Aboriginal Australia: An Introductory Reader in Aboriginal Studies*, St. Lucia, University of Queensland Press, 1994, 123–39.

31   Flood, *Archaeology of the Dreamtime*, 241.

32   Howard Morphy, "Colonialism, History and the Construction of Place: The Politics of Landscape in Northern Australia," in Barbara Bender, ed., *Landscape Politics and Perspectives*, Oxford, Berg Publishers, 1993, 205–44, at 234.

33   Bill Neidjie, *Speaking for the Earth: Nature's Law and the Aboriginal Way*, Washington, Center for Respect of Life and Environment, 1991, 53. Reprinted from Big Bill Neidjie, Stephen Davis, and Allan Fox, *Kakadu Man*, Brisbane, Prestige Litho, 1985.

34   James L. Kohen, *Aboriginal Environmental Impacts*, Sydney, University of New South Wales Press, 1995; Åke Hultkranz, "The Owner of the Animals in the Religion of the North American Indians," in Christopher Vecsey, ed., *Belief and Worship in Native North America*, Syracuse, NY, Syracuse University Press, 1981, 135–46; J. Donald Hughes, *American Indian Ecology*, El Paso, Texas Western Press, 1983, 32–4.

35   Gostin and Chong, "Living Wisdom," 124.

36   Kohen, *Aboriginal Environmental Impacts*, 86–9.

37   John Dargavel, *Fashioning Australia's Forests*, Melbourne, Oxford University Press, 1995, 16.

38   Stephen J. Pyne, *Burning Bush: A Fire History of Australia*, New York, Henry Holt, 1991, 124.

39   Tim Flannery, *The Future Eaters*, New York, George Braziller, 1995, 180–6.

40   Eric Rolls, "The Nature of Australia," in Tom Griffiths and Libby Robin, eds, *Ecology of Empire: Environmental History of Settler Societies*, Seattle, University of Washington Press, 1997, 35–45, at 37.

41   Alexander MacGregor Stephen, *Hopi Indians of Arizona*, Southwest Museum Leaflet No. 14, Los Angeles, Southwest Museum, 18.

42   Theophrastus, *De Historia Plantarum*, 1.3.6.

43   Stephen Budiansky, *The Covenant of the Wild: Why Animals Chose Domestication*, New York, William Morrow, 1992.

44   Purushottam Singh, *Neolithic Cultures of Western Asia*, London, Seminar Press, 1974, 208–12; Erich Isaac, *Geography of Domestication*, Englewood Cliffs, NJ, Prentice-Hall, 1970.

45 N.I. Vavilov, *Origin and Geography of Cultivated Plants*, New York, Cambridge University Press, 1992, 430; Bruce D. Smith, *The Emergence of Agriculture*, New York, Scientific American Library, 1994, 5–8.

46 Michael A. Hoffman, *Egypt Before the Pharaohs: The Prehistoric Foundations of Egyptian Civilization*, New York, Knopf, 1979, 85–90.

47 Mark Nathan Cohen, *Health and the Rise of Civilization*, New Haven, CT, Yale University Press, 1989, 116–22.

48 Anne H. Ehrlich and Paul R. Ehrlich, *Earth*, New York, Franklin Watts, 1987, 59.

49 John T. Hack, "The Changing Physical Environment of the Hopi Indians of Arizona," *Papers of the Peabody Museum of American Archaeology and Ethnology, Harvard University* 35, 1, 1942, 1–85, at 19–38.

50 C. Daryll Forde, "Hopi Agriculture and Land Ownership," *Journal of the Royal Anthropological Institute of Great Britain and Ireland* 41, 1931, 365.

51 Scott Rushforth and Steadman Upham, *A Hopi Social History*, Austin, University of Texas Press, 1992, 71.

52 Gordon B. Page, "Hopi Land Patterns," in *Hopi Indian Agriculture and Food*, Flagstaff, AZ, Museum of Northern Arizona, Reprint Series 5, 1954, 8–15.

53 Muriel Porter Weaver, *The Aztecs, Maya, and Their Predecessors: Archaeology of Mesoamerica*, 3rd edn, San Diego, CA, Harcourt Brace Jovanovich, 1993, 13–15.

54 Ernest Beaglehole, *Notes on Hopi Economic Life*, Yale University Publications in Anthropology 15, New Haven, CT, Yale University Press, 1937, 33.

55 Maitland Bradfield, *The Changing Pattern of Hopi Agriculture*, Royal Anthropological Institute of Great Britain and Ireland, Occasional Paper 30, 1971, 5.

56 Elsie Clews Parsons, *Pueblo Indian Religion*, Chicago, University of Chicago Press, 1939, 198.

57 Ibid., 31.

58 Ernest Beaglehole, *Hopi Hunting and Hunting Ritual*, Yale University Publications in Anthropology 4, New Haven, CT, Yale University Press, 1936, 11.

59 Richard O. Clemmer, *Continuities of Hopi Culture Change*, Ramona, CA, Acoma Books, 1978; Alfred F. Whiting, "Hopi Indian Agriculture: I, Background," *Museum Notes, Museum of Northern Arizona* 8, 10, April 1936, 51–3.

60 Richard O. Clemmer, *Roads in the Sky: The Hopi Indians in a Century of Change*, Boulder, CO, Westview Press, 1995, 88–90.

3   The great divorce of culture and nature

Cities are not separate from the natural world on which they depend. In a north Indian city, men and women build a new apartment structure largely by hand. They carry tiles in wooden hods on their heads, tiles formed of earthen clay that have been baked by burning wood and charcoal, which is partially oxidized wood, from the shrinking forests on hillsides far to the north. The scaffolding is of bamboo that grew in the same forests, tied with ropes of hemp from fields that can be seen in the hazy distance from the top of the building. Like all cities, this one uses resources transported from the land near at hand or far away.

In Shanghai, I visited a marketplace where an astonishing variety of stalls lined the lane, stocked with every staple for the kitchen: vegetables and fruits from gardens just outside the city, live ducks from a nearby lake, lotus, water chestnuts, and snails that country people brought to sell. What is often described as a series of economic transactions also can be seen as human beings manipulating and using other species of animals and plants.

An easy walk from the center of Ávila in Spain took me along the crowded streets of a thriving provincial capital, through a gate in the massive walls, through wheat fields, vineyards, and olive orchards, to a viewpoint where, looking back over the city, I could glimpse the pineclad heights of the Sierra. In this short distance, I saw examples of many different ways in which the land is used to meet the preferences and needs of an urban population.

Each of these scenes has something important in common with the early cities that arose in the river valleys of Mesopotamia, the Nile, and the Indus, or on the loess plains of north China. The state with its religious and political institutions, the specialization of human occupations, the stratification of society into classes, and the development of arts such as monumental architecture, writing, and the measurement of space and time, appeared first and developed most fully in these large, densely populated human centers. The city is a structured human relationship with the natural environment. Although it is an artificial creation of human culture, it can also be seen as an ecosystem related to other ecosystems. Every activity of human beings in it requires some resource from the surrounding environment. The city is not a truncated phenomenon, but has a natural context consisting of the many cycles of organic and inorganic substances that constantly affect it. Cities are part of the ecosystems within which they exist, although they make extensive changes within them and reorganize nature for their own benefit. Too often cities are studied only as a series of human social relationships and economic arrangements, and their intimate, constant, and necessary connections with the natural processes of the Earth are forgotten.[1]

A more productive agriculture was the necessary condition for the genesis of cities, since they were larger, more densely populated, and organized in a more complex way than the villages that preceded them. They required an agrarian economic base that could produce a food surplus. This was done in part by expanding cultivated land at the expense of forests, wetlands, and arid country. But in order to feed large numbers of men and women engaged in activities that did not produce food, such as rulers, priests, military commanders, and scribes, it was necessary to have a system in which the labor of a farm family could provide food for others besides itself. This was often achieved through large-scale water management aimed at controlling floodwaters, or providing waters to fields through canals.

The staple food of city dwellers consisted of grains such as wheat, barley, and millet. Rice was cultivated in China as early as the Shang dynasty (*c.*1750–1100 BC),[2] and was possibly present in the Indus Valley at approximately the same time.[3] The plow, a technological innovation, helped to create an agricultural surplus, and thus to make cities possible. Seed selection, fertilization techniques, and crop rotation also made contributions.

The effects of flood control and irrigation on the environment were among the impacts of urbanization. Rivers carry sand, silt, and suspended organic matter, all of which settle out when the water slows. Where a river was contained between levees to prevent flooding, as in Mesopotamia and north China, it caused the riverbed to rise above the surrounding land and made floods worse when dikes finally broke. Siltation also occurs in canals and, unless people undertake the heavy burden of removing silt to adjacent spoil banks, shortens the useful life of these works. Eventually it may overwhelm them in spite of these efforts. Another effect was salinization, the gradual increase of salts in waterlogged soil as a result of evaporation. Flowing water dissolves salt, and more of it after deforestation exposes salt-bearing rocks to rainfall. When the water is spread on the fields and evaporates, the salt accumulates. High salt concentrations obstruct germination and impede the absorption of water and nutrients by plants, or prevent growth. Salinization can be serious wherever irrigation is practiced in dry climates on poorly drained soils, which was the situation in much of Mesopotamia and the Indus Valley.

The rise of cities created demand for materials and fuels. Architecture became extensive, complex, and massive. The need for building materials was immense, considering residences and places of business, temples, palaces, and tombs, along with walls and citadels. Most materials for construction came from the Earth, consisting of clay dried in the sun or baked into brick, and stone. Stone quarries scarred many a hillside. Fuel for brick kilns required huge quantities of wood and charcoal, which came from the forests. The cities of the Indus Valley, for instance, were constructed of baked brick. Timber was also of major importance in building, being used to support ceilings and roofs, and for scaffolding during construction, adding to pressures on woodlands.

An improved metallurgy produced tools, weapons, and ornaments of copper and then of bronze. Cities were often centers of metallurgy, or spawned such centers in their vicinity, and the demands for fuel threatened forests. The ore had to be dug out of the ground, leaving pits and tunnels; and had to be raised to a high temperature for smelting (2012°F or 1100°C for copper), which required the burning of wood or charcoal for fuel. It required roughly 15–25 tons[4] of charcoal to produce one ton of copper. The effect of cutting vegetation for this one use alone at a major center of production would have involved the divestiture of hundreds of thousands of hectares of trees.[5] Copper compounds are poisonous, so workers were at risk, and pollution from the wastes of manufacture was dangerous to humans and other organisms.

The people were divided into a number of new occupations, some new and unique to the city. As a result, many citizens of urban centers belonged to groups whose jobs meant they were insulated from the land and no longer worked intimately with living animals and plants. They spent their time indoors or in marketplaces, manufacturing or selling products, or working in government, the law, and religion. Their food was obtained through trade, not directly from the sources. Those who were most urbanized included the leaders and political decision-makers.

Warriors attempted to defend the agricultural lands and other resource interests of the city, and strove to seize those of neighboring cities. Before the adoption of horses and chariots, success depended on amassing numbers of foot soldiers on the battlefield. Commanders demanded the service of almost all able-bodied men and made warrior status the requirement of citizenship, a fact that prevented women from formal participation in political life. Acts of war destructive of the environment were often done deliberately to deprive rival cities of the means of support and resistance. Armies set fires, trampled crops, cut trees, and disrupted water supplies.

Merchants formed an important occupational group in the early cities,[6] and the marketplace evolved, usually an open square near the center, surrounded by a sheltered walkway where stalls or shops could be erected. There the produce of local farms, handicrafts of artisans, and items of clothing were offered in trade. In addition, services like haircuts and dentistry were available, as well as prepared food and alcoholic drinks. The marketplace had notable environmental effects from the start, facilitating the exchange of resources and increasing demand for them.

Growth in numbers and density of population produced problems of pollution, waste disposal, and the spread of diseases, affecting the health, stature, and longevity of the inhabitants. Drinking water was drawn from wells, rivers, and canals subject to contamination. Mesopotamian documents mention the danger of death from drinking bad water.[7] To pollution from sewage and offal were added wastes from industrial activities such as metallurgy, leather tanning, and pottery kilns. These accumulated until rain washed them into rivers and ground water. A few early cities arranged for removal, or built sewers and latrines, such as are found in the ruins of Knossos on Crete. Wastes as well as the concentration of human bodies and stored foodstuffs attracted opportunistic organisms.

Human health suffered by every measure. Neolithic villagers were less healthy than hunters and herders, but city dwellers showed further decline; studies of their skeletal remains show that they were shorter in stature, lived briefer lives, suffered more from bad teeth and bones, and were subject to communicable diseases.[8] To these dangers must be added warfare, slavery, and human sacrifice. An unconscious tradeoff had been made which forfeited quality of life for quantity of human numbers and security for the community. For individuals, urban life was rarely an improvement over earlier societies.

Just as important as the transformation of the environment where the city stood was the way in which urban demands affected the surrounding area at greater distances as the city grew. Cities could exploit resources at a distance, directly and indirectly, becoming dependent on trade routes vulnerable to hostile disruptions or natural calamities.

The most damaging effect of cities on the environment was deforestation as a result of the demands for building material and fuels. It began close to the centers and spread outward along lines of transportation such as rivers, coastlines, and roads. Forest products are heavy and bulky, and were exploited as much as possible over the shortest and easiest routes. However, many cities had to reach out further. The cities of the Indus Valley, for example, brought deodar cedar wood from the Himalaya. The cedar forests of the Lebanon mountain range furnished timber to the Sumerians and the Egyptians, who had to transport it long distances. Later, King Hiram of Tyre gave cedar and cypress timber to King Solomon to build the temple in Jerusalem; it was shipped in the form of sea-going rafts.[9]

Warfare meant destruction in ancient times, and the natural environment was not safe from it. Crops were destroyed when armies marched over them, or fought battles on the fields, although blood and fallen bodies might briefly fertilize the soil. Longer-lasting damage was done by cutting down groves of fruit trees, an action forbidden by the Hebrew Bible.[10] Armies also set forests on fire, diverted rivers, and deliberately polluted sources of water.

Many of the regions where cities first appeared are today arid and sparsely vegetated. It is a thought-provoking sight to see the remains of a great city like Ur, with its massive ziggurat (a step-pyramid-like structure topped by a temple), or the citadel of Mohenjo-Daro, surrounded by desiccated and largely deserted landscapes.[11] Desertification is intensified by salinization, and both processes went on in the neighborhood of cities. Another factor demonstrated by hydrologic studies is the altering of watercourses. The channels of the Yellow River, the Tigris and Euphrates, and the Indus and its tributaries, shifted over significant distances. Although many of these often disastrous displacements occurred spontaneously, others were occasioned by the construction of canals and other water control structures, and by deforestation and subsequent floods and desiccation. Thus early cities had a hand in creating the deserts that later enveloped them. Progress, it appears, is not inevitable.

The city cannot be understood properly unless it is seen as an ecosystem, that is, as a series of ecological relationships. It does not exist in isolation, but interacts with other ecosystems and functions as part of a larger ecosystem. A study of the city, therefore, must see human social factors as operating within a complex series of ecological processes that impact and affect them. The city-dwellers, like their Neolithic ancestors, depended on a natural system for survival. But this fact was less immediate for them. Feedback from natural systems was less instantaneous. Therefore it could seem to them that culture and nature were two separate realms, and that culture, representing order and security, should be dominant over chaotic nature.

Such a viewpoint was mistaken. The city of the Afro-Asiatic Bronze Age (*c.*3000–1000 BC) was no less a part of the larger ecosystem than the Neolithic village or Palaeolithic hunters' camp. It was more populous, though, and more complex. Its decision-making had a wider impact on the environment, and needed to be informed by better knowledge of the workings of surrounding ecosystems. This knowledge was not always available. Mistakes made in urban economic arrangements were more far-reaching than before, and might mean that a city imposed demands on the environment at a level that was not sustainable. This actually happened many times. Cities shrank, or their sites were abandoned completely. But before this happened, or while it was happening, they depleted their environment, and they did so over an extensive landscape, sometimes including distant places from which they drew resources. Culture acted as though it were divorced from nature only at its own peril.

The Uruk Wall: Gilgamesh and urban origins

The rooms of museums do not always transport their visitors in spirit to far places and distant times, particularly if the visitors are tired and have walked through dozens of rooms just previously. So it was with few expectations that I stepped into the room in the Pergamon Museum in Berlin that contains a large number of objects from Uruk, a site in Iraq that was once a great Sumerian city. One side of the room was occupied by the patterned baked clay bricks of a city wall. I read the label. This was a piece of the wall of Uruk, the city where Gilgamesh was king! Suddenly, a thrill of excitement seized the back of my neck, and fragments of the most ancient epic poem that survives on Earth came into my mind.

> I will proclaim to the world the deeds of Gilgamesh ... In Uruk he built walls, a great rampart ... Look at it still today: the outer wall where the cornice runs, it shines with

the brilliance of copper; and the inner wall, it has no equal. Touch the threshold, it is ancient ... Climb upon the wall of Uruk; walk along it, I say; regard the foundation terrace and examine the masonry: is it not burnt brick and good?[12]

This passage is not a mere poetic boast; the circuit of walls around Uruk measured 9.6 km (6 mi), with 900 towers. Archaeology presently indicates that the earliest civilized societies arose as a series of city-states in Sumeria, located in the alluvial land along the lower courses of the Euphrates and Tigris rivers, near where they empty into the Persian Gulf. This area would be desert were it not for the rivers and the irrigation they make possible. The walls of the old cities of Sumeria, built as they were of clay brick, have more or less eroded into the rounded mounds of the cities they formerly protected. But the restored wall of Uruk reflects something of its ancient glory to the inner eye.

The wall is the symbol not of the city alone, but also of a new view of the world, which entailed a "Great Divorce," a sense of separation between culture and nature that came about with the origin of cities. Walls were meant to keep enemies out, but they stood also as tangible signs of a division between what was inside and what was outside: within was the ordered lifestyle of a city; and without was a comparatively chaotic world. In the day of Gilgamesh the wall-builder this distinction was recent, and it indeed marked a divorce between civilization and nature. The psychological separation was much more marked than anything experienced by hunters, herders, or village farmers. The distinction between the crowded centers of human civilization, the productive countryside, and the lands beyond where wild creatures lived, was clearly recognized.

The motif of human struggle against hostile nature is prominent in the mythologies of Mesopotamia, where the first cities arose. In the Old Babylonian epic of creation, *Enuma Elish*, which is patterned on much earlier Sumerian texts, the world is shown to be the result of a battle between Tiamat, the female monster of chaotic nature, and Marduk, the champion

Figure 3.1 Mostly abandoned, these terraces in the arid country near Petra, Jordan, illustrate the impression of environmental deterioration over large areas of the Near East. Photograph taken in 1976.

of the new order of the gods. Marduk captures Tiamat in a net, drives the wind into her mouth to distend her belly, shoots an arrow through her heart, and "split[s] her like a shell-fish into two parts,"[13] making the sky of the upper half and the sea of the lower half. He then proceeds to build Esharra, a city of the gods, in the sky. This is an instance of the prevalent idea among urban folk that the city, with its straight streets, monuments, and walls, is an earthly copy of a model of divine order, the heavenly city.

At the same time, the wild and its inhabitants became enemies and game. When Gilgamesh's kingship in his city of Uruk becomes oppressive, the gods fashion a wild rival for him, a hairy man named Enkidu, who lives in the wilderness with the animals, running with them and warning them away from hunter's traps. He is a man among wild creatures, learning from them and protecting them. Then Gilgamesh sends a woman to seduce Enkidu. Besides sex, she offers him bread and wine, foods transformed from cultivated plants by human art. After that, the animals fear and flee from him, and Enkidu has to enter the city. One of the labors of Gilgamesh and Enkidu is an expedition to a sacred cedar forest in the mountains, where they kill the wild guardian, Humbaba, cut down the trees, and take the logs back to his city to use in building a palace. Undomesticated animals were extirpated, especially if they threatened crops and herds; Gilgamesh is portrayed killing lions simply because he sees them "glorying in life."[14] In destroying forests, the inhabitants of cities were also destroying the habitats of many species of animals.

The theme of struggle between culture and nature seems to have been a masculine rather than a feminine attitude to many recent writers. That it was so should not be surprising, since battle is a warriors' metaphor, warriors are usually men, and warriors eventually controlled the political and economic structure of the early cities. In pre-urban societies, women and men seem to have had complementary roles, with neither completely dominant. The tasks of the sexes were not rigidly defined; women sometimes hunted and men often gathered, for example, while planting or weaving were done by either sex, or both.[15] In the cities of the Urban Revolution, roles were more strictly divided, and male warriors tended to fill those that were dominant. Men wrote most of the literature, too, although not exclusively so, and the warrior images of combat and conquest are prominent as a result. It is important not to make too hard and fast a rule of this; there were warrior goddesses and male earth gods in the myths of cities. But the thought that attitudes to nature might have been more sympathetic if women had continued to be as balancing a force in urban societies as they had been in earlier ones does not seem unreasonable.

Indeed, the art and literature of the early civilizations of Mesopotamia repeat an unmistakable note of pride in human triumph over nature, and often this note resounds to glorify human technological achievements. Flood control and irrigation are the basis of a well-ordered state. Kings are portrayed armed with bows and arrows, using nets, and riding in chariots, engaged in the ceremonial hunting and killing of wild animals, particularly powerful ones such as lions, ibexes, and wild bulls.

The creation of cities in Mesopotamia was an aspect of a changed relationship between human beings and the environment, based on a more intensive agriculture using two new inventions: the plow and systematic large-scale irrigation. Early cities were not large by modern standards; Uruk had a population of perhaps 25,000. But for a human aggregation of this size, it was necessary for agriculture to produce a surplus. This happened with the invention of the ox-pulled plow and irrigation. The fertile, sandy soil of Mesopotamia was easily turned by the ox-drawn plow. The rivers provided the needed water, but their flow was so undependable that control by major irrigation works was demanded. These works of irrigation conquered sections of the land and won rich sustenance from its basic fertility.

Figure 3.2 The wall of the ancient city of Uruk in Sumeria, more than 5,000 years old, is decorated with a colored pattern of painted terracotta cones. It symbolizes the separation of the urban center from the rural and wild areas outside the city. Photograph taken in the Pergamon Museum, Berlin, in 1991.

Thus a Mesopotamian king felt justified in listing the construction of a new canal, along with the defeat of his enemies in battle, as the major events of his reign. The systems of canals that brought water to the fields constituted the Sumerians' most extensive and labor-consuming achievement. This new agriculture enabled a much larger human population to live in expanded settlements, and many people no longer had to work on the land, so that specialized occupations flourished in the cities. It was necessary for society to create the institutions that would organize food production and distribution, the import of useful materials, and the defense of one city against the appropriation of its lands and goods by another.

The urban dwellers raised mighty works of baked and unbaked clay bricks: temples, shrines raised on lofty ziggurats, palaces, and thick city walls. But the lack in Sumeria of some important building materials, especially timber and stone, meant that cities like Uruk had to import them from far away. In the flat, alluvial land where agriculture flourished there were few substantial forests and little stone or metallic ore. Native materials such as reeds and clay could be used in ordinary domestic construction, but roofs of ever larger temples and palaces required long, straight timbers that the treeless Mesopotamian plain could not supply, and sculptured images and other decorations had to be made from stone and metal that the alluvial soil did not contain. Mountains to the north and west had abundant supplies of stone and timber. These products, as well as luxury goods, were obtained by the far-ranging merchants. Merchants were an important segment of Sumerian society, but it must not be imagined that they represented free enterprise. Their activities were managed by the rulers, and when they traveled to other cities, their status was that of quasi-ambassadors. These merchant-venturers traveled by land, river, and sea. To the east, they traded as far as the Indus Valley for timber, ivory, and precious stones.[16] In the west, they brought fine woods from Lebanon, copper from Cyprus, and were in touch with Egypt almost continuously by way of the Red Sea. Every Sumerian city had a marketplace where the items of trade as well as those of local manufacture were available. In Mesopotamia, women engaged in trade in many commodities, and were the proprietors of bars.[17]

In order to support the growing trade, cities needed to increase production of items to exchange, principally grain, ceramics, and textiles. This led to additional pressure on the land. Farmers shortened the fallow period, overplanted, plowed marginal lands, and intensified irrigation – practices which led to salinization.[18]

Copper and bronze metallurgy appeared around 3000 BC as the Sumerian cities flourished, which justifies the name "Bronze Age" for the period that followed. The early metallurgists undoubtedly adopted some of the methods used by the manufacturers of pottery to achieve the high temperatures that were necessary. Both processes required considerable volumes of fuel, mainly wood and charcoal, which increased the demands of the Sumerians on the vegetative cover of the region.[19]

This was unfortunate, because flooding was a continual danger for the Mesopotamian cities, and deforested mountain slopes higher up in the drainage of the two rivers allowed faster and more silt-laden runoff to swell the inundations. The Tigris and Euphrates sometimes rose high enough to break the levees, destroying villages and fields. Cities tried to raise themselves above the flooded plains by adding to the accumulated mounds upon which they were built. They placed the temple dwellings of the gods on platforms, and then even higher on ziggurats. The system of canals and dikes was in constant danger of disruption by flooding and erosion. The silt and mud carried by the waters settled out wherever they slowed, and constant dredging was required to keep the canals open. The "spoils," or excess material, piled up along their banks until the canals were 10 m (30 ft) or more above the surrounding fields. This hampered their ability to drain the land and was a danger in time of flood.

Salinization is a danger wherever irrigation is practiced in warm, dry climates, and was disastrously prevalent in lower Mesopotamia. When irrigation water raised the natural water table and evaporated, salts accumulated. Poor drainage, made worse by the silt that had been deposited, made it hard to correct the situation by leaching salt from the fields with fresh water. Ground water became more and more saline. Farmers increasingly turned from the sensitive wheat to barley, which is more salt-tolerant. Over large areas the ground became so saline that white salt crystals could be seen on the surface and cultivated plants were unable to grow. Those fields had to be abandoned, and it became more difficult to find new areas for irrigation and cultivation. A survey by Thorkild Jacobsen and Robert Adams found evidence of increased salinity and declining yields in southern Mesopotamia between 2400 and 1700 BC. Speaking of the area where the first cities arose, the investigators concluded, "That growing soil salinity played an important part in the breakup of Sumerian civilization seems beyond question."[20]

The once flourishing cities of ancient Sumeria – Uruk, Ur, and the others – are now abandoned mounds in a desert environment. Satellite photographs indicate that the fertile land of Mesopotamia today has shrunken significantly from the extent it covered in Sumerian times. This is not the result of climatic change alone, although both rainfall and temperature have varied from one period of time to another. They represent an ecological disaster caused by overuse and eventual exhaustion of the land. In Mesopotamia, of all regions studied by ancient historians, there is the clearest relationship between environmental devastation caused by humans and the decline of cities and their civilizations. But it is, unfortunately, not the only example.

The Nile Valley: ancient Egypt and sustainability

Ask a fairly well-read person about labor conditions during the construction of the pyramids, and you will probably be told that the workers were slaves toiling under the lash. In fact, they were agricultural laborers whose work was commandeered during the off-season, and they were provided with lodging and food – what amounted to wages in the days before coinage.[21] Inscriptions record the pharaoh's boasts at how well he treated the workers. The laborers' own graffiti show they were organized into teams that competed to fill their quotas. Conditions were not always to their liking, however; laborers on tombs in the Valley of the Kings went on strike for reasonable wages.

The technology Egyptian workers had at hand was rudimentary. With the pyramids, it was a matter of stone on stone, supplemented by wooden mauls and wedges. Later the Bronze Age came to Egypt, and stoneworkers could use metal tools, but since granite is harder than bronze saws, they needed powdered quartz as an abrasive. A relief from the Middle Kingdom shows 172 workers moving a huge stone statue, pulling it with ropes.[22] It is on a sled (no wheels), and a man stands on the runner pouring lubricating liquid on the ground. With methods such as these, the Egyptians constructed what the Romans, thinking of their own useful aqueducts, would call "the idle pyramids,"[23] and decorated tombs hidden in the desert, intended never to be seen again by human eye.

But the Egyptians also had a useful, sophisticated technology that kept their civilization operating well. That was the system of water management that used the natural flooding of the Nile River, with irrigation works and careful planning, to keep the agricultural base functioning. It was an appropriate technology for the ecological situation of a rainless land watered by an exotic river flowing from East Africa. No other ancient civilization lasted so long while maintaining a relatively stable economic pattern. Some historians talk about the constancy of Egyptian culture through so many centuries in a disparaging tone, attributing lack of change to absence of creative thought. But the long-lasting stability of Egyptian civilization may have come from the sustainability of Egypt's ecological relationships. Karl Butzer said that a history of flood-plain civilization in the Nile Valley offers a test case of human–land relationships, adding, "It has become difficult to ignore the possibility that major segments of ancient Egyptian history may be unintelligible without recourse to an ecological perspective."[24]

The Egyptians lacked science in the modern sense. But they expressed an understanding of the workings of nature in religious images, and they explained technology in terms of the sacred. In this perspective, irrigation was an activity originated by the gods. Sacred geometry, sacred astronomy, and sacred records were marshaled to assure what we would call sustainability. Geometry, elaborated through trial and error to reestablish boundaries between fields when markers had been swept away in the flood, was regarded as a hallowed occupation devised by the wise god Thoth and entrusted to trained priest-scribes. Temples were oriented to keep watch on the revolutions of the sun and stars, which would tell when to open canals. Papyri containing these arcane branches of knowledge were kept in temple libraries.

Indeed, early pieces of art show that irrigation was practiced by the pharaoh himself. The first-dynasty Scorpion-King mace head shows the king digging a canal, and "Canal-digger" was an important administrative title. Canal building was believed to be a major occupation of those in the blessed world beyond death. Some scholars think that the monarchy of the pharaoh was an outgrowth of the need to direct hydro-engineering on a country-wide scale,[25] although most irrigation work was supervised by local officials in the nomes, districts the size of American counties. Butzer, believing that they evolved as local irrigation units, maintained, "These nomes, as basic territorial entities, originally had socioeconomic as well as ecological overtones, but then became increasingly administrative in nature."[26]

Irrigation works extended cropland area beyond the area naturally flooded. The two types of land were kept distinct: *Rei* fields were those ordinarily covered by flood; *Sharaki* land required artificial irrigation. Laborers dredged channels, dug ditches, built dams, constructed dykes and basins, and used buckets to raise water. These activities were considered parts of a holy occupation. Major projects sponsored by pharaohs were commemorated as good works; Pepi I (2390–2360 BC), for instance, cut a canal to water a new district. Inscriptions boast, "I made upland into marsh, I let the Nile flood the fallow land," and "I brought the Nile to the upland in your fields so that plots were watered that had never known water before."[27] Kheti I (2100 BC) announced, "I initiated a channel ten cubits [5.2 m; 17 ft] wide ... I caused the water of the Nile to flood over the ancient landmarks."[28] The flow of water from the Nile into the great oasis of Fayum was controlled, and the level of Lake Karun was regulated to permit irrigation above its shores.

Technological inventions were made, such as the shaduf, a bucket on a long counterbalanced arm. Nilometers were installed near the First Cataract and elsewhere to measure the height of the river and to help predict the extent of the annual flood. Egypt incorporated such advances into the system of environmental regulation.

Egypt remained an agrarian rather than an urban society. As Adolf Erman put it, "Agriculture is the foundation of Egyptian civilization."[29] It is necessary to look at agriculture in order to understand the ecological relationships of the Egyptians. Sustainability was provided by the deposition of fertile alluvial soil containing mineral material and traces of

Figure 3.3 Salinization in the Fayum Oasis, Egypt. Evaporation of water used in irrigation in this basin below sea level has left crystals of salt in the soil in the foreground. Photograph taken in 1981.

organic debris brought down in the flood from the mountains and swamps further south. The Greek historian Herodotus, observing that the soil of Egypt had been formed by the river's sediment, pronounced Egypt to be the "gift of the Nile."[30] The Egyptians were aware of this: an early monument reads, "The Nile supplies all the people with nourishment and food."[31] Their environment encouraged them to think of processes of nature as operating in predictable cycles. The Nile flooded its banks at almost the same time every year (beginning in late July or early August). The only fertile land was what the river watered in the long, narrow valley floor of Upper Egypt and the broad Delta of Lower Egypt.

The flood was not totally predictable: a high Nile might wash away irrigation works and villages, or a low Nile might fail to water the land adequately.[32] In some periods when the river failed, rebels or invaders took advantage of weakness and unrest. As a result, Egyptian history was punctuated by times when pharaonic government collapsed. But traditional patterns of environmental relationships reappeared with phenomenal tenacity. As John Wilson expressed it, "The Nile never refused its great task of revivification. In its periodicity it promoted the [Egyptians'] sense of confidence; in its rebirth it gave [them] a faith that [they], too, would be victorious over death and go on into eternal life. True, the Nile might fall short of its full bounty for years of famine, but it never ceased altogether, and ultimately it always came back with full prodigality."[33] The natural regime, channeled by technology that was adapted to it, provided the environmental insulation necessary for a sustainable society.

Some difficult environmental problems appeared in spite of Egypt's record of success. A reliable food supply allowed overpopulation. When population increased to near the highest level that could be supported in a year of good harvest, any abnormally low harvest would bring the danger of famine. Reliefs on the causeway of Unas at Sakkara show people starving, their ribs conspicuous. The biblical story of Joseph's interpretation of Pharaoh's dream, and his advice to build granaries to prepare for hard times, is a reflection of the actual situation in Egypt.[34] Fat years were interspersed with lean ones, and population had ups and downs as a result. The pharaoh and governmental officials tried to even out fluctuations of supply and demand by storing a surplus in good years and distributing it when the harvest failed. Granaries have been excavated; one tomb at Amarna records forty granaries with a total capacity of 1,120 cubic meters (39,580 cu ft). Prices fluctuated in difficult periods: in the fifty-five years between the reigns of Ramses III and VII (1182–1127 BC), for example, the price of emmer wheat rose from eight to twenty-four times base price, and then fell under Ramses X, XI, and XII (*c.*1100 BC).[35]

The Egyptians' joy in their work was captured in pictures of plowing, hunting, and building. Active as these portrayals are, they show no realization that the environment was being altered. Egyptian art has little feeling of progress, decay, or the destruction of nature. For them, time ran in cycles, not along an inexorable line. But destructive changes nonetheless occurred.

Egypt suffered less from salinization than Mesopotamia because the regular flood leached salt from the soil. Salinization did occur in irrigated areas above flood line, and was serious in the Fayum, which is below sea level.

Although Egypt is seldom thought of as tree-clad, deforestation was a problem. The desert is more than 90 percent of Egypt's area, but the watered land had sections full of trees.[36] Tomb paintings show trees being cut. Egypt had plenty of firewood and fine woods for carving and cabinet-making, but few tall, straight trees, hence it imported timber from Phoenicia, where thick forests of conifers flourished on the slopes of the Lebanon mountains. Egyptian ships reached Byblos and other Phoenician ports as early as the reign of Snefru, first pharaoh of the Fourth Dynasty (*c.*2650 BC), to obtain cedar, juniper, fir, pine, and other timber trees for construction. In the Middle Kingdom, Egyptian influence was dominant on the Phoenician coast; in the New Kingdom, the area was conquered. In Egypt itself, after cutting for fuel and other purposes, the destruction of forests was made permanent by grazing of domestic animals, especially goats, which nibbled all the small trees that could form new forest.

The need for wetlands, plants, wildlife in sustaining the ecology of this land threatened by desert should be evident. But the habitats of wild animals, birds, and aquatic creatures gradually shrank and then disappeared, perhaps so slowly that few Egyptians were aware of what was happening. Eventually "the almost total disappearance of large game from the [Nile] valley, with increasing importation of captured animals for symbolic hunts by the nobility, argues for eradication of the natural vegetation."[37]

The worship that the Egyptians accorded to animals did not prevent wild animals from being hunted; still less did it save them from the effects of habitat destruction. In predynastic times, as petroglyphs and other works of art attest, Egypt possessed a variety of species as rich as that now found in East Africa. By the end of the Old Kingdom, however, elephant, rhinoceros, giraffe, and gerenuk gazelle were missing or rare north of the First Cataract, and the wild camel was extinct. Barbary sheep, lion, and leopard survived, but in reduced numbers. Some of this depletion was due to climatic change, but possibly more was due to habitat reduction and deliberate destruction. Amenhotep III boasted on one scarab that he had killed 102 lions with his own hand; lions were so honored that only kings could take them as prey, but kings gained glory by killing them.[38] By the Middle Kingdom, the ranges of some of the antelope species had been limited and their numbers decimated.[39]

As a result of these processes, Egypt at the end of the ancient period was environmentally changed, but still an abundant land. The Nile continued to bring sufficient water and sediment in most years to guarantee good crops. Grain, other foodstuffs, and crops such as flax for linen and papyrus for paper, were usually abundant enough to meet Egypt's needs and to be exported. Egypt was in most respects self-sufficient, so that the Egyptians were content with their land. Some modern writers have interpreted this contentment as an attitude that was "insular and self-satisfied."[40] That this was not the case is clear from the vigorous way in which they pursued the timber trade to obtain a resource in which they were not well-supplied at home. Although later subjected to foreign conquest, the land of Egypt, with the time-honored technology of irrigation, continued to be productive. It was the breadbasket of the Roman Empire.

That there might come a time when there would be no crocodiles or wild papyrus in their land was unimaginable to the ancients. But such a time was to come; indeed, it has come. Bird life is now at an ebb. The ibis is scarcely seen in Egypt, and of the fourteen species of duck in ancient Egyptian art, only one now breeds there.[41] A similar fate awaited the fish of the Nile. Today, unfortunately, the natural cycles that assured Egypt's sustainability have been disrupted. The Aswan Dams, which finally eliminated the annual flood, are described in Chapter 7.

Tikal: the collapse of classic Maya culture

Before modern settlement had spread in the Petén region of Guatemala, the Maya site of Tikal was isolated in a vast jungle. I flew there with my wife and son in an airplane that had survived many flights; its cracked windows had never been repaired. Clouds gathered as we winged northward; when we neared Tikal they were solid beneath us. The landing strip had no guidance system, so we feared the pilot would turn back, but a narrow opening appeared, neatly framing the dirt airstrip, and the plane dove through it. Suddenly, we saw the towering pyramids of the Maya city rising above the canopy of rainforest trees that surrounds them. It is one of the truly spectacular archaeological sites. Then we were safely down. The central temples have been restored, but others are still covered by vegetation and look like high, abrupt natural outcroppings. The Temple of the Giant Jaguar is tallest, lifting its limestone crest 44 m (144 ft) above its base. All around are other buildings, including palaces less impressive only by comparison with the steep pyramids beside them.

Tikal was one of the independent Maya city-states of the southern lowlands, numerous and populous, that elbowed one another for room across the moist tropical land between the Gulf of Mexico and the Caribbean Sea. They flourished in a Classic Age from the third to ninth centuries AD, and then were abandoned. A brilliant culture disappeared, and the population shrank to a tiny remnant. As John Lowe put it in *The Dynamics of Apocalypse*, "In dealing with the Maya collapse, we are not describing socio-political eclipse, but rather a profound social and demographic catastrophe."[42]

Before the effects of clearing for agriculture and building, the Southern Maya Lowlands were covered by tall rainforest interrupted by scattered savannas and wetlands. The forest had a multistoried canopy, with emergent trees rising as high as 40 m (130 ft) above the ground. The number of tree species was astounding, among them chicozapote, ramón, mahogany, strangler fig, sapodilla, breadnut, logwood, avocado, mamey, and the sacred ceiba.[43] The dry season lasts from January to May, when it becomes increasingly hot. The wet season, from May to December, produces rainfall as high as 3,000 mm (120 in).[44] The water that collects flows towards the Caribbean or the Gulf in river systems, the largest being the Usumacinta, which runs northwestward. A wide diversity of animals lived in the ecosystems, including deer, peccary, tapir, paca, agouti, rabbit, and other herbivores, and predators such as jaguar, puma, smaller cats, coatimundi, and fox. Reptiles and amphibians including frog, iguana, crocodile and snake were numerous. Various monkeys and scores of species of birds including the quetzal, parrot, toucan, curassow, quail and wild turkey also thronged the forests. Insect diversity was staggering; only a small proportion has been given scientific names.[45]

The impression of Classical Maya culture (AD 200–900) held by scholars has changed radically since the mid-1950s. Previously, the dominant idea[46] was that Maya cities consisted of huge but sparsely populated ceremonial centers run by peaceful priests concerned with calendars and astronomy, supported by peasants laboring at slash-and-burn agriculture. This picture emerged because earlier Mayanists concentrated on the elite and their spectacular structures, and because the only hieroglyphs that could then be understood recorded dates and time periods. But the archaeologists of the 1960s turned to remains left by the lower class.[47] The University of Pennsylvania conducted a scientific survey of Tikal and its environs which discovered that occupation was far wider and denser than had been suspected.[48] A much larger population, and more intensive agriculture, had to be postulated. Similar results came from other sites. It appeared that Maya settlement was not sporadic, but was a dense blanket of habitation with population densities close to those of rural modern China.

At the same time, scholars achieved the decipherment of Maya hieroglyphics after a breakthrough by Yurii Knorozov, a Leningrad linguist who had never visited the Maya sites. This work was forwarded by the energetic Linda Schele of the University of Texas, who convened seminars of leading experts. The inscriptions spoke not only of gods, planets, and the calendar; they also recorded events in the reigns of Maya kings, including wars, conquests, capture and sacrifice of enemy leaders, and the letting by kings of their own blood as an offering and a way to obtain visions.[49] The Maya elite emerged through the glyphs as a flamboyant and bloodthirsty set, more human if less likable than before.

As wider-ranging surveys, including sophisticated aerial photography, looked at the southern lowlands, evidence came to light for agricultural methods other than slash-and-burn, a method that requires long fallow and cannot support high population densities. Increasingly intensive agriculture utilized virtually all the available soils. Drainage and irrigation canals had been constructed in wetlands. For example, in the oddly-named Pulltrouser Swamp, crisscrossing channels were dug, with raised fields or platforms between them where crops could be grown.[50] Other surveys found evidence of terracing on hillsides, as the Maya attempted to use varying soils in marginal situations.[51]

The principal crops were maize, beans, and squash, supplemented by amaranth, manioc, and chili peppers, and cotton and sisal for fiber. Tree crops such as breadnut and cacao, growing in longer-lasting orchards, were also utilized. It appears that the variety of crops increased as the Maya, faced with an increasing population and limited land, experimented in an attempt to increase production.

The dense and increasing population is evident in the size of the great centers. The population of Tikal in the eighth century is estimated at between 40,000 and 90,000, comparable with the 50,000 in Shakespeare's London.[52] Water was supplied to the city from enormous reservoirs and catch basins.[53] There were extensive residential areas; thousands of mounds locate houses of the lower class, who labored in agriculture, but also in quarrying, stonemasonry, woodworking, and ceramics manufacture.

Figure 3.4 The Temple of the Giant Jaguar (Temple I) rises above the plaza in the ancient Maya city of Tikal, Guatemala. This represents the Classical Phase of Southern Lowland Maya civilization, with high population and intensive use of environmental resources. Photograph taken in 1974.

Tikal was not the capital of a great empire; it and its rivals were city-states with limited areas of political control. They entered into alliances and engaged in frequent warfare. Surveys found fortifications, a surprise for believers in Maya pacifism. The monuments declare that these wars glorified rulers, but the underlying reason for them was a desperate struggle for limited resources of food, fuel, and fiber. The first war recorded in presently readable texts took place between Tikal and Uaxactún in AD 378.[54] As the Maya reached the peak of population and material culture in 750, warfare increased, becoming more celebrated – and more destructive.[55]

The collapse, when it came, was relatively sudden. It happened within fifty to a hundred years. The last date on a stela at Tikal is 869. No dates anywhere in the classical calendrical system are recorded after 889. These facts are symptoms of a cessation of every aspect of Maya elite culture.[56] No more monuments were built; no elaborate tombs, no temples, no palaces, no offices for the bureaucracy. No fine polychrome pots were thrown, no beautiful jade jewels carved. The classical writing systems disappeared. Where were the ceremonies? Where the ball games, processionals, and visits of rulers? All gone, with the elite class that had performed them. What was lost? "An entire world of esoteric knowledge, mythology, and ritual."[57]

There is more, however. It was not just the decapitation of a culture. By AD 850, two-thirds of the population were gone, and the eventual loss is estimated at 75–85 percent.[58] In Tikal for a time, only one-tenth of the residential platforms were occupied. Then not only the classical centers were abandoned, but the countryside as well. Second-growth forest invaded exhausted farmlands. Millions of people disappeared from the southern lowlands. All social, economic, and political systems collapsed. Eventually, rainforest returned. It was one of the greatest demographic disasters in history.

What were the reasons for the collapse? This question was the subject of a seminar at the School of American Research in Santa Fe, New Mexico, in 1970. The proceedings were edited in a landmark volume by T. Patrick Culbert.[59] At the time, there were many competing explanations, and Richard E.W. Adams gave a review of previous theories, ranging from earthquakes and hurricanes through diseases to invasion. With the new information from surveys and decipherment, an opportunity to move toward consensus appeared. Almost a quarter of a century later, an explanation had emerged that, in broad outline, the majority of Mayanists could accept. Culbert stated it: "Most concur that centuries of uninterrupted growth put the Maya in a perilous position from which almost any disaster – drought, erosion, or social disorder – could have triggered a decline."[60] What were the elements of peril? Insights came from a team that studied the valley of Copán.[61] They found that under the pressure of population growth, agricultural intensification resulted in deforestation and catastrophic soil erosion. The society degraded its environment in the attempt to increase production of food and fuel. Of course, scholars still disagree about the weight to be given the various factors that helped to bring about the collapse, and, as often in history, it is probably a case of multiple causation.

Population increase is mentioned by virtually all Mayanists as a factor contributing to the collapse. The evidence of dense occupation over a wide area found at Tikal is paralleled in sites all over the southern lowlands. As Peter D. Harrison remarked in regard to Pulltrouser Swamp, "The Late Classic period exhibits, here as elsewhere, the same explosion of occupation that has come to be expected in all parts of the Maya lowlands. There is yet to be found a site that will adequately disappoint in this regard."[62] In an investigation of sediment cores in Lake Salpetén, researchers found an increase in phosphorus loading from human excreta of 9.6 times.[63] Similar results were obtained from other lakes.[64] Don S. Rice estimates that the Maya population increased by an order of magnitude from AD 300 to 800, with most of the surge after AD 650, and that the population density reached 250/sq km (650/sq mi), a figure comparable to that of rural China in the twentieth century.[65] Culbert's figures, slightly under Rice's, would yield a total population of 21,600,000 if applied throughout the region.[66] Undoubtedly that figure is too high, but the impression of a large, dense population is correct. "Population was spread thickly over the countryside as well as in proximity to urban centers," indicates Adams.[67] Such levels were attained by exponential growth during the Classic Period, and were followed by an appalling crash. The Maya used admirable ingenuity in attempting to sustain their expanding population, but it proved nonetheless to be unsustainable.

It is the classic economic problem of increasing food supply to feed a burgeoning population. The Maya extended agriculture into every part of their landscape that could be used for food production, using new methods to farm swamps and hillsides. Lacking animals that could be harnessed for plowing, they used human energy to develop production on increasingly difficult soils. In the process, they degraded the environment, exhausting nutrients and exposing the soil to erosion. Food production per unit population dropped as human numbers increased and yields declined. Importing food was not a viable solution, since with humans as the only burden bearers, maize can be efficiently imported only up to 90 km (56 mi), and the entire region was suffering shortage.[68] The land was unable to support the numbers of people that were living there by the mid-ninth century, and was even less able to provide for further increase. Culbert calls it "an exemplary case of [ecological] overshoot."[69]

At the same time, there was an increase in monumental construction as temples were built, pyramids enlarged, and stelae carved. Perhaps the elite, aware of the crisis, had decided to increase their use of the technology of sacrifice to gain the aid of the gods. This certainly was counterproductive, since it took workers away from food production, demanding more energy from common people who were receiving less food per capita.[70] It also placed demands on resources, not least on forests, since wood was required as scaffolding in construction and as fuel for making plaster from limestone.[71]

These demands, along with clearance for agriculture, meant near disappearance of rainforests throughout the Maya landscape. Pollen studies show a regressive loss of rainforest plants and an increase in grassland species, maize, and weeds from the time of Maya occupation to about AD 1000.[72] Additional evidence comes from studies of lake-bottom sediments, which show that sedimentation rates increased greatly during the Classic Maya Period, and the character of the sediments indicates that they resulted from erosion caused by deforestation.[73] Regeneration of the forests would have been inhibited if, as some studies suggest, the climate in this period was unusually dry.[74] Recent studies, however, have not found archaeological evidence for drought in the southern lowland Maya heartland.[75] The forests were largely gone by the Late Classic Period. This was a subsistence crisis, because the average householder in the tropics consumes over 900 kg (about one ton) of wood per person per year in food preparation and other domestic uses.[76] The search for wood meant forest loss beyond the immediate environs. The effects of deforestation include erosion, salinization, loss of water-retaining ability, and decline in transpiration with consequent decrease in humidity and rainfall. Even with a marginal decline in rainfall, torrential rains would have occurred from time to time, perhaps as the result of hurricanes, washing away soil that had been deprived of forest protection. The removal of forests caused restriction or extinction of forest animals; wild animal foods to supplement the Maya diet may have decreased.

Physical anthropologists have detected disease and nutritional impoverishment among the Maya.[77] Skeletons show deterioration in health through the Late Classic.[78] Average height decreased. Children's teeth show caries and enamel hypoplasia, signs of fasting and of a diet in which proteins decreased and carbohydrates increased. Diet-related diseases such as scurvy and anemia appeared. Diseases characteristic of high population densities such as Chagas' disease, *Ascaris* worms, and diarrhea, became common.[79] Average life span was in the thirties, with infant mortality of 40 percent.

After the collapse of the Classic Maya in the southern lowlands, the great cities were abandoned and the rainforest returned, although for a time there were squatters in places like Tikal; traces of their fires, garbage, and graffiti remain. A few cities survived with decimated populations, especially those located near water trade routes and with something to export, such as cacao or cotton.[80] But to the north, in the Yucatán, cities expanded and prospered for a time, and new cities appeared. Whether people fled from the south and swelled numbers in the north is unclear.[81] Later on, there were disasters in the north too; resources there, including water, may have proved inadequate. Puuc Maya centers, such as Uxmal, Sayil, Kabah, and Labná, gave way around AD 1000, and Chichén Itzá fell 200 years later. When the conquistadors arrived, they encountered a shadow of Maya civilization. The last independent Maya kingdom, the Itzá of Tayasal, resisted the Spaniards bravely and was defeated only in 1697, 175 years after Cortez first encountered the Mayas.

The Maya as a people did not disappear. They tried to assert independence in wars against Spanish, Mexican, and Guatemalan governments that continued sporadically into the 1990s. They number in the millions today, and their population is increasing. Non-Maya immigrants from outside are moving into the rainforest, felling it to make farms, and hunting the animals. Rich landowners have taken over huge ranches, and poorer people are forced onto hillsides where the soil is less rewarding. The Maya, who have lived there for millennia, have little choice but to participate in the destruction of their landscape. Is it a repeat of the historical tragedy of the Classic Maya collapse? The people who suffer most directly from tropical deforestation, and have the most to lose from it, are local forest communities. If they had the power to act in their own interests, they might provide an impetus for conservation and sustained use, since they have a tradition of knowing the forest and how to live with it. Unfortunately, they are seldom allowed to participate in the management of their forests by plantation owners, governments, wood products businesses, multinational banks and corporations, and sometimes even international conservation organizations.

For untold generations the ancient Maya lived within and made cultural adaptations to the rainforest environment. They invented agricultural methods to use differing parts of their landscape. With all their genius and civilization, however, they suffered an ecological collapse. Within the context of an ecosystem, no one species can succeed indefinitely by monopolizing as many of its energy streams as possible, while increasing its numbers without limit.

To the west 250 km (155 mi) in the Mexican state of Chiapas, we visited another Maya city, Palenque. It is not as large as Tikal, and its buildings not quite as high, but it is set in front of jungle-clad hills as if on a stage, calculated to impress. Its palace boasts a unique tower, and the Temple of the Inscriptions gives weight to the architectural assemblage. That temple gained distinction as the first Maya pyramid known to have a tomb beneath it when Alberto Ruz Lhuillier found the intact burial chamber of a Maya ruler there in the early 1950s. In the floor of the temple atop the pyramid is a rectangular opening, once hidden by a flagstone slab. We entered it and followed the stairway, under complex corbel vaulting, that slants downward toward the west side of the pyramid and then turns sharply east, still downward, to the burial chamber. It took us a few minutes, but Ruz and his team had labored for four seasons to clear the passageway of stone rubble and concreted lime.[82] There, underneath a beautifully carved basalt slab weighing almost 6 tons, they found a sarcophagus containing the jade-ornamented body of Pacal (Shield) the Great, ruler of Palenque in AD 615–83. Carvings on the sarcophagus sides show ten of his ancestors as personified trees, two each of five different species (cacao, avocado, sapote, guayaba, and nance), symbolically placing the king in the center of a sacred grove – the Maya protected certain groves of cacao and other trees as sacred places.[83] Pacal himself is shown in exquisite relief on the lid at the moment of death, being transformed into a ceiba, most sacred of trees, the symbolic World Tree.[84] The evident identification of the Maya monarch with tree, grove, and rainforest is poignant. When he ruled, neither he nor his people could have predicted that they and their descendants would bring down the living web on which they depended, and that when the city was emptied of human inhabitants, trees would return to fill the plazas of Palenque. But today, outside the archaeological zone, around the raw new agricultural fields visible from the Temple of the Inscriptions, the rainforest is falling again.[85]

Conclusion

When cities appeared in the landscape, a new split between culture and nature entered human minds. City and countryside were still parts of an ecosystem that embraced both, but it was a reorganized ecosystem in which forms of energy such as food and fuel flowed toward the urban center. Agriculture produced a surplus beyond the amount needed to feed the peasants who labored on the land, and this surplus fed the rulers, priests, soldiers, and workers in specialized occupations. When food supply increased, population also tended to expand, and the demand for resources rose proportionately. This cycle of growth continued until it approached the limits of the local ecosystem. The early cities had ways of postponing the inevitable crash – conquering neighboring lands and cities, engaging in trade over longer distances, importing metals and timber, and adopting more intensive agricultural technologies such as irrigation. But the basic problem remained. That is, an exponentially expanding population and economy within a finite ecosystem. Conquest could deplete as well as expand resources, lengthening trade routes reached the point where the effort to bring in resources required more energy and wealth than was brought in, deforestation made flooding more serious and unpredictable, and intensive agriculture introduced erosion, salinization, and other factors that reduced production. Limits were exceeded, the food supply declined, and the fall of a civilization was typically more sudden than its rise had been. The same basic problem, in various guises, returned in later historical periods, and the following chapters contain examples to illustrate it.

Notes

1   J. Donald Hughes, "An Ecological Paradigm of the Ancient City," in Richard J. Borden, ed., *Human Ecology: A Gathering of Perspectives*, Baltimore, MD, Society for Human Ecology, 1986, 214–20.
2   Kwang-chih Chang, *The Archaeology of Ancient China*, New Haven, CT, Yale University Press, 1977, 218, 289.
3   Vishnu-Mittre and R. Savithri, "Food Economy of the Harappans," in Gregory L. Possehl, ed., *Ancient Cities of the Indus*, Durham, NC, Carolina Academic Press, 1979, 207.
4   One metric ton equals 1.1 English/US tons. Since the two units are relatively close in scale, only the weights in metric tons will be given throughout this book.

5   Theodore A. Wertime, "The Furnace versus the Goat: The Pyrotechnologic Industries and Mediterranean Deforestation in Antiquity," *Journal of Field Archaeology* 10, 1983, 445–52, at 451.
6   Morris Silver, *Economic Structures of Antiquity*, Westport, CT, Greenwood Press, 1995, 97–9.
7   H.W.F. Saggs, *Civilization Before Greece and Rome*, New Haven, CT, Yale University Press, 1989, 123.
8   Mark Nathan Cohen, *Health and the Rise of Civilization*, New Haven, CT, Yale University Press, 1989, 116–22.
9   Marvin W. Mikesell, "The Deforestation of Mt. Lebanon," *Geographical Review* 59, 1969, 1–28; 1 Kings 5. 1–10.
10 Deuteronomy 20. 19–20.
11 R.E. Mortimer Wheeler, *Civilizations of the Indus Valley and Beyond*, New York, McGraw-Hill, 1968, 72.
12 N.K. Sandars, ed., *The Epic of Gilgamesh: An English Version with an Introduction*, London, Penguin Books, 1972, 59.
13 James B. Pritchard, ed., *The Ancient Near East: An Anthology of Texts and Pictures*, Vol. 1, Princeton, NJ, Princeton University Press, 1973, 35.
14 Sandars, *The Epic of Gilgamesh*, 94.
15 Sandra Lin Marburg, "Women and Environment: Subsistence Paradigms, 1850–1950," *Environmental Review* 8, 1, 1984, 7–22.
16 Bridget and Raymond Allchin, *The Rise of Civilization in India and Pakistan*, Cambridge, Cambridge University Press, 1982, 188–9; Shereen Ratnagar, *Encounters: The Westerly Trade of the Harappan Civilization*, Delhi, Oxford University Press, 1981, 99–156.
17 Silver, *Economic Structures of Antiquity*, 54–5.
18 Charles L. Redman, *Human Impact on Ancient Environments*, Tucson, University of Arizona Press, 1999, 185–6.
19 John Perlin, *A Forest Journey: The Role of Wood in the Development of Civilization*, Cambridge, MA, Harvard University Press, 1991, 35–43.
20 Thorkild Jacobsen and Robert M. Adams, "Salt and Silt in Ancient Mesopotamian Agriculture," *Science* 128, November 1958, 1251–8, at 1252.
21 Mark Lehner, *The Complete Pyramids*, London, Thames & Hudson, 1997, 224–5.
22 This is an often-reproduced scene from the tomb of Thothhotep (Djehutihotpe) at Deir el-Bersha. C.R. Lepsius, *Denkmäler aus Ägypten und Äthiopien*, Berlin, Nicolai, 1849–56, Part 2, Plate 134; Ahmed Fakhry, *The Pyramids*, Chicago, University of Chicago Press, 1969, 13.
23 Frontinus, *Aqueducts* 1. 16.
24 Karl W. Butzer, *Early Hydraulic Civilization in Egypt*, Chicago, University of Chicago Press, 1976, 56.
25 Karl Wittfogel, *Oriental Despotism: A Comparative Study of Total Power*, New Haven, CT, Yale University Press, 1957.
26 Butzer, *Early Hydraulic Civilization in Egypt*, 105; Michael A. Hoffman, *Egypt Before the Pharaohs*, New York, Alfred A. Knopf, 1980 (reprint, London, Ark Paperbacks, Routledge & Kegan Paul, 1984), 30–2.
27 First Intermediate Period tomb inscription, Siut.
28 Hoffman, *Egypt Before the Pharaohs*, 313, from J.H. Breasted, *Records of Ancient Egypt*, Vol. 1, Chicago, University of Chicago Press, 1906, 188–9.
29 Adolf Erman, *Life in Ancient Egypt*, London, Macmillan, 1894 (reprint, New York, Dover, 1971), 425.
30 Herodotus, *Histories* 2. 5.
31 Erman, *Life in Ancient Egypt*, 425, from R. Lepsius, *Denkmäler aus Ägypten und Äthiopien*, Vol. 3, Berlin, 1858, 175.
32 Butzer, *Early Hydraulic Civilization in Egypt*, 51–6.
33 John A. Wilson, *The Culture of Ancient Egypt*, Chicago, University of Chicago Press, 1951, 13.
34 Genesis 41. 1–37.
35 Butzer, *Early Hydraulic Civilization in Egypt*, 55–6.
36 Ibid., 25.
37 Ibid., 86–7.
38 Herrmann Kees, *Ancient Egypt: A Cultural Topography*, Chicago, University of Chicago Press, 1961, 20.

39   Butzer, *Early Hydraulic Civilization in Egypt*, 26–7.
40   Hoffman, *Egypt Before the Pharaohs*, 24.
41   Kees, *Ancient Egypt*, 93–4.
42   John W.G. Lowe, *The Dynamics of Apocalypse: A Systems Simulation of the Classic Maya Collapse*, Albuquerque, University of New Mexico Press, 1984.
43   A more complete list and description may be found in Arturo Gómez-Pompa, "Vegetation of the Maya Region," in Peter Schmidt, Mercedes de la Garza, and Enrique Nalda, eds, *Maya*, New York, Rizzoli, 1998, 39–51.
44   Don S. Rice, "Eighth-Century Physical Geography, Environment, and Natural Resources in the Maya Lowlands," in Jeremy A. Sabloff and John S. Henderson, eds, *Lowland Maya Civilization in the Eighth Century A.D.: A Symposium at Dumbarton Oaks, 7th and 8th October 1989*, Washington, DC, Dumbarton Oaks Research Library and Collection, 1993, 20.
45   John S. Henderson, *The World of the Ancient Maya*, Ithaca, NY, Cornell University Press, 1997, 27–32.
46   This paradigmatic view is associated with the names of the respected Mayanists, Sylvanus G. Morley and J. Eric S. Thompson. See Sylvanus G. Morley, *The Ancient Maya*, Stanford, CA, Stanford University Press, 1946 and subsequent editions, and J. Eric S. Thompson, *The Rise and Fall of Maya Civilization*, Norman, Oklahoma University Press, 1954 and subsequent editions.
47   Jeremy A. Sabloff, *The New Archaeology and the Ancient Maya*, New York, Scientific American Library, W.H. Freeman, 1990, 167–8.
48   See the comprehensive Tikal Reports series, including Coe's *Excavations in the Great Plaza, North Terrace and North Acropolis of Tikal*, 6 volumes, Tikal Report 14, Philadelphia, University Museum, University of Pennsylvania, 1990.
49   Linda Schele, "The Maya Rediscovered: The Owl, Shield, and Flint Blade," *Natural History* 100, 11, 1991, 6–11.
50   Peter D. Harrison, "Settlement and Land Use in the Pulltrouser Swamp Archaeological Zone, Northern Belize," in Scott L. Fedick, ed., *The Managed Mosaic: Ancient Maya Agriculture and Resource Use*, Salt Lake City, University of Utah Press, 1996, 177–90.
51   Sheryl Luzzadder-Beach and Tim Beach, "Wetlands as the Intersection of Soils, Water and Indigenous Human Society in the Americas," in J.R. McNeill and Verena Winiwarter, eds, *Soils and Societies: Perspectives from Environmental History*, Isle of Harris, UK, White Horse Press, 2006, 91–118.
52   Sabloff, *New Archaeology*, 79.
53   Vernon L. Scarborough, "Flow Of Power: Water Reservoirs Controlled the Rise and Fall of the Ancient Maya," *The Sciences* 32, 2, March 1992, 38–43.
54   Schele, "The Maya Rediscovered," 6.
55   David Stuart, "Historical Inscriptions and the Maya Collapse," in Sabloff and Henderson, eds, *Lowland Maya Civilization* , 321–54.
56   Richard E.W. Adams, "The Collapse of Maya Civilization: A Review of Previous Theories," in T. Patrick Culbert, ed., *The Classic Maya Collapse*, School of American Research, Albuquerque, University of New Mexico Press, 1973, 22.
57   Michael D. Coe, *The Maya*, 4th edn, London, Thames & Hudson, 1987, 128.
58   T. Patrick Culbert, *Maya Civilization*, Washington, Smithsonian Books, 1993, 118; Richard E.W. Adams, *Ancient Civilizations of the New World*, Boulder, CO, Westview Press, 1997, 65.
59   T. Patrick Culbert, ed.,*The Classic Maya Collapse*, School of American Research, Albuquerque, University of New Mexico Press, 1973.
60   T. Patrick Culbert, "The New Maya," *Archaeology* 51, 5, 1998, 48–51, at 50.
61   Eliot M. Abrams, AnnCorinne Freter, David J. Rue, and John D. Wingard, "The Role of Deforestation in the Collapse of the Late Classic Copan Maya State," in Leslie E. Sponsel, Thomas N. Headland, and Robert C. Bailey, eds, *Tropical Deforestation: The Human Dimension*, New York, Columbia University Press, 1996, 55–75.
62   Harrison, "Settlement and Land Use," 181–2.
63   Don S. Rice, "Paleolimnological Analysis in the Central Petén, Guatemala," in Fedick, ed., *Managed Mosaic*, 193–206.
64   Don Stephen Rice and Prudence M. Rice, "Lessons from the Maya," *Latin American Research Review* 19, 3, 1984, 7–34. The Rices list other studies with similar results, at 25 and references.
65   Don S. Rice, "Paleolimnological Analysis," 196.

66   Culbert, "The New Maya," 49. Rice's figures, similarly treated, would yield a peak total population for the southern lowlands of 23,400,000. In 1964, the population of the Petén was only 25,910, but had risen to 200,000 by 1979.

67   Adams, *Ancient Civilizations of the New World*, 56.

68   Lowe, *Dynamics of Apocalypse*, 120.

69   T. Patrick Culbert, *The Lost Civilization: The Story of the Classic Maya*, New York, Harper & Row, 1974, 116.

70   Elliot M. Abrams, *How the Maya Built Their World: Energetics and Ancient Architecture*, Austin, University of Texas Press, 1994.

71   Don S. Rice, "Eighth-Century Physical Geography, Environment, and Natural Resources in the Maya Lowlands," in Sabloff and Henderson, eds, *Lowland Maya Civilization in the Eighth Century A.D.*, 11–63, at 51.

72   Don S. Rice, "Paleolimnological Analysis," 198; Julian C. Lee, "Creatures of the Maya: The Impact of Pre-Columbian Agriculture Can Still Be Seen on Many of the Yucatan's Frogs, Lizards, and Snakes," *Natural History* 90, 1, January 1990, 44–51, at 48.

73   Don S. Rice, Prudence M. Rice, and Edward S. Deevey, Jr., "Paradise Lost: Classic Maya Impact on a Lacustrine Environment," in Mary Pohl, ed., *Prehistoric Lowland Maya Environment and Subsistence Economy*, Cambridge, MA, Peabody Museum of Archaeology and Ethnology, Harvard University, 1985, 91–105.

74   David A. Hodell, Jason H. Curtis, and Mark Brenner, "Possible Role of Climate in the Collapse of Classic Maya Civilization," *Nature* 375, 1 June 1995, 391–4.

75   Nicholas Dunning and Timothy Beach, "Stability and Instability in Prehispanic Maya Landscapes," in David L. Lentz, ed., *Imperfect Balance: Landscape Transformations in the Precolumbian Americas*, New York, Columbia University Press, 2000, 179–202, at 198.

76   Don S. Rice, "Eighth-Century Physical Geography," 28.

77   Ibid., "Eighth-Century Physical Geography," 43.

78   Adams, *Ancient Civilizations of the New World*, 64; Frank P. Saul, "Disease in the Maya Area: The Pre-Columbian Evidence," in Culbert, ed., *Classic Maya Collapse*, 301–24.

79   Demitri B. Shimkin, "Models for the Downfall: Some Ecological and Culture-Historical Considerations," in Culbert, ed., *Classic Maya Collapse*, 269–300, at 279, 282.

80   Jeremy A. Sabloff, "Ancient Civilization in Space and Time," in Schmidt, Garza, and Nalda, eds, *Maya*, 53–71, at 70. Just how "thriving" the remnant towns may have been is questionable; they certainly lacked monumentality and many aspects of classic Maya culture.

81   Jeremy A. Sabloff, "Interpreting the Collapse of Classic Maya Civilization: A Case Study of Changing Archaeological Perspectives," in Lester Embree, ed., *Metaarchaeology: Reflections by Archaeologists and Philosophers*, Boston, MA, Kluwer Academic Publishers, 1992, 99–119, at 108.

82   Alberto Ruz Lhuillier, "The Mystery of the Temple of the Inscriptions," *Archaeology* 6, 1, Spring 1953, 3–11.

83   Arturo Gómez-Pompa, José Salvador Flores, and Mario Aliphat Fernández, "The Sacred Cacao Groves of the Maya," *Latin American Antiquity* 1, 3, 1990, 247–57.

84   Linda Schele and Peter Mathews, *The Code of Kings: The Language of Seven Sacred Maya Temples and Tombs*, New York, Scribner, 1998, 113–25.

85   Philip Howard, "The History of Ecological Marginalization in Chiapas," *Environmental History* 3, 3, July 1998, 357–77, indicates that a disaster similar to the classic Maya collapse is happening today, accompanied by civil strife along ethnic and economic lines.

4 Ideas and impacts

In the later phase of ancient history, two important processes of change transformed human roles within the natural world. One occurred in the sphere of attitudes to nature, and the other occurred in the sphere of human actions impacting the natural environment. The processes were simultaneous and influenced each other.

Human impacts on nature increased in scale in this period, which for Eurasia and north Africa is roughly the last eight centuries BC and the first eight centuries AD, due to an unsteady but general increase in human populations and the appearance of great empires which rose, conquered large territories and populations, flourished, declined, and fell. They had the ability to organize numbers of people in vast projects that transformed the landscape, such as irrigation schemes, road building, terracing of hillsides, mining, and logging.

Among the empires that occupied segments of the Earth's surface were those of the New Babylonians and the Persians, Alexander the Great and his Hellenistic Successors, the Mauryans of India, the Qin and Han of China, the Carthaginians, Parthians, Sassanids, and Romans. The Romans will serve as the quintessential example of an ancient empire in the last section of this chapter, but there were a number of others whose population, technology, and resource use had impacts on the natural environment that resulted in damage and possibly also contributed to their downfall. Among the impacts that must be mentioned are deforestation, depletion of wildlife, overgrazing, soil erosion, salinization, additional forms of agricultural exhaustion, air and water pollution, noise pollution, and various other urban problems affecting health.

This age also saw the origin or reformation of several great systems of thought and rules of behavior. Indeed, the term "Axial Age" is often used for the early part of it because so many of these systems which deeply formed and changed humanity's worldviews appeared during that time, often as the result of the work of figures such as Zoroaster, Confucius, Lao Tzu, Pythagoras, Buddha, Mahavira, and the Jewish prophets. Some of these systems are religions or philosophies; others might best be described as ways of life that are generally accepted in societies. They were embraced by large numbers of people, and in some cases continued their influence through every subsequent period of human history down to the present. These systems had important effects on human behavior in regard to ecosystems, but to varying degrees. How far can we praise or blame these widely accepted and often contrasting systems for the maintenance or damage of a sustainable human relationship with the rest of the community of life?

The first section of this chapter discusses the partial failure of the Greek polis, Athens in particular, to adapt its economy to natural systems although great philosophers of the fifth and fourth centuries BC considered the problem and offered advice. The second section focuses on China from the Warring States period through the Qin Dynasty to the beginning of the Han, touching on the sages who strove to persuade the emperors, and the bureaucracies that served them, to embrace their social programs. The third section addresses the complicated problem of the decline of the Roman Empire and asks whether environmental problems constituted a major cause.

There were several great systems that formed human ideas about the natural world in the last few centuries BC and the first few centuries AD. They can be placed in three general categories: (1) traditional, evolving systems that included earlier ideas along with new ones; (2) systems created by reformers who taught the oneness of life including humans and nature; and (3) monotheistic religions that made humans God's stewards with dominion over and responsibility for the rest of creation.

The first group evolved within the context of the traditional views and rites inherited from earlier periods, and to some extent never rejected them. Examples include Hinduism and Shinto. In both India and Japan, however, there was in early times the intrusion of a different people who, while introducing new beliefs, also absorbed some of the ideas and practices of the aboriginal groups. In India, the Sanskrit literary tradition, beginning with the *Vedas*, was in large part the work of an invading population of herding folk, who worshipped gods including many who personified the striking phenomena of nature, such as Indra, a sky-god, and Surya, the sun, with hymns and sacrifices.[1] Undoubtedly the great reverence of Hindus for the cow can be traced back to the herdsfolk. Less positive attitudes of the cattle-herders can also be found. In the epic *Mahabharata*, Krishna and Arjuna burned the great Khandava forest with all the creatures in it as an offering to Agni, god of fire. They patrolled the edge of the forest, driving back any living thing that tried to escape. This incident is symbolic of pastoral clans clearing the forest and removing forest tribes to make room for their grazing animals.[2] But Hinduism also preserved elements of pre-Vedic India; Shiva, a god with both hunting and agricultural affinities, has been traced back to the Indus Valley civilization. Venerable groves and sacred tree species such as the pipal (*Ficus religiosa*) originated among forest-dwellers and swidden farmers, as did the popular worship of animal-shaped gods such as Hanuman, the courageous monkey, and elephant-headed Ganesha. The Indian heritage contains a conservation ethic reflected in another incident of the *Mahabharata*. The heroic Pandavas were living in exile in the forest and hunting for food, when the wild animals appeared to the eldest of them in a dream and asked him to move to another part of the forest so the animal populations could recover.[3]

Within the traditional systems, the symbols and practices of ancestral hunter–gatherer, agrarian, and pastoral societies and the attitudes toward nature typical of them tended to persist, although they were also transmuted. The process of change in these systems was typically evolutionary rather than revolutionary. They tended to retain the worship of natural entities, including nature gods, and to teach practices that embody conservation. They perceived that spirits inhabit and animate the natural world including animals, plants, and objects such as rivers and mountains. Visitors to India are often impressed by the toleration shown by ordinary Indians to the presence not only of cattle, but of many wild species as well, in fields, villages, and even in cities. This is in part due to the belief in reincarnation, which implies a common destiny of different forms of life, since human souls after death may be reborn as animals or plants.

Traditional people venerated powers of life, such as the fertility of Mother Earth, and sought to cooperate with them by participating in rituals. In its origins, the traditional Shinto of Japan was a form of nature worship rooted in the forest.[4] The forest had many *kami* (spirit beings) including gods, powerful animals, and the dead, and these were often associated with sacred places. The religion contained underlying shamanistic and animistic elements including tree worship and the belief that humans must respect plants and animals, which share life equally with them.

The beliefs of traditional religions were often accompanied by practices and taboos that tended to support conservation, but such practices were not always effective. Madhav Gadgil theorizes that a human society will favor ecological prudence when it is in its interest to do so, and this is the case when it occupies a relatively stable environment, is sedentary and at a population level close to the carrying capacity of its local resources, has a closed group structure, and is not experiencing rapid technological change.[5] The so-called caste system in India, by limiting the use of specific resources to designated hereditary groups, may have helped to preserve communities of this type, and therefore to establish a pattern of conservation.

Traditional systems also produced more universalistic ideas, sensing that one great spiritual reality lies beyond all lesser spirits, and nature itself. Hindu philosophy, with its vision of the oneness of all beings, supports a positive attitude to nature. Chapter 7 of this book gives further consideration to Hinduism and the environment.

A second group of world systems originating in the ancient world taught the oneness of life, and an ethic based on respect for all living things, and therefore seems to have encouraged the preservation of nature. These religious philosophies had great reformers such as Mahavira (Jainism), Siddhartha Gautama (Buddhism), and Lao Tzu (Taoism) as their exponents.

Among the Greek philosophers, Pythagoras and the Pythagoreans especially belong to this group. They viewed the universe (*cosmos*) as an organism of which humans and other creatures are parts.[6] Phercydes, teacher of Pythagoras, held that the world is a single living being, and Empedocles described it as animate, ensouled, and intelligent.[7] Plato followed the Pythagoreans when he said that the cosmos is "that living creature of which all living creatures, severally and generically, are portions ... [and] are by nature akin to itself."[8] Like mortals, the cosmic animal is alive; it has a body and is "endowed with soul and reason." Such a view of the universe led these thinkers to posit a cyclical interplay of elements and living beings within the organic unity of the cosmos. Empedocles described an endless recycling. He insisted that because all things are composed of the same four elements, there is a constant process of reassemblage in which there is no creation "out of nothing," nor annihilation. These ideas offered the rudiments of a philosophical grounding for ecological thought.

Aristotle also presented an image of the cosmos as an organic whole, of which creatures are parts.[9] "All things are ordered together somehow, but not all alike – both fishes and fowls and plants; and the world is not such that one thing has nothing to do with another, but they are connected."[10] Since the cosmos is in motion, natural cycles take place, such as the circulation of the elements, including air and water.[11] Animals also approximate the cycles of the cosmos; their biological processes follow the periods of the sun and moon, and reproduction is their imperfect imitation of the eternity of the cosmos.[12]

These conceptions of the cosmos present a world in which plants and animals, including human beings, are not simply individual entities, but are related to a system whose nature they share. This means that ecological thought, including questions of the relationships of living creatures to one another and to the environment, was at least theoretically possible. While it cannot be said that ecology was a major theme in Greek philosophy – "ecology" as a word is derived from Greek roots, but does not occur in ancient Greek literature[13] – some philosophers did inquire into the subject that would later receive the name, among them Aristotle and his most brilliant student, Theophrastus.

The ethical teaching of some philosophers such as the Pythagoreans and Plato forbids hurting or damaging any living being. This precept seemingly would favor the protection of nature. The Pythagoreans taught that because all living things, including humans, have a common origin and natural ties, and are formed of the same components, including the soul, they are all related and should be treated with respect. They forbade taking the lives of animals or plants, as well as eating food that required killing an organism. They banned beans and many other plant foods along with meat. Many foods could be consumed without killing, so far as they knew, such as milk, cheese, honey, wine, oil, fruits (as long as one did not eat the seeds), and leafy vegetables. Plato argued for a benevolent attitude by human beings toward other living beings, since as parts of nature, humans find their welfare dependent upon that of nature as a whole. [14] "To pursue one's own good is the very same thing as to contribute to the good of the whole natural organism. To attain one's own good, one must contribute to the good of the whole of which one is a part, that is, one must be just."[15] If human beings disrupt the balance of the system, therefore, they harm themselves.

The first principle of Buddhist ethics is "do no harm," neither to human beings, to animals or plants, or even to things usually considered inanimate. Buddhist literature recognizes the importance of preserving the forest as a habitat for animals such as tigers. The Buddha banned his followers from throwing their wastes or leftover food into rivers, lakes, or the sea, and urged them to "guard the lives of all living beings abiding there."[16] All told, Buddhists were counseled to treasure and conserve nature, of which human beings are part.

The Jainas were, if anything, even more insistent on this principle. One of their strongest teachings is *ahimsa*, or non-violence to any living thing, a doctrine that influenced Mahatma Gandhi. Occupations like warfare or farming, which required killing fellow beings, were forbidden to them. The strictest Jainas sweep their paths to avoid stepping on insects, and wear gauze masks to keep from inhaling the tiniest creature. This is perhaps the most extreme teaching in any religion of denying oneself in order to preserve other forms of life, and it was understandably followed by few. A handful of Jaina ascetics hastened death by starving themselves rather than eating anything, animal or plant, that had been killed.

These religious philosophies counseled their followers to live the simplest lives possible, making few demands on the environment. According to Lao Tzu's book, the *Tao Te Ching*, for example, a wise person will be at one with *Tao*, the way of nature, will use as few resources as possible, and will not damage the natural environment.

Ideally, a Buddhist man would spend at least part of his life as a monk. Rules forbade monks to cut down trees. Monks also were not allowed to eat ten kinds of meat, mostly from forest animals. Right living implies for all Buddhists a non-wasteful mode of life.

Jainas believe that individual souls must save themselves through heroic asceticism. The truest Jaina is a mendicant monk who sacrifices bodily comforts, perhaps going without clothing ("sky-clad"), in order to purify soul from matter.

This group of religious philosophies also fosters the appreciation of nature. The Buddha was born in a grove of *sal* trees, according to legend.[17] He achieved enlightenment while meditating beneath a sacred pipal tree. Buddhist temples often have sacred groves of trees that are protected, along with the creatures that live in them. Taoism's emphasis on nature strongly influenced Chinese painting and other arts, including garden arrangement. More will be said about Taoism in the section below on China.

A third group of worldviews that arose before the end of the ancient world are the monotheistic faiths, Zoroastrianism, Judaism, Christianity, and Islam. Zoroastrianism, the religion of ancient Persia, will be discussed below in connection with the Persian Empire. The three "Abrahamic" religions – Judaism, Christianity, and Islam – agreed that there is only one God, who created the universe, including the Earth and all the creatures within it. Among God's creations were human beings, to whom God gave dominion over the Earth and all its creatures.[18] But God did not relinquish sovereignty over the natural world. Therefore humans are his representatives, and their dominion over and stewardship of the Earth and its non-human inhabitants must be exercised within ultimate responsibility to the Creator. This is the doctrine of stewardship, which is recognized in the three religions as a preventative against heedless exploitation of the created world. Whether it worked that way in actual practice is open to question – a similar division of doctrine and deeds can, of course, be observed in all religions.

Commandments of the Torah, the books of the Jewish law, indicated what Jews, and to a lesser extent humans in general, were directed to do or forbidden from doing. Many of the commandments given specifically to the Jews have important environmental applications. For example, a Jewish army besieging an enemy city may not destroy trees:

> When in your war against a city you have to besiege it a long time in order to capture it, you must not wantonly destroy its trees, wielding the axe against them. You may eat of them, but you must not cut them down. Are trees of the field human beings who can withdraw before you into the besieged city?[19]

But there is more. In their commentaries the rabbis regarded the words, "you must not wantonly destroy," as a general principle, to be applied not only during a military action; a tree may never be cut down with destructive intent. In a positive sense, planting a tree is a good work and a sign of peace.[20] Further, the command applies not only to trees, but further to all things from which humankind may benefit, such as food, clothing, and water. Nor does scripture give license to destroy the non-useful part of nature, since enjoyment is a value, too, and even to destroy what one does not enjoy would damage one's humanity.

The rabbis applied this principle of "do not destroy" in numerous specific cases that prohibit wastefulness. For example, one should not adjust a lamp to burn too quickly, for this would be wasteful of the fuel.[21] One is not allowed to throw bread or to pass a cup of liquid over bread at table; in both cases the bread (symbolic of all food) could be ruined.[22] Several commandments require kindly treatment of animals; for instance, the weekly Sabbath day of rest applies to them.[23] Beyond that, there is divine providence for each species; God desires them to be perpetuated. A mother bird may not be taken along with her eggs or young.[24] The Talmud has the raven rebuke Noah, who is about to send him out of the ark, where there are only two ravens, over the flood waters to search for dry land, "If sun or rain overwhelm me, would not the world be lacking a species?"[25]

There are also statements of concern for environmental conditions. Builders of the cities of the Levites were commanded to provide the amenity of an open pasture 1,000 to 2,000 cubits (500 to 1,000 meters or yards) around cities, free of construction and cultivation.[26] Thus the covenant constrained the Jews to control their appetites, to respect the rights of other living things, and to work in ways that enhance the landscape and prevent its misuse. According to one text, just after the creation, God spoke the following words to Adam:

> See my works, how fine and excellent they are! Now all that I have created, for you have I created. Think upon this, and do not corrupt or desolate my world, for if you do corrupt it there will be none to set it right after you.[27]

Christianity began within Judaism and kept the Hebrew Bible among its sacred books, so it is not surprising that many important Christian ideas about the natural world were inherited directly from Judaism. Jesus, as portrayed in the gospels, spent much of his life outdoors, retiring for prayer and spiritual renewal to the mountains, sea, and desert wilderness. He taught in such places as well. His teachings were framed as parables using images from the natural world such as trees, birds, seeds and growing grain, vines, and sheep. God provides for all creatures; he said, "look at the birds of the air, they neither sow nor reap ... and yet your heavenly Father feeds them."[28] Not a single sparrow "is forgotten before God."[29] He told people to "Consider the lilies of the field: they neither toil nor spin; yet I tell you, even Solomon in all his glory was not arrayed like one of these."[30] He urged them to lead a simple life, accepting in gratitude the necessities of life from nature as gifts of God.

Paul wrote, "Ever since the creation of the world, [God's] invisible nature, namely his eternal power and deity, has been clearly perceived in the things that have been made."[31] The cosmos is a natural demonstration, visible to everyone, of "a living God who made Heaven and Earth and the sea and all that is in them."[32] Therefore nature is not evil: "Everything created by God is good, and nothing is to be rejected if it is received with thanksgiving."[33] God has "put everything in subjection to humankind,"[34] according to the New Testament as well as *Genesis*, and Christianity affirms human stewardship. It makes an ethical statement about the way in which humans handle the mineral, vegetable, and animal realms, teaching that creation should be treated kindly and responsibly, protected, and preserved. The character of human beings is revealed in the way they treat what has been entrusted to their care. Sin, in environmental terms, consists of injuring creation by using it in ways that affront God's purposes. All ecological damage is a sin, a rupture in creation itself. The broken state of nature is the fault of humankind. The salvation of the world includes all creation. In Christ, God entered into human life, a human body, and therefore the natural world. Paul says that God was in Christ reconciling to himself "all things, whether on Earth or in heaven."[35] The whole creation desires its restoration, and the vision of the future is not the destruction of the world, but its renewal.[36]

Islam shares attitudes toward nature that are characteristic of the monotheistic faiths, but has its own unique ethos, much of which derives from roots in Arabic culture. The Quran, revealed through the prophet Muhammad, contains ethical principles that require the good treatment of the natural environment. The scholar Seyyed Hossein Nasr provides a compelling picture of the role of nature in Islam:

> The Islamic view of the natural order and the environment, as everything else that is Islamic, has its roots in the Quran, the very Word of God, which is the central theophany of Islam ... The Quran addresses not only men and women but the whole of the cosmos. In a sense, nature participates in the Quranic revelation ... The soul which is nourished and sustained by the Quran does not regard the world of nature as its natural enemy to be conquered and subdued but as an integral part of its religious universe sharing in its earthly life and in a sense even in its ultimate destiny.[37]

Allah, God, is the one creator of all things, and Islam teaches that humankind is God's steward (*al-khalifah*): the Quran states, "I am setting on the Earth a steward."[38] Humans in Islam are the central creatures of the earthly sphere, but can exercise power over things only in obedience to God's laws. And Divine Law (*al-Shari'ah*) specifically includes duties to the natural environment. Laws forbid pollution and instruct the planting of trees and gentle treatment of animals. Since Islamic governments are delegated the authority to enforce the laws of God, they have the responsibility to protect nature within their realms and to establish environmental justice.[39] Unfortunately, not many of them as yet have exercised this responsibility in a creditable manner.

The three monotheistic faiths have been blamed for motivating ecological damage because they seem to separate God from nature, leaving nature without spiritual worth, and because by placing human beings above the rest of creation, they apparently give them permission to use other creatures and the Earth itself without considering that they have any value of their own.[40] But in Judaism, Christianity, and Islam as represented in their scriptures safeguards exist against the misuse of creation. If those who professed belief in them had always acted according to their precepts, they would have cared for nature. It is to other historical factors that we must look for an explanation of the fact that the homelands of the monotheistic faiths became environmentally devastated even in early times.

This recalls the subject of impacts, and the great empires mentioned above that caused many of them. Some of these empires tried to apply the teachings of a religion or philosophy. The Indian ruler Ashoka, after a bloody campaign, turned to the peaceful tenets of Buddhism and promulgated laws that commanded the planting of trees and kind treatment for animals. To what extent did the ideas of the accepted system affect the impacts of the society on the natural environment in a positive or negative way? It seems that the answer must be equivocal. There were positive effects, but empires have never been able to control the actions of all those over whom they hold sway. Officials become corrupt, and economic forces are in most times stronger motives than moral suasion. Individuals and communities choose what seems to be in their short-term interest above what would be best in the long term for the Earth, society, or themselves.

The Persian emperors adopted Zoroastrianism as their official religion. The dates of the prophet Zoroaster (Zarathustra) are uncertain, but tradition placed his birth around 660 BC. He had a vision which led him to teach that there is only one good God, Ahura Mazda, the Wise Lord of fire and the sun. Opposing God is Angra Mainyu (Ahriman), evil spirit of darkness and the lie, who will eventually be overcome. Every individual, said Zoroaster, must choose between good and evil. After death, each soul will be judged; those who have habitually done evil will be cast into Hell, but those who have chosen good will successfully cross a bridge into Paradise. Among good acts recommended by Zoroaster were kindness to living things and keeping the elements fire, water, and earth free from pollution. He taught that "trees are among the good creatures of [Ahura Mazda, and that] tending them is an act of reverence."[41] The epic *Shah-Nama* records that King Goshtâsp planted a cypress tree in honor of Zardosht (Zoroaster), as a sign that he had adopted the "good religion."[42] Modern Zoroastrians in Iran make pilgrimages to numerous sacred groves growing near springs, waterfalls, and streams, a custom of great antiquity.[43] The Zoroastrian priesthood erected "towers of silence" where the bodies of the dead were exposed to birds of prey so as not to pollute the elements. The dualism of this faith, however, also cut through the natural world. Creatures were classified in two groups, good and evil. Dogs, cattle, and trees were good, but creatures such as wolves, snakes, flies, and demons of disease were in Ahriman's camp, and it was thought a virtuous act to kill them.

The Persian emperors sought to practice Zoroastrianism including reverence for nature and the ritual maintenance of the purity of the elements fire, water, and earth, so that under Persian law pollution was forbidden. The Greek writer Xenophon says that when the king of the Persians traveled through any of his wide-flung provinces, he pointedly observed the condition of the land. Where a landscape was well cultivated and planted with trees, he rewarded the local governor with honors; but where he found deforestation and deserted fields, he replaced the miscreant with a better administrator. So the king judged the worth of his appointees by the care they gave to the land and its inhabitants, believing this just as important as maintaining a garrison for defense or a flow of taxes.[44] The principles seem clear: a governor who cares for the Earth can be trusted to govern well, and the quality of an administration can be judged by the state of the environment in its territory. It seems certain that Xenophon did not imagine this story; he knew the Persian Empire at first hand since he had marched through much of it with a mercenary army and had faced its king on the battlefield. He observed that the Persian king and nobles had paradises, large enclosed tracts of land with forests where wild animals were kept. Unfortunately, in spite of these good efforts, the Persian landscape deteriorated. The hillsides were divested of trees and open to erosion. The irrigation of the Persians, often through underground aqueducts called qanats, was accompanied by salinization like that described for Mesopotamia in Chapter 3. Wildlife was hunted out, in part because of the belief that many fierce wild animals were "evil," and should be exterminated. Agriculture expanded at the expense of forest – not always the best idea in an arid land. The experience of the Persians illustrates that in establishing a sustainable natural landscape, good ideas are not enough.

Athens: mind and practice

Herodotus, the ancient Greek historian, describes a famous tunnel that runs under a mountain on the island of Samos.[45] Built to carry an aqueduct to the city from springs in the hills, it is about 1,000 m (3,300 ft) long. It was designed by Eupalinus at the order of the tyrant Polycrates in 530 BC, a remarkably early date for such a structure. I went to Samos mainly to see it. On arrival, I discovered that it and the other archaeological sites on the island were closed by a strike of the guards. I sympathized with them – they are paid little – but when would I have another opportunity to see the ancient work? Fortunately Greek hospitality prevailed, and when I explained my interest to the leading archaeologist on the island, she arranged for a knowledgeable colleague to get me, along with my wife and daughter, unseen into the tunnel. The remarkable story of its construction is that the builders began at both ends and met in the middle. This required a level of mathematical and engineering sophistication on the part of the designers that is astonishing, but may be explained by the fact that one of the citizens of Samos then was Pythagoras, who devised the right triangle theorem. He left the island to escape the tyranny of Polycrates, but may have stayed there long enough to consult with Eupalinus concerning the tunnel. We followed it as it ran, straight as a laser, into the mountain. Suddenly, the bore swerved, first one way, then the other, as if the excavators had been searching for the other section. Obviously they found it, but ironically, my guide told me, a modern survey showed that if the builders had continued the original straight line, they would have joined almost perfectly, with an error of centimeters. Eupalinus had designed the work correctly, but the contractors did not rely on the theoreticians, and resorted to trial and error. Philosophers had marvelous ideas about building cities and their associated works, but the people who actually built and operated them did not always trust or follow the philosophers.

Ancient Greek cities had to contend with a land where water is a limiting factor. Plato advised assuring a supply of unpolluted water, associating it with tree plantations and sanctuaries, which had sacred groves.[46] Aristotle advised that drinking water should be kept separate from irrigation water, and that temples be built on high places, where their sacred groves would presumably help with the water supply. Athens actually took water from nearby streams and from springs and wells close at hand. Solon the lawgiver had encouraged well-digging in the city, and dozens were excavated in or near the marketplace. As the city grew, local supplies became inadequate, so to tap sources in the country, aqueducts were constructed. These ran at ground level as covered canals, or were raised or buried to maintain a working grade. Athenian engineers devised an underground tunnel from Mount Pentelicus to Athens which was provided with a vertical airshaft every 15 m (50 ft). Meton built another aqueduct to Piraeus. Once in the city, the water was conducted into fountain houses such as the Enneakrounos (Fountain of Nine Spouts). Necessary as this water was to the city, it had been taken from an arid countryside where it was critical to the biota.

The city government supervised water supply, and the office of water commissioner was a prominent one. The Superintendent of Fountains was elected by show of hands. This places the office in the category of those who, like the generals, needed special skills and therefore could not be selected by lot, as most Athenian magistrates were. Water theft, a fairly common offense, was punished by fines. Themistocles, when he was water commissioner, used the fines to pay for a statue of a female water carrier.[47] Sources of water varied greatly in purity, and physicians stressed the need for clean water. Athenaeus of Attaleia wrote a work, "On the Purification of Water," discussing filtration and percolation.

Athens was not isolated from its hinterland.[48] The impact of ancient cities like Athens on the natural environment – on the land and its resources, on air and water, and on animal and plant populations – produced environmental problems prefiguring modern ones, including modification of ecosystems in the surrounding countryside. The citizens had to face decisions regarding land use and urban planning. Athens suffered from crowding, noise, air and water pollution, accumulation of wastes, plagues, and additional dangers to life and limb.

Greek writers from early times assigned many environmental changes to human agency. From Hesiod on, they lauded agriculture, holding that through it humankind was improving the Earth. Human beings were the natural caretakers of the Earth, and its creatures were placed in their custody. Well-planned efforts make the landscape more beautiful and serviceable for human purposes. Humankind improves plants and animals through domestication; in the same way, the extension of civilization amended a defect of the wilderness, a barren waste that was a haunt of beasts. The most moving statement from classical Athens of our species' ability to control other creatures and change the Earth comes from the playwright Sophocles, in the hymn he gave the chorus in *Antigone*. "Many wonders there are, but none so wonderful as Humankind," it begins.[49] This creature can cross the sea and plow the soil, snare birds and beasts, and tame the horse and mountain bull, knows speech and thought, and how to avoid frost and rain – but does not know how to escape death or to prefer justice to evil.

One question of practical environmental import that emerged in ancient Greek thought was: how ought humans to arrange their homes and provide for their own lives? Can a city, or nation, be planned so as to harmonize with the natural environment? A book entitled *Airs Waters Places* bears the name of Hippocrates, the great medical writer of Athens' Golden Age. The author discusses the ways in which a city's setting and natural amenities determine the health and predominant psychology of its inhabitants. He stresses the importance of such factors as exposure to winds and sunlight, the quality of the water supply, the type of soil, and the presence or absence of forests and marshes. The prevailing changes of the seasons, he says, affect various cities in different ways. "Such cities as are well situated with regard to sun and wind, and use good waters, are less affected by such changes."[50] Hippocrates is concerned with description, not city planning, but his ideas have prescriptive value, and later authors such as Aristotle applied them in a normative way.[51]

Athens had grown organically but planlessly around the defensible height of the Acropolis. Streets were a jumble of narrow passages yielding only to the Sacred Way, a straight ceremonial road, and to the open space of the Agora, where there were facilities for trade and political affairs. The philosophers thought population must be limited, not because resources are limited, but in order to keep social control. "A very populous city can rarely, if ever, be well governed."[52] Plato's ideal state would have had 8,000 adult male citizens (although the *Republic* envisions women among the state's educated guardians), while Aristotle would have allowed no more citizens than could see and hear each other at the same time; in those days, of course, there were no public address systems. But Athens was larger than that. Within the walls, which were no further than 1.6 km (1 mi) from the Acropolis, perhaps 100,000 inhabitants lived during the middle of the fifth century BC. City-dwelling Athenians had little space. No wonder Socrates sought the tree-shaded banks of a stream outside the city for his philosophical conversation with Phaedrus![53]

The arrangement of Athens' buildings and streets was not so much chaos as an adjustment of human habitation and movement to the shape of the natural site, its topography and drainage, and to structures that had been established in the past. Rational city planning found its sphere in the expansion of the port and the establishment of colonies. The first city planner to work in Athens was Hippodamus of Miletus, a "metrologist," according to Aristotle, who "discovered the method of dividing cities"[54] by applying principles he observed in celestial phenomena to urban design. He "divided the land into three parts: one sacred, one public, the third private: the first was set apart to maintain the customary worship of the gods, the second was to support the warriors, the third was the property of the husbandmen."[55] He is chiefly noted for the "Hippodamian" plan, in which regularly spaced straight streets cross one another at right angles to make rectangular blocks, some of which are designated as locations of public buildings and a marketplace. He created new plans for Athens' port, Piraeus, and Thurii, Pericles' panhellenic colony in Italy. The Hippodamian rectilinear plan was copied by other cities.

Meton, another Athenian urban planner, also based designs on celestial phenomena. He is caricatured by Aristophanes in the *Birds* as the would-be architect of Cloudcuckooland, where the plan suggested is radial: "In its center will be the market-place, into which all the straight streets will lead, converging to this center like a star."[56] Plato's model city in the *Laws* had twelve equal quarters centered on an acropolis.[57] Modern Greek architect Constantine Doxiadis detected a radial arrangement of the monumental buildings on the Athenian Acropolis when viewed from its ceremonial entrance.[58]

A city is more than the built-up area, however. Its ecosystem includes the surrounding lands on which it depends for food and other resources. Atistotle, describing his model city, turned as a necessary preliminary to an examination of natural features such as the sea, mountains, and forests.[59] In the Greek city-state, town and country were a unit. No political distinction between them existed in classical Athens; peasants of distant villages were co-citizens of the democracy along with urban residents. Land use decisions were made by the sovereign people and applied throughout the territory. Categories of use and ownership were recognized, with specific laws and administrative arrangements. The result was a radical transformation of the countryside.

Philosophers thought a city should be "autarchic," finding the natural resources it needed in its own territory. Plato advised that a city should have essential resources near at hand so that it can be as self-reliant as possible.[60] Aristotle, while believing that the ideal city should produce necessities for itself, recognized that practical considerations would occasion trade. The city should be situated in a place "suitable for receiving the fruits of the soil, and also for the bringing in of timber and other products that are easily transported," preferably from its own territory.[61] Autarchy was never achieved in classical Athens. A city could be self-sufficient only if it could establish a sustainable mode of subsistence with its local eco-systems. The Athenians would not accept such limitations; consequently the economic needs of a militarily powerful city, populous by ancient standards, could only be met by reaching outward through trade and conquest.

The productive rural area around the city consisted of cropland, gardens, orchards, and grazing land. Each category was governed by specific laws and supervised by a city official. For example, owners of olive groves had to pay a tax of three-quarters of a pint of oil annu-ally for each tree to the city, and were forbidden to uproot more than two producing trees in any given year.[62] Xenophon offered opinions on how humans should relate to the natural environment in his work, *Economics*. He concentrated on a subject he knew well as a land-owner: farmers and their relationship to the Earth. The principle he advocates is fundamen-tal: humans should learn from the Earth herself, as from a goddess. "The earth willingly teaches righteousness to those who can learn; for the better she is served, the more good things she gives in return."[63] As a teacher, Earth is not mysterious, but clear and open. "For the Earth never plays tricks, but reveals plainly and truthfully what she can and what she cannot do. I think that because she conceals nothing from our knowledge and understand-ing, the Earth is the surest tester of good and bad men."[64] As an example, Xenophon points out that "to farm correctly, one must first know the nature of the earth ... For you are not likely to get a better yield from the land by sowing and planting what you want instead of the crops and trees that the soil prefers."[65] The way to find out what the Earth prefers, besides observing what crops one's neighbors have planted successfully, is to look at what grows on untilled land with the same type of soil: "Being uncultivated it reveals its own nature. For if the wild growth it bears is good, then by being well treated, it will be able to bear good cultivated crops."[66] Even individual plants can give instruction; the grapevine shows by climbing trees that it needs support, and by spreading and then dropping its leaves indicates when its fruit is ready to gather.[67] The Earth herself reveals when she has been well

Figure 4.1 The Parthenon and other buildings in Athens, Greece, crown the Acropolis, a limestone outcropping around which the city grew. The stone used in their construction is marble brought from Mount Pentelicus, 13 km (8 mi) distant. Photograph taken in 1966.

cared for. A good farmer should buy neglected land and practice what might be termed "agricultural therapy" to heal it. Xenophon advised ways to do this, including the use of green manure and the treatment of saline soils.[68] "Nothing improves more," he maintains, "than a farm that is transformed from an unworked state to fruitful fields."[69] He carefully laid out and took scrupulous care of his own estate.[70]

On the mountains near Athens were forests used as sources of fuel, timber, fodder, and other products. Wood was even more a necessity than in modern cities. It was used to build ships for the navy that maintained the city's sea power, for merchant ships and other means of transport, and in construction (even marble temples had wooden roof supports). As fuel, usually converted into charcoal, it smelted the ore from the famous silver mines at Laurium, fired the popular ceramic vases, cooked food, and heated buildings in winter. A man could make a comfortable living by bringing firewood into the city on a donkey.[71]

By the mid-fifth century BC, the surface of Attica – Athens and its environs – was largely deforested. Forests were prevented from regrowing by the constant grazing of goats and the unrelenting demand for firewood. Erosion depleted the mountain soils, deposited silt along the coastlines, and dried many springs.[72] The result was declining agriculture and a chronic shortage of wood and other forest products. One of the most interesting statements of environmental change in ancient literature is Plato's description of the deforestation and erosion of Attica in the *Critias*. He mentions that in his own day, one could see large beams in buildings from trees that had been cut from hillsides where he could find only low-growing flowering plants, "food for bees."[73] The result was that the rains, instead of being held by the forests, rushed into the sea, leaving former springs dry and carrying away the soil, so that what was left was like the bones of a man wasted by disease.[74] I once took the road up Mount Parnes and saw the rocks, "the bones of the land," laid bare by erosion, with a small remnant forest that suggested what the whole mountain might have looked like before the devastation. Plato had some practical suggestions to avoid deforestation: goats, which damage trees and crops, should be watched by keepers;[75] indiscriminate gathering of firewood should be forbidden, and its supply should be regulated by district foresters;[76] and fire must not be allowed to spread.[77] These positive regulations seem designed to prevent the deforestation he had observed in Attica.[78] Aristotle further counseled that the resources of the landscape surrounding the city should be kept safe by "Inspectors of Forests" and "Wardens of the Country" provided with guardhouses and mess halls.[79] Conservation was, therefore, part of Aristotle's idea of the good city, as it was with Plato.

In practice, however, the exhaustion of local resources forced the Athenians to search for timber abroad. Much of the aggressiveness of Athens can be explained in this way. Diplomats sought advantageous timber deals in treaties with forested lands like Macedonia.[80] Groups of Athenian colonists were dispatched to tree-bearing Chalcidice and Italy. Timber towns like Antandros were dragooned into the Athenian empire, and the timber trade became an issue in conflicts with other maritime cities such as Corinth. As a major argument in favor of his ill-fated expedition to Sicily, Alcibiades mentioned access to the island's forests.[81] By the end of classical times, these woodlands had been depleted. Aristotle's student Theophrastus listed sources of wood in the lands around the Aegean Sea, making clear that Athens was dependent on supplies from places such as Macedonia and Asia Minor.[82] The decline of Athens can be correlated with the failure of the city to maintain the forest ecosystem.

The city government asserted sovereign ownership of unoccupied land within its territory and supervised its use. The living and nonliving components of open land interacted with other parts of the urban ecosystem. Wildlife could be hunted or ores could be mined. Ten officers called *polêtai* negotiated and recorded three- or ten-year leases for mines in the public domain.

One category of land use remains: sacred space, areas dedicated to gods and goddesses. Theoretically these were untouched, with economic activities including hunting and wood gathering forbidden, but in practice the precincts could be leased to private persons and revenue collected. In Athens, the magistrate with jurisdiction over this activity was the Royal Archon, whose duties had to do with religion.

Roads are a means of reaching outward to tap the resources of the countryside and other cities through trade and exploitation. While Athenians preferred travel by sea if they were going far, they also constructed roads. The one used to bring marble from quarries on Mount Pentelicus, and the Sacred Road to Eleusis, were paved with limestone slabs. But the geographer Strabo complained that Greek roads were bad, poorly drained, and often steep.[83]

Garbage and sewage presented a considerable problem. Laws enforced by city commissioners directed that waste matter be carried outside the walls for ten stadia (2 km or 1.25 mi) before it was dumped. Sewers were often covered, and excess water kept them flushed out. Athens had a sewer that provided fertilizer for her fields, but not every house was connected; many had their own cesspools.

Temperature inversions, natural occurrences as common then as in Athens today, held smoke and dust in suspension above the city. Many people are surprised to discover ancient references to air pollution, but although the volume of pollution was undoubtedly less then, it should be remembered that in more recent times, in large non-industrial cities with few cars, fires and the dust of human activities produced a heavy pall.

Cities were a haven for opportunistic organisms that share human habitations, and a number of these were vectors of disease that spread to country districts. Treatment of the dead was a concern; Athenians often buried the deceased rather than cremating them. Corpses were a potential source of disease, so a strict law forbade burial within the city walls. Outside the gates, tombs lined the roads.

Athens had an overwhelming effect on the environment where it was located, and beyond. This was true both in the built-up area and in the immediate vicinity. Cities are, after all, habitats constructed by humans for human occupation. Many problems found in modern cities are not new; ancient cities knew them to a greater or lesser extent. But the impacts of urbanism were by no means limited to the area covered by dwellings and fields, or even to the greater territory over which a city exercised political authority. Athens exploited the resources of the land it could dominate along its frontiers, and its tentacles of trade and economic power reached outward to draw valued materials of many kinds from lands located overseas or across mountain barriers. Places around the Aegean Sea were deforested in response to Athens' demands for fuel and shipbuilding timber. Nowhere was the economic influence of Athens more evident than in the lengths to which its leaders were willing to go to obtain grain for its hungry population. Athens reached into the hinterland of the Black Sea and to trading colonies in Egypt and Syria.

Aristotle gave attention to the role of the environment and resources in economics. This seems appropriate, since "economy" and "ecology" share the same Greek root, and their concerns cannot be separated without danger. "The perfect state," he declares, "cannot exist without a due supply of the means of life."[84] In the *Politics*, he makes a distinction between "natural economy," that is, activities such as agriculture, pastoralism, and hunting, which turn resources into products with intrinsic value, and "unnatural economy," activities using experience and art, such as retail trade, which makes money from exchange. "The

Figure 4.2 Between the columns of the Temple of Apollo, Corinth, the height of Acrocorinthos, bare of forests, is visible. The juxtaposition of the achievements of civilization with a depleted environment is instructive. Photograph taken in 1959.

means of life must be provided beforehand by nature ... Wherefore the art of getting wealth out of fruits and animals is always natural." But "Retail trade is not a natural part of the art of getting wealth," he maintains, since it is a mode where humans extract riches from each other, and these riches are of the spurious kind; one cannot eat coins. He commends natural economy and censures its unnatural counterpart. "The most hated sort" of trade, he continues, "is usury, which makes a gain out of money itself, and not from the natural object of it ... of all modes of getting wealth this is the most unnatural."[85] Aristotle places extractive industries, such as cutting timber and mining, in a third category; they are "partly natural." They are natural because they depend on resources taken from the Earth, but they are unnatural because, without "bearing fruit" (perhaps referring to their sustainability) they are sold for profit.[86] This certainly applies to mining, and it probably applies to timber cutting in the way it ordinarily was practiced in Aristotle's Greece.

Philosophical thought such as Aristotle's did not determine actual practice in ancient Athens. Even when they agreed with each other, which was seldom, the philosophers were a tiny minority. Of course they had influence beyond their numbers, and there were attempts to carry philosophical systems into practice in a few utopian communities. Pythagoras' followers controlled Crotona for a while, and Syracuse briefly accepted the guidance of Plato's philosophy. But although some aspects of philosophy could well have provided positive environmental attitudes, these would not have been effective in conservation without dependable knowledge of the workings of nature and the effects of human actions upon it. In some places traditional knowledge survived, the result of centuries of trial and error.

Figure 4.3 An Orthodox church in Messenia, Greece, its walled cemetery filled with cypresses, suggests the appearance of an ancient temple and walled sacred grove, whose trees and animals were protected from woodcutting, hunting, and other uses. Photograph taken in 1959.

There were subsistence farmers with the kind of respect for the Earth that Xenophon described, and their practices were successful adaptations to the ecosystems they had to live within or perish. They took good care of the land as long as their lives were not disrupted by the frequent wars.

Science in general, and ecology in particular, had at most a beginning among the Greeks. It would have been difficult for them to decide which practices were likely to bring the best results when an environmental problem appeared for the first time, or which intensified in the course of time from a tolerable level to an intolerable one.

It must be concluded that the course environmental problems took in the landscape was not chiefly the result of the concepts of the natural world held by the Greeks. It was also, and probably more importantly, the result of the technology they used, the population levels they reached, the economic measures they took to feed, clothe, and shelter themselves, and the common patterns of their rural and urban lives. Only through studying the interaction of all these factors is it possible to gain an understanding of the ecological processes that underlie the history of ancient Athens.

The failure of Athens to adapt its economy to natural systems in harmonious ways is a cause of its decline, and is one reason why the power it exercised in the Golden Age did not persist very long into the Hellenistic period that followed. The citizens placed too great a demand on available resources, depleted them within their sphere, and then went as far as they could to gain access to additional resources, including imperial expansion throughout the Aegean lands and beyond, until this effort collapsed. They faltered because they failed to maintain the balance with their own environment that is necessary to the long-term survival of any human community. They treated nature as an apparently inexhaustible mine rather than as a living system, as an exploitable resource rather than as part of a community that included them as well. Ecological failures interacted with social, political, and economic forces to assure that Athens suffer a disastrous decline in the level of civilization.

Xian: Chinese environmental problems and solutions

Xian, an ancient capital, is redolent of the Chinese past. The central district is contained within a mighty square of walls that escaped the fate of so many city walls during the Mao Tse Tung era – they were removed and replaced by ring roads, but Xian's was so venerable that it was spared. The city's museum, called the Stone Forest because it has a staggering collection of inscribed tablets, evokes the achievements of several dynasties. From the top of the Great Goose Pagoda, I could look to the east and see the huge earthen tumulus of Qin Shi Huangdi, the first Qin dynasty emperor, which rises above the river valley 35 km (22 mi) away. Now 50.5 m (166 ft) high, it originally towered almost 150 m (500 ft). It is the mausoleum of a man who united China in the late third century BC and controlled its economy. Beyond the tomb lie vaults containing thousands of terracotta warriors, horses, and chariots that were buried to guard the emperor for eternity. Surrounding these was a royal park 56 sq km (22 sq mi) in area, containing animals such as deer, chamois, and rare birds.

Qin Shi Huangdi inspired awe among his people, but little love. He commanded the construction of the Great Wall, an intimidating barrier against northern barbarians. The wall also interfered with the migrations of wild and domestic animals and the people who depended on them.[87] Hundreds of laborers died in that project. Along the wall, a belt of elm trees was planted as an environmental amenity and a barrier against the desert.[88] He established a precedent in the suppression of dissent, ordering that all books be burned except those dealing with useful subjects such as medicine, pharmacy, agriculture, and arbo-riculture, as well as divination by tortoise shell and yarrow. He also spared the records of his own state of Qin. He commanded scholars to desist from discussing the past, and buried alive several hundred who protested.[89] In terms of the natural environment, he showed what an efficient bureaucratic autocracy with a consistent economic policy could accomplish, even in so vast and varied a country as China.[90]

The purpose of the book burning was to suppress "the discussions of the various philoso-phers," which were used by opponents of the regime "to discredit the decrees of laws and instructions," "to cast disrepute on their ruler," and to "lead the people to create slander."[91] There had been a "hundred schools" of philosophy in the decades before the Qin conquest of China. Qin Shi Huangdi followed one of them, Legalism, and wanted to wipe out the others, but his enmity was directed mainly at the most prominent and influential of them, the Confucianists, and to some extent also toward the Taoists. What were the teachings of these schools concerning nature and the environment?

The author of the *Tao Te Ching*, the basic text of Taoism, is given the name Lao Tsu. Little is known of him, but he possibly lived in the sixth century BC. The book portrays the Tao (Way) as the principle of Nature, which underlies all existence. A human being lives wisely by following nature:

> Man follows the earth's law.
> Earth follows heaven's law.
> Heaven follows the Tao's law.
> Tao follows what is natural.[92]

The *Tao Te Ching* advises people to live a "good and simple life."[93] This would entail doing as little to interfere with the natural order as possible; indeed, the principle of action is *wu wei*, "working without doing,"[94] allowing nature to indicate the path. The Taoist ideal state would be a very small village with few or no laws, where people live at peace and do not seek their fortune elsewhere.[95] Such a community would live lightly on the earth, preserving life and making few changes, none if possible, in the natural order. It would be hard to imagine any philosophy more unlike that of the Qin emperor.

For Confucianists, the Tao was found in proper social relationships. Their founder, Confucius, lived in the late sixth and early fifth centuries BC. The individual, for him, was an uncarved block capable of being shaped by education. He taught that persons of all social levels can learn, through knowledge of past examples and practice of ritual, to become participants in a harmonious social hierarchy. A child is born into definite relationships, such as those with parents, rulers, friends, and fellow beings, and learns the duties that result from these relationships. This philosophy is radically anthropocentric. But it also teaches that humankind and nature must be in proper relationship.[96] The animosity of Qin Shi Huangdi to the Confucianists is not based on their attitude to nature, but on the Confucianist tendency to criticize a ruler who failed to follow what they considered the right customs.

In the days of Mencius, a philosopher who lived in the fourth century BC and wrote one of the four classics of Confucianism, China was in a period of rapid economic and environmental change. Businesses involved in markets and trade were active, with expanding use of coinage. A burgeoning population was making demands on agricultural production, and forests were being felled to make room for fields of wheat and rice, as well as to provide fuel and building materials. Officials of the states into which China was divided needed expertise in handling political and environmental affairs.

A section of Mencius' book that has impressed historians is his description of Ox Mountain, an outstanding demonstration of the sage's acuteness in observing environmental change and its causes:

> There was a time when the trees were luxuriant on the Ox Mountain. As it is on the outskirts of a great metropolis, the trees are constantly lopped by axes. Is it any wonder that they are no longer fine? With the respite they get in the day and in the night, and the moistening by the rain and dew, there is certainly no lack of new shoots coming out, but then the cattle and sheep come to graze upon the mountain. That is why it is as bald as it is. People, seeing only its baldness, tend to think that it never had any trees. But can this possibly be the nature of the mountain? Can what is in man be completely lacking in moral inclinations? A man's letting go of his true heart is like the case of the trees and the axes. When the trees are lopped day after day, is it any wonder that they are no longer fine? ... Others ... will be led to think that he never had any native endowment. But can that be what a man is genuinely like? Hence, given the right nourishment there is nothing that will not grow, and deprived of it there is nothing that will not wither away ... [97]

Mencius saw a mountain that had been stripped of its forest by logging, and observed the way in which grazing can make deforestation permanent by preventing the growth of new trees.[98] This passage is remarkably similar to the description in the *Critias* by Plato, in the same century, of the deforestation of mountains near Athens.[99] In both cases, the philosophers report processes of which they were eyewitnesses. Mencius wrote that Confucius had climbed two mountains, and made similar ascents himself.[100] Undoubtedly many highlands in China were suffering the fate of Ox Mountain.

Figure 4.4 The historic city of Xian, which was the capital during the Qin dynasty, is one of very few in modern China that retains its wall. Photograph taken in 1988.

Mencius considered land management one of the most important responsibilities of the state. He advised rulers to make periodic inspection tours of their domains, and to observe the condition of the land as evidence of the quality of stewardship, or lack of it, among their subordinates. He told them that officers should be rewarded if the land is well cared for, but "on the other hand, on entering the domain of a feudal lord, if he finds the land is neglected, ... then there [must be] reprimand."[101] The same observation was made by his near contemporary Xenophon, concerning the king of the Persians, as mentioned above.[102] In both cases, the principle that the authorities must rule on behalf of the inhabitants was recognized. Mencius insisted that it is inadequate for a ruler to wish his people well; he must show his benevolence by instituting economic programs to advance their welfare.[103] He insisted that "the people are of supreme importance; the altars to the gods of earth and grain come next; last comes the ruler."[104] Rulers were not exempt from labor on behalf of the people. A landlord had to plow the soil to grow grain for the sacrifices,[105] and it was the duty of the ruler to care for the land so that it would provide an environment that nurtured native human goodness. The condition of the environment in a country offered telling evidence concerning the merit of its government.

In theory, the ruler owned the land and allotted it to those who used it. A benevolent ruler must pay close attention to land and labor, since peasants would flee from the territory of a malevolent lord to that of a provident one.[106] The benevolent ruler will begin by resurveying the land. Mencius favored a traditional method of distribution called the well-field system, after the character *jing* for a water well, which looks something like a tic-tac-toe board, or the Western sign for "number" (#). A square of land was divided in this manner into nine smaller squares, each of the eight outer plots being assigned to one farm family, and the center plot being a public field cultivated by all eight families with the produce going to the government.[107] In this way, labor could be shared and the tax would be lighter in a year of crop failures. Mencius opposed fixed taxes that required the same payment by farmers whether crops were heavy or light.[108] It is interesting to note that the well-field system is based on the "nine divisions" scheme of cosmography, which subdivided the world, the continent, and China itself into nine sections, similarly arranged, with Mount Kun-lun, the *axis mundi*, in the center.[109] Symbolically, such an arrangement would make each nine-field unit a microcosm reflecting the structure of the universe. This should not be taken to mean that the system itself is mythical, however. Recent geographical study of landscapes in north China reveals rectilinear patterns consistent with the hypothesis that the well-field system was not a mere ideal, but was extensively used.[110] Theoretically, family plots throughout China would have been equal, and the size mentioned by Mencius was about 1.8 ha (4.5 acres). But in practice, peasant allotments varied greatly.[111]

Can the well-field system be regarded as a primitive form of socialism?[112] Some of its elements would make it seem so. The land, which was the means of production in this agricultural society, in theory belonged to the state and was parceled out equally among the farmers. Strictly speaking, there was no private property, since the land could not be sold. The arrangement was something like a commune, since Mencius obviously intended that the "eight families would form a community with close relations of friendship and mutual aid."[113] Every farmer made a contribution of labor on common land, and therefore gave a portion of produce which was set at a reasonable percentage of annual yield. The pattern was ordained and managed from above, which is perhaps more like socialist practice than socialist theory. It is interesting, though, that decisions on practices of cultivation in the plots were left to those with the most practical experience, that is, the farmers themselves. Other aspects of the system seem feudal, since labor on the public field was obligatory, farmers were bound to their assigned plots in the fashion of serfs, and the structure supported a class of landowning nobles.

A distinctive emphasis of Mencius is his recommendation of conservation practices. He said, "Earth is more important than Heaven, and Man more important than Earth."[114] His grasp of the principle of the wise use of renewable resources can scarcely be faulted. His advice to King Hui of Liang is notable:

> If you do not interfere with the busy seasons in the fields, then there will be more grain than the people can eat; if you do not allow nets with too fine a mesh to be used in large ponds, then there will be more fish and turtles than they can eat; if hatchets and axes are permitted in the forests on the hills only in the proper seasons, then there will be more timber than they can use.[115]

Here it was assumed that regulations governing economic activities would be promulgated and enforced. The people should be allowed to work in the fields at seedtime and harvest, presumably not being marched off to war. The nets with wide mesh to be used in fishery would allow the small fish and turtles to escape and grow to catchable size. A form of sustained-yield forestry would assure a supply of wood in succeeding years. Mencius' advice concerning forest conservation was particularly sound. In the Ox Mountain passage, he observed the advance of deforestation and its causes. In this section, he advised careful practices of timber harvesting and the planting of trees; in other places, he objected to the building of huge mansions and indicated the wisdom of preventing the waste of cut logs.[116]

If the Chinese rulers and people had heeded Mencius' advice, the environment of China would not have been so badly degraded.[117]

Chinese governments in this period asserted authority in the area of forest and fishery management. Land surveys included attention to forests, lakes, and coastal zones.[118] Mencius referred to foresters, gamekeepers, and managers of lakes stocked with fish as ordinary positions on the staff of a ruler.[119] Methods of cultivating timber trees were well known and practiced.[120] In certain times and places, these measures no doubt resulted in conservation, but the overall impression one gains from Chinese history is of the uneven but inexorable advance of deforestation.

A major force both in deforestation and removal of wildlife in Mencius' time was the expansion of agriculture into undeveloped land. In the two centuries before, the ox-drawn iron plowshare had come into use, supplementing human labor with a major new source of energy. Other tools and methods of fertilizing had been invented.[121] Thus it is not surprising that Mencius spoke of the increase of cultivated land at the expense of the wild. He sometimes did this by referring approvingly to the deeds of mythical kings who had originally cleared the land for human habitation.[122] The contemporary Legalist, Shang Yang, urged the rulers to take measures to cultivate waste lands as a deliberate policy to increase population.[123] Rulers often ordered the cultivation of wasteland to combat famine. Mencius compared the expansion of fields at the expense of forest to the military conquest of neighboring lands. He opposed opening up new lands for tyrants, saying that those who do so deserve punishments, including death, equal to those imposed on men who make war or secure alliances for the same evil rulers. The land base, he believed, should be increased only for rulers who practiced benevolent government.[124]

A measure that inhibited agricultural expansion but added to environmental amenities was the establishment of gardens, parks, and reserves by the rulers. These were not wilderness; in Chinese gardens every bit is designed, and art exhausts itself to be indistinguishable from nature. Mencius thought that inappropriately large enclosures of land would deprive the people of their livelihood, but one needed to consider also whether it would be open to the people to enjoy and to use in customary ways such as gathering firewood and hunting small game. Ordinary folk would resent even a small park if they were kept out, but would take pleasure in an extensive one if their ruler shared it with them.[125] A park should be created not only for the ruler's private enjoyment, nor only to preserve animals, plants, and land, but most importantly for the benefit and enjoyment of the people.

Do animals have value for Mencius, and should they be conserved? Mencius saw the human feeling of compassion for an animal as positive and ennobling, although less so than a similar feeling for other human beings.[126] When Mencius told King Xuan of Qi that he knew that the king could bring peace to his people, Xuan asked how Mencius could tell. Mencius said it was because King Xuan had seen an ox being led to sacrifice, and could not stand to see it shrinking with fear, so he had spared it and ordered a sheep slain in its place.[127] As the sage observed, "Even the devouring of animals by animals is repugnant to men."[128] Since the king felt empathy for an animal, Mencius was certain that he could feel similarly for his people. He had no opposition to animal sacrifice as such, since Confucius himself had taken part in sacrifices.[129] But it was strange that the king, moved by the suffering of the ox, should have ordered a sheep sacrificed in its place. Mencius explained that it was because the king had seen the ox, but not the sheep. His conclusion is not that the sheep should also be spared, but that a gentleman should stay away from the kitchen to spare his own feelings! Mencius himself was not a vegetarian, since he once remarked that his favorite dishes were fish and bear's paw.[130]

Mencius' advice on the treatment of animals must be considered in light of his distinction between human nature and animal nature. Mencius used the word *xing* for "nature" in two senses, that is, specifically human nature and nature in general.[131] One of the best examples is in the famous passage about Ox Mountain,[132] where Mencius compared the nature of the mountain, which is that it would be forested, with the nature of a human being, which is good. Mencius' purpose is to argue for his doctrine of the original goodness of every individual, but he is not speaking only of human nature. The mountain, too, has a basic nature which is good, and when it is violated it becomes "no longer fine." It is best if everything in the world, human person and mountain alike, can develop in accord with its own nature. In the same sense, animal nature is good. Indeed, there is nothing at all wrong with it for animals. "In that case," as Mencius once asked, "is the nature of a hound the same as the nature of an ox and the nature of an ox the same as the nature of a man?"[133] Obviously not. Every animal has its own nature, distinct from that of other species, and human beings have their own nature, different from the general "nature" that they share with animals and the universe. The distinct nature of humans is the recognition and honoring of appropriate relationships to other human beings.[134] Human beings, if they abandon their relationships and duties to other humans, which constitute their own proper nature, become false animals, which is unnatural to them.[135] But it is laudable for human beings to behave like animals if that means to follow their own inherent good nature, as animals follow theirs: "The people turn to the good as water flows downwards or as animals head for the wilds. Thus the otter drives the fish to the deep; thus the hawk drives birds to the bushes."[136] For humans, the good is natural. One who wants to follow the right path can do so simply by looking into one's inmost nature, because the knowledge of right and wrong is inborn.

Mencius criticized the extravagant life of many nobles which was wasteful of resources, since he believed that nothing is "better for the nourishing of the heart than to reduce the number of one's desires."[137] Despising "men of consequence," he said,

> Their hall is tens of feet high; the capitals are several feet broad. Were I to meet with success, I would not indulge in such things. Their tables, laden with food, measure ten feet across, and their female attendants number in the hundreds. Were I to meet with success, I would not indulge in such things. They have a great time drinking, driving, and hunting, with a retinue of a thousand chariots. Were I to meet with success, I would not indulge in such things. All the things they do I would not do ... Why, then, should I cower before them?[138]

Mencius advised a middle way, not the ostentation of these men of consequence, nor the absolute self-sufficiency, amounting to poverty, espoused by Hsu Hsing and some Taoists.[139] But he admired the natural man who lived in the wilderness, sensing that he was closer to the truth and more amenable to education in wisdom than many who have been raised in more civilized surroundings.[140] Some modern Chinese commentators have criticized Mencius as being too strongly in favor of preservation and, consequently, "anti-development."[141]

Mencius' advice was not taken seriously enough by the rulers of China. Too often they squandered their states' resources on ostentatious new palaces, tombs, indulgences, and, above all, military adventures. The increased use of bronze and the introduction of iron required fuel wood for metallurgy. More hillsides were denuded and exposed to erosion. Economic crises and famines persisted after the fourth century BC, and set the scene for a takeover of the entire country by the militaristic, centralizing state of Qin and the creation of the imperial system that would dominate China for the centuries following.

The Legalist philosophy favored by the first Qin emperor held that human nature is unruly and must be constrained by law. To bring population into balance with resources, Legalists advised allowing private ownership of farms so that self-interested farmers who worked hard would increase their holdings and their production. They did not apply the same principle to commerce, since they thought prosperous entrepreneurs would accumulate resources that the ruler ought to command. They thought that the only activities the state should encourage were agriculture and the military, and that both of these should be kept under the ruler's autocratic scrutiny. The Qin emperor severely restricted business and supposedly encouraged agriculture, although the private farmers were subjected to a grain tax of 50 percent. To attack the problem of a short food supply, he used two methods. First, production was augmented by bringing more land under cultivation and by constructing irrigation systems. Second, an "ever-normal granary" stored grain in years and districts of abundant crops, to be distributed in cases of famine.[142] His military conquest of all China meant that there could be empire-wide policies including a single currency, standardized weights and measures, trade regulations, and a centrally directed corps of officials. For a time, it must have seemed that he had brought China's ecological crisis under control.

His own lifestyle, it seems, was not guided by conservation. In 212 BC, Qin Shi Huangdi began the construction of a huge palace for himself in Xian, which required so much fine timber in its construction that entire mountains were stripped of trees to build it, as a poet remarked:

> When the Six Kingdoms came to an end
> When the Four Seas were unified
> When the mountains of Sichuan were denuded
> Then the Apang palace appeared.[143]

It was not the deforestation, but the ostentation, that became an issue for the opposition, and the Qin dynasty proved short-lived. Although Qin Shi Huangdi may have died of overwork, his successor was not so dedicated, and his regime collapsed. The Qin palaces were burned by rebels. In the Han dynasty that followed, the teachings of Confucius and Mencius returned to favor. The new leaders decided that the ideas that should form Chinese culture were those of the Five Classics that had, fortunately, survived the Burning of the Books, and the skills government employees should have would be based upon an understanding of the classics. They set the foundations of the famous Chinese examination system that provided able administrators for centuries afterwards. Some Qin policies continued, however, especially those dealing with agriculture, and the time to come would see an uneven but inexorable rise in population, a lamentable loss of forests, and continued danger from famine in spite of an expansion of agricultural land. The Yellow River Valley suffered severe deforestation and resultant soil erosion and flooding, a process that continued for centuries, spreading through hilly areas in the south with Han Chinese settlers and affecting the Yangtze River Valley.[144] China would become one of the most severely deforested centers of civilization.

Rome: environmental reasons for the decline and fall

Trajan's Column is one of the most imposing monuments that survive from ancient Rome. Built of Carrara marble, it is 3 m (10 ft) or so in diameter and rises 35.1 m (115 ft) above the pavement, counting a base that is 5.4 m (17.6 ft) high.[145] When it was dedicated in AD

113, it bore in addition a 5.5 m (18 ft) bronze statue of the emperor in whose honor it was erected; the statue fell at an unknown time, and was replaced in 1588 by another of St. Peter. The most interesting feature of the column is the marble relief more than 1 m (3 ft) in height, which spirals upward around the column twenty-three times and contains, among many other things, 2,500 human figures. It celebrates Trajan's conquest of Dacia, a territory included in modern Romania, and is regarded by some experts as a principal source of information about Roman military equipment and operations.[146] But it also reveals a lot about the effects of the Roman army on the environment.

More than two hundred trees are represented in the relief. Usually their species can be identified; oaks are the most common, their leaves and acorns clear in many cases; there are also pines, cypresses, and olives. Sometimes they serve as background or as dividers between scenes – the sculptor's indication that the action was taking place in forested Dacia. Some of them stand alone, and others in groups as a kind of "shorthand" for forest. Many are in the foreground, often being chopped down vigorously by ax-wielding Romans or Dacians. Sometimes the military axmen are clearing roads through thick woodland to allow passage for the legions. More often, they can be seen carrying away logs and using them to make siege terraces, catapults, battering rams, and beacon fires. One such beacon, not yet ablaze, is made of 144 logs.[147] There are many structures that demanded timber in their construction: camps, forts, palisades and other defense works, warships, boats, and barges loaded with barrels. Then there are the bridges of boats, huge assemblages of wood. Two of them, shown near the beginning of the relief, cross the Danube: "Each boat carries, amidships, a stout pier of logs firmly held together by horizontal slats. In between every pair of boats there is a pontoon of closely fitted planks; and the piers and pontoons carry the timber roadway structure of the bridge, with railings at the sides."[148] Each of the soldiers crossing the river carries a wooden stake. The emperor offers sacrifice on a fire altar.

The work to supply the huge amounts of wood necessary for military operations was done by *classiarii*, technical support units for the army, directed by "ax masters." If necessary, these men could fight with their axes, as the column relief shows. The transformation of the landscape by these operations was massive. Toward the end of the relief, a scene in northern Dacia where a forest god contemplates a little lake among the woods, rich in game such as deer and boars, is followed by a tame pastoral landscape where sheep and cattle graze around a single tree that bears only two meager tufts of leaves above a trunk whose branches almost all have been lopped.[149]

Most historians have given up trying to find one all-conquering cause for the decline and fall of the Roman Empire, and have retreated to the safer ground of multiple causation. What brought down Rome was a number of processes that interacted.[150] One of these was the Roman mistreatment of the natural environment, including overexploitation of scarce natural resources such as forests and soil, and failure to find sustainable ways to interact with the ecosystems of Italy and the many other lands, including Dacia, which they conquered.

Study of Roman writings, archaeological reports, and scientific studies of deposits of silt from erosion and ancient pollen grains have led me, and others, to the conclusion that environmental factors were important causes of the decay of Roman economy and society, and that the most important of these factors were produced by human activities. The result of the process of deterioration is evident in the landscape, where impressive ruins are often surrounded by desolate, desert-like environments.

Some of the wisest Romans were aware that humans often abuse the natural world. Seneca remarked, "If we evaluate the benefits of nature by the depravity of those who misuse them, there is nothing we have received that does not hurt us. You will find nothing, even of obvious usefulness, such that it does not change over into its opposite through man's fault."[151]

Among the many ecological problems suffered by the Roman Empire, most were caused by the Romans themselves. Deforestation and its consequence, soil erosion, leads the list of these disasters. Early in the twentieth century, Vladimir Simkhovitch suggested that these were the main causes of the calamity of agricultural exhaustion.[152] Agricultural crises were responsible for rising prices, food shortages, and labor shortages. The extinction of many species of animals and plants affected agriculture in unsuspected ways. B.D. Shaw, the perceptive historian of Roman North Africa, averred: "the tens of thousands of animals purposefully hunted down for the arena were, of course, a small proportion of the total that yielded to more mundane processes such as the systematic destruction of their habitat by the expansion of agricultural settlements."[153] With each species that is extirpated, the ecosystem verges closer to collapse, so by hunting and capturing animals for slaughter in the

Figure 4.5 Trajan's Column, in the Forum of Trajan, Rome, bears a spiral marble relief celebrating the emperor's conquest of forested Dacia (Romania), illustrating the effects of military activity on the ancient European Environment. Photograph taken in 1994.

arena, the Romans were weakening their economies in the long run.[154] They were unaware of this, because they thought that by killing off animals that sometimes raided their herds, they were doing a good thing. But predators ate a far greater number of rodents and other animals that devour crops, and increase in numbers of the latter reduced agricultural production.

Industry in the Roman Empire did not make up as large a segment of the economy as it does today, but it had significant environmental consequences. One can still see scars of ancient mining and quarrying, although they are often eclipsed by modern operations. Demands on forests for timber and fuel for mining, smelting, metallurgy, and firing of ceramics were even more destructive.[155] Pollution may not have been produced on the modern scale, but the Romans lacked technology to reduce effluents to the air or water, except for construction of chimneys to disperse noxious smoke high in the air, as in the silver smelters of Spain mentioned by Strabo.[156] Studies of ice cores drilled in Greenland have shown that lead in the atmosphere increased during Roman times.[157]

Air pollution due to smoke, dust, and odors from urban activities made life unhealthy. Water was polluted by sewage that fouled ground water and made wells unsafe, especially in cities. Not every town had well-maintained aqueducts like those that supplied Rome. But in that great city, the *Cloaca Maxima*, or "main drain," discharged into the Tiber River, threatening not only those living downstream, but Rome itself when the river flooded and untreated effluent invaded the streets. Toilet and garbage pails were emptied out windows, attracting vermin and rotting into sludge so deep that, in places like Pompeii, stepping stones were provided for pedestrians. Such conditions provided breeding grounds for diseases which did not spare Rome.

Why did the Romans fail to maintain a sustainable balance with the Mediterranean ecosystem within which they lived?[158] Among the answers to that question is the general Roman attitude to the natural world. It would seem that the way they regarded nature would help to determine their decisions and actions that affected it. The early Romans saw the landscape as the sacred space of the gods. They avoided actions that would anger their deities, such as killing deer in temple forests, and tried to please the gods by planting trees. These traditions contained ecological wisdom, but there was always the danger that they would deteriorate into automatic rituals and lose their intimate connection to natural processes. Romans tended to cut corners for economic expediency, as when Cato the Elder advised the use of a handy prayer, "to the god or goddess whom it may concern," whenever a Roman farmer wanted to cut trees or plow ground in a sacred grove, where it was ordinarily forbidden to do so.[159]

Whatever gave short-term profit was the rule in developing natural resources. The Romans had turned the nations of the Mediterranean basin into provinces; they seem to have treated nature, too, as a conquered province.[160] The Romans had made citizens of other lands into slaves; they appear to have assumed they could do the same with the Earth and all her creatures. Their pragmatism, however, was short-sighted.

In order to survive and prosper in the long run, the Romans needed knowledge of how nature operates and what the effects of their actions might be. It would be unfair to minimize the amount of practical knowledge that Roman farmers and herders possessed, accumulated through trial and error over many centuries. These came down to each generation in the form of wise instructions, sayings, and agricultural writings such as those of Cato, Varro, and Columella. However, the Romans lacked anything that could be called science in the modern meaning of that term. Experiment and hypothesis were almost unknown disciplines. Philosophers asked some questions that might be termed ecological, but their answers were based on the doctrines of the schools to which they belonged, and were of limited application to environmental problems. Even as incisive a philosopher as Lucretius believed that agriculture was declining because the Earth was getting old.[161] So neither accepted wisdom nor philosophy was a guide to ecological sustainability.

The Romans are deservedly admired for their technology. Frontinus boasted that their aqueducts were better built than Egyptian pyramids or Greek monuments, and certainly more useful.[162] Technology will aid in the long-term survivability of a civilization only if it is appropriate. It might be supposed that Roman technology was environmentally less damaging than its modern counterparts, since it was simpler, utilizing human and animal power for the most part and waterpower to some extent. However, the Romans brought their efforts to bear during centuries, and even simple technologies can be destructive when they are used over large territories for long periods of time, as the inroads into forests resulting from dependence on wood and charcoal for energy demonstrate. Ironically, the technological achievements of the Romans we most admire most are the very ones that show clearly their ability to damage the environment.

No other ancient empire combined large size with social control as effectively as did Rome. A civilization can direct its effects on the environment efficiently only if it can use positive and negative methods to get its people to act in ways that are considered to be social goods. To achieve goals desirable for society requires individual sacrifices. For instance, a goatherd will not keep his animals off a hillside where tree seedlings are growing just because it would be good for shipbuilders to have a forest there in a few years. Roman

Figure 4.6 A Roman mosaic shows an African elephant being taken aboard ship to be brought to Rome, possibly to be exhibited in the amphitheater. Image from the "Corridor of the Great Hunt," Roman Villa of Casale, near Piazza Armerina in Sicily, dating from the fourth century AD. Photograph taken in 1994.

efforts were geographically far-reaching, accessing resources located at great distances. Roman roads and ships brought timber from the Alps and Lebanon. Tin came to the Mediterranean from beyond the Strait of Gibraltar. Gigantic projects like the Roman road system, which had enough total length to reach the moon, show that rulers had ways of getting cooperation. People evaded regulations when they could, however, if it seemed in their interest. Rome at its most efficient, say in the days of the autocratic emperor Diocletian, could accomplish more and invade its citizens' lives more thoroughly than any other ancient empire, but could not approach the ability of a computer-equipped modern state to keep informed about its citizens and make sure that they perform social duties.

Much of Roman social organization was occupied by military preparation and war effort, a fact abundantly illustrated by Trajan's Column. War was damaging to the ancient landscape. The structure of the Roman government was designed like an army, and periods of peace, when the priests symbolically closed the gates of the Temple of Janus, were so rare as be recorded with wonder by historians. The well-known *Pax Romana* lasted for almost two hundred years, but it was not uninterrupted and it did not end wars on the frontiers. In fact the Dacian conquest, which illustrates how far war's environmental devastation extended beyond Rome's neighborhood, occurred during the "Roman Peace." The military anarchy of the Third Century followed close on its heels: fifty years of war that left no province untouched. War hurt agriculture in many ways. Taxes for the military were collected mainly from farmers, reducing their ability to invest in producing crops. Military campaigns devastated the countryside, slaughtered farmers and their families, and requisitioned or destroyed crops and buildings. Army agents conscripted farmers, who often spent years fighting instead of caring for the land, inevitably neglecting terraces and irrigation works. The passage of armies living off the country and trampling crops was a calamity noted often in literature, and Roman generals used deliberate environmental warfare that demolished an enemy's natural resources and food supplies. The reliefs of Trajan's Column show soldiers setting fire to villages and rounding up peasants as prisoners and slaves.

Figure 4.7 The Tiber River served as water supply and means of transportation for the imperial city of Rome. Photograph taken in 1959.

The trend of the Romans' actions affecting the environment over the centuries was destructive. They exploited renewable resources faster than was sustainable, and consumed nonrenewable resources as rapidly as they could. They failed to adapt their economy to the environment in sustainable ways and placed an insupportable demand on the natural resources available to them. Thus they failed to maintain the balance with nature that is necessary to the long-term prosperity of a human community. They depleted the lands they ruled, and in so doing undermined their own ability to survive. Environmental changes as a result of human activities must be judged to be one of the causes of the decline and fall of the Roman Empire.[163]

Conclusion

Systems of ideas have power to shape human action. Individuals have at times behaved according to the principles of a doctrine they have accepted, consciously or unconsciously, although the behavior would seem to have been counter to self-interest and may even have resulted in death. When they have commanded the allegiance of human societies, systems of ideas have built institutions and monuments, stimulated conquest as well as resistance to conquest, and changed the face of the Earth. The empires of the ancient world were quick to realize the potential for wider social control offered by religions and philosophies, and sometimes turned them into state doctrines. Their importance in environmental history therefore should not be underestimated.

It should be kept in mind, however, that the use of systems of ideas by states is far more common than the attempt of states to follow those systems whenever they conflict with what the states consider to be their own interests. Therefore the fact that environmentally positive teachings can be found in ancient religions and philosophies does not always surely indicate that they were put into practice. The desire to maintain power and the search for resources to maintain it undoubtedly took precedence in most cases.

Many of the systems of ideas that flourished in the ancient world continued to be active in subsequent periods, including modern times. Sometimes it is suggested that the wider observance of one or another philosophy or religion would improve humankind's relation to the Earth. But the ecological process of the relationship of human societies to the rest of nature is dynamic. A rule of behavior that produces a positive result in one time and place may do the opposite under other circumstances. New occasions teach new duties. The religions and philosophies that wish to help our species adapt, survive, and at the same time preserve the community of life must be able to change, discarding outmoded formulations and recognizing the need to respond to ever-new environmental challenges.

Notes

1   Peter Marshall, "Hinduism: The Way of Understanding," in *Nature's Web: An Exploration of Ecological Thinking*, London, Simon & Schuster, 1992, 24–40.
2   Madhav Gadgil and Ramachandra Guha, *This Fissured Land: An Ecological History of India*, Berkeley and Los Angeles, University of California Press, 1992, 79.
3   Madhav Gadgil, "The Indian Heritage of a Conservation Ethic," in Bridget Allchin, F.R. Allchin, and B.K. Thapar. eds, *Conservation of the Indian Heritage*, New Delhi, Cosmo Publications, 1989, 15–16.
4   Takeshi Umehara, "The Civilization of the Forest," *New Perspectives Quarterly* 16, 2, 1999, 40–8.
5   Gadgil, "Indian Heritage of a Conservation Ethic," 13–21, at 14–15.
6   J. Donald Hughes, "Gaia: Environmental Problems in Chthonic Perspective," *Environmental Review* 6, 2, Fall 1982, 92–104, at 98–9.

7   Hermann S. Schibli, *Pherekydes of Syros*, Oxford, Clarendon Press, 1990, 69–77; Empedocles fragments 8–9.

8   Plato *Timaeus* 30d.

9   Mohan Matthen, "The Organic Unity of Aristotle's World," in Laura Westra and Thomas M. Robinson, eds, *The Greeks and the Environment*, Lanham, MD, Rowan & Littlefield, 1997, 133–48.

10 Aristotle *Metaphysics* 107a17–20.

11 Aristotle *Meteorology* 342a29–33, 346b18–347a12.

12 Aristotle *Generation of Animals* 731b30–33, 777b17–20.

13 My repeated searches of the corpus of ancient Greek literature in the *Thesaurus Linguae Graecae*, now available as a CD containing virtually everything extant which was written in Greek between 800 BC and AD 800 have failed to find *oikologia* or any variants. It has occurred to me, however, that the founders of modern ecology may have borrowed the "eco-" ("oeco-") portion of "ecology" not from *oikos* ("house"), but from *oikeios* ("appropriate"), a word used by Aristotle and Theophrastus to designate the favored habitat of an animal or plant species.

14 Plato *Laws* 10.903b–d.

15 Timothy A. Mahoney, "Platonic Ecology, Deep Ecology," in Westra and Robinson, eds, *Greeks and the Environment*, 52.

16 Chatsumarn Kabilsingh, "Early Buddhist Views on Nature," in Allan Hunt Badiner, ed., *Dharma Gaia: A Harvest of Essays in Buddhism and Ecology*, Berkeley, CA, Parallax Press, 1990, 8–13, at 11.

17 Another story says that he had a past life as a sal tree. Christopher Key Chapple, "Animals and Environment in the Buddhist Birth Stories," in Mary Evelyn Tucker and Duncan Ryûken Williams, eds, *Buddhism and Ecology: The Interconnection of Dharma and Deeds*, Cambridge, MA, Harvard University Press, 1997, 131–48, at 140.

18 *Genesis* 1. 26–30.

19 *Deuteronomy* 20. 19–20.

20 Genesis 21. 31–33.

21 *Shabbat* 67b.

22 *Berakhot* 50b

23 *Deuteronomy* 5. 12–15.

24 *Deuteronomy* 22. 6–7.

25 *Sanhedrin* 108b. See Jonathan Helfand, "The Earth is the Lord's: Judaism and Environmental Ethics," in Eugene C. Hargrove, ed., *Religion and Environmental Crisis*, Athens, GA, University of Georgia Press, 1986, 38–52.

26 *Numbers* 35. 2–5.

27 *Ecclesiastes Rabbah* 8. 28.

28 *Matthew* 6. 26.

29 *Luke* 12. 6.

30 *Matthew* 6. 28–9.

31 *Romans* 1. 20.

32 *Acts* 14. 15.

33 *First Timothy* 4. 4.

34 *Hebrews* 2. 8.

35 *Colossians* 1. 17, 20.

36 *Revelation* 21. 5.

37 Seyyed Hossein Nasr, "Islam and the Environmental Crisis," in Steven C. Rockefeller and John C. Elder, eds, *Spirit and Nature: Why the Environment is a Religious Issue*, Boston, MA, Beacon Press, 1992, 86–108, at 88.

38 *Quran* 2. 30.

39 Iqtidar H. Zaidi, "On the Ethics of Man's Interaction with the Environment: An Islamic Approach," in Hargrove, ed., *Religion and Environmental Crisis*, 107–26.

40 The most widely known critic of Christianity in ecological terms is Lynn White, and of monotheism generally, probably Arnold Toynbee. See Lynn White, "The Historical Roots of Our Ecologic Crisis," *Science* 155, 1967, 1203–7, and Arnold Toynbee, "The Religious Background of the Present Environmental Crisis," *International Journal of Environmental Studies* 3, 1972, 359–67.

41 R.C. Zaehner, *The Teachings of the Magi: A Compendium of Zoroastrian Beliefs*, New York, Oxford University Press, 1976, 21.

42 Ferdowsi, *The Epic of the Kings: Shah-Nama, the National Epic of Persia, by Ferdowsi*, tr. Reuben Levy, Chicago, University of Chicago Press, 1967, 192.

43 Mary Boyce, *A Persian Stronghold of Zoroastrianism: Based on the Ratanbai Katrak Lectures, 1975*, Oxford, Clarendon Press, 1977, 52, 137, 156, 254, 255–6, 262–3, 269.

44 Xenophon, *Oeconomicus* 4. 8–9.

45 Herodotus, *Histories* 3. 60.

46 Plato, *Leges* 761b–d, 845d–e, 947e.

47 Plutarch, *Themistocles* 31.

48 Moses I. Finley, "The Ancient City from Fustel de Coulanges to Max Weber and Beyond," *Comparative Studies in Society and History* 19, July 1977, 305–27.

49 Sophocles, *Antigone* 332–75.

50 Hippocrates, *Airs Waters Places* 10.

51 Aristotle, *Politics* 1330a–b.

52 Aristotle, *Politics* 1326a–b.

53 Plato, *Phaedrus* 229–30.

54 Aristotle, *Politics* 1330b.

55 Aristotle, *Politics* 1267b.

56 Aristophanes, *Birds* 1004–9.

57 Plato, *Critias* 112a–e, 115a–c; *Laws* 5.745B–E.

58 Konstantinos Apostolou Doxiadis, *Architectural Space in Ancient Greece*, tr. and ed. Jaqueline Tyrwhitt, Cambridge, MA, MIT Press, 1972.

59 Aristotle, *Politics* 1326b–1327a.

60 Plato, *Leges* 704d–707d.

61 Aristotle, *Politics* 1327a.

62 Demosthenes, *On the Dead* 71.

63 Xenophon, *Oeconomicus* 5. 12.

64 Ibid. 20. 14.

65 Ibid. 16. 2–3.

66 Ibid. 16. 5.

67 Ibid. 19. 17–19.

68 Ibid. 16. 12, 20. 12.

69 Ibid. 20. 22–3.

70 Xenophon, *Anabasis* 5. 3. 7–13; *Oeconomicus* 4. 13–14, 20–5.

71 Demosthenes, *Against Phaenippus* 42. 7.

72 Plato, *Critias* 111b–d.

73 Ibid.

74 Owen Goldin, "The Ecology of *Critias* and Platonic Metaphysics," in Westra and M. Robinson, eds, *Greeks and the Environment*, 73–80, presents a very careful analysis of this passage.

75 Plato, *Leges* 639a.

76 Ibid. 849d.

77 Ibid. 843e.

78 Plato, *Critias* 111b–d.

79 Aristotle, *Politics* 1331a–b.

80 Marcus Niebuhr Tod, *A Selection of Greek Historical Inscriptions*, Vol. 2, Oxford, Clarendon Press, 1933, 111.

81 Thucydides, *Peloponnesian War* 6. 90.

82 Theophrastus, *History of Plants* 5. 8.

83 Strabo, *Geography* 5. C235.

84 Aristotle, *Politics* 1325b.

85 Ibid. 1256b–1258b.

86 Ibid. 1258b.

87 John R. McNeill, "China's Environmental History in World Perspective," in Mark Elvin and Liu Ts'ui-Jung, eds, *Sediments of Time: Environment and Society in Chinese History*, Cambridge, Cambridge University Press, 1998, 31–52, at b47.

88 Richard Louis Edmonds, *Patterns of China's Lost Harmony: A Survey of the Country's Environmental Degradation and Protection*, London, Routledge, 1994, 30.

89   John K. Fairbank and Edwin O. Reischauer, *China: Tradition and Transformation*, Boston, MA, Houghton Mifflin, 1978, 57.

90   Lester J. Bilsky, "Ecological Crisis and Response in Ancient China," in Lester J. Bilsky, ed., *Historical Ecology: Essays on Environment and Social Change*, Port Washington, NY, Kennikat Press, 1980, 66.

91   Derk Bodde, *China's First Unifier: A Study of the Ch'in Dynasty as Seen in the Life of Li Ssu*, Leiden, E.J. Brill, 1938, 23.

92   *Tao Te Ching*, 25. Quotations from the *Tao Te Ching*, unless otherwise noted, are from Lao Tsu, *Tao Te Ching*, tr. Gia-Fu Feng and Jane English, New York, Random House Vintage, 1972.

93   Ibid. 57.

94   Ibid. 47.

95   Ibid. 80.

96   Wenhui Hou, "Reflections on Chinese Traditional Ideas of Nature," *Environmental History* 2, 4, October 1997, 482–93.

97   Mencius, 6. A. 8. Quotations from Mencius, unless otherwise noted, are from the translation by D.C. Lau, *Mencius*, London, Penguin, 1970. This passage is on 164–5.

98   Philip J. Ivanhoe, "Early Confucianism and Environmental Ethics," in Mary Evelyn Tucker and John Berthrong, eds, *Confucianism and Ecology: The Interrelation of Heaven, Earth, and Humans*, Cambridge, MA, Harvard University Press, 1998, 59–76, at 68–9.

99   Plato, *Critias* 111b–d.

100 Mencius, 7. A. 24, 187.

101 Ibid. 6. B. 7, 176.

102 Xenophon, *Oeconomicus* 4. 8–9.

103 Herrlee G. Creel, *Chinese Thought from Confucius to Mao Tse-tung*, Chicago, University of Chicago Press, 1953, 82.

104 Mencius, 7. B. 14, 196.

105 Ibid. 3. B. 3, 108.

106 Ibid. 3. A. 3, 99 (that men should not leave their villages), 4. A. 9, 121–2 (that people flock to benevolent rulers).

107 Ibid. 3. A. 3, 97–100; Fung Yu-lan, *A Short History of Chinese Philosophy*, New York, Macmillan, 1950, 75; J.J.L. Duyvendak, *The Book of Lord Shang*, Chicago, University of Chicago Press, 1928, 41–4.

108 Mencius, 2. A. 5, 82.

109 John S. Major, "The Five Phases, Magic Squares, and Schematic Cosmography," in Henry Rosemont, Jr., ed., *Explorations in Early Chinese Cosmography*, Chico, CA: Scholars Press, 1984, 133–5.

110 F. Leeming, "Official Landscapes in Traditional China," *Journal of the Economic and Social History of the Orient* 23, 1980, 153–204, cited in Joseph Needham, *Science and Civilization in China*, Vol. 6, Cambridge, UK, Cambridge University Press, 1984, 101–3.

111 Mencius, 3. A. 3, 97–8; Francesca Bray, "Agriculture," in Needham, *Science and Civilization in China*, Vol. 6, 429.

112 This question is raised by several authors, including Benjamin I. Schwartz, *The World of Thought in Ancient China*, Cambridge, MA, Harvard University Press, 1985, 280–2, and Arthur Waley, *Three Ways of Thought in Ancient China*, London, George Allen & Unwin, 1939, 120. This analysis substantially agrees with that of Schwartz.

113 Creel, *Chinese Thought from Confucius to Mao Tse-tung*, 82.

114 Mencius, 2. A. 1, 85.

115 Ibid. 1. A. 3, 51. See also 7. A. 22, 186, which repeats another part of the same passage with small variations.

116 Ibid. 7. B. 34, 201; 1. B. 9, 68.

117 Creel, *Chinese Thought from Confucius to Mao Tse-tung*, 82.

118 Bilsky, "Ecological Crisis and Response in Ancient China," 66.

119 Mencius, 3. B. 1, 106; 5. A. 2, 140.

120 Ibid. 6. A. 13–14, 167–8. The trees mentioned in these sections are definitely ones used for their wood, but the identification of the species is argued. They perhaps include *Paulownia, Rottera, Catalpa*, and *Sterculia*. See James Legge's translation of *The Works of Mencius*, Oxford, Clarendon Press, 1895 (reprint, New York, Dover, 1970), 415–17n.

121 Bilsky, "Ecological Crisis and Response in Ancient China," 67–8; Tuan, *China*, Glenside, PA, Aldine, 1969, 64.

122 Mencius, 3. A. 4, 102.

123 Duyvendak, *The Book of Lord Shang*, 175ff., 214–15.

124 Mencius, 4. A. 14, 124.

125 Ibid. 1. B. 2, 61–2; 3. B. 9, 113.

126 John Berthrong, "Motifs for a New Confucian Ecological Vision," in Tucker and Berthrong, eds, *Confucianism and Ecology*, 237–63, at 257.

127 Mencius, 1. A. 7, 54–6.

128 Ibid. 1. A. 4, 52.

129 Ibid. 5. B. 4, 154.

130 Ibid. 6. A. 10, 166.

131 I.A. Richards, *Mencius on the Mind: Experiments in Multiple Definition*, London, Kegan Paul, Trench, Trubner & Co., 1932, 4–5.

132 Mencius, 6. A. 8, 164–5.

133 Ibid. 6. A. 3, 161.

134 Ibid. 4. B. 19, 131.

135 Ibid. 3. A. 4, 102; 3. B. 9, 114; 4. B. 28, 133–4.

136 Ibid. 4. A. 9, 122.

137 Ibid. 7. B. 35, 201.

138 Ibid. 7. B. 34, 201.

139 Ibid. 3. A. 4, 100–4.

140 Ibid. 7. A. 16, 184–5.

141 Edmonds, *Patterns of China's Lost Harmony*, 25.

142 Robert B. Marks, *Tigers, Rice, Silk, and Silt: Environment and Economy in Late Imperial South China*, Cambridge, Cambridge University Press, 1998, 226–39, explains the idea of "ever-normal granaries" and traces its history.

143 *Fan chou wen chi*; quoted by L.S. Yang in *Les aspects économiques des travaux publics dans la Chine impériale*, Paris, Collège de France, 1964, 37. See Tuan, *China*, 40.

144 Edmonds, *Patterns of China's Lost Harmony*, 30.

145 Frank Lepper and Sheppard Frere, *Trajan's Column: A New Edition of the Cichorius Plates*, Gloucester, UK, Alan Sutton, 1988, 14.

146 Lino Rossi, *Trajan's Column and the Dacian Wars*, Ithaca, NY, Cornell University Press, 1971.

147 Lepper and Frere, *Trajan's Column*, Plate IV.

148 Rossi, *Trajan's Column and the Dacian Wars*, 132–3.

149 Lepper and Frere, *Trajan's Column*, Plates CIX, CXIII.

150 J. Donald Hughes, "Environmental Problems as Factors in the Decline of the Greek and Roman Civilizations," in *Pan's Travail: Environmental Problems of the Ancient Greeks and Romans*, Baltimore, MD, Johns Hopkins University Press, 1994, ch. 11, 181–99.

151 Seneca, *Quaestiones Naturales* 5. 18. 5.

152 Vladimir Grigorievitch Simkhovitch, "Rome's Fall Reconsidered," in V.G. Simkhovitch, ed., *Toward the Understanding of Jesus and Other Historical Studies*, New York, Macmillan, 1921, 84–139.

153 Brent D. Shaw, "Climate, Environment, and History: The Case of Roman North Africa," in T.M.L. Wigley, M.J. Ingram, and G. Farmer, eds, *Climate and History: Studies in Past Climates and Their Impact on Man*, Cambridge, Cambridge University Press, 1981, 382.

154 J. Donald Hughes, "Hunting in the Ancient Mediterranean World," in Linda Kalof, ed., *A Cultural History of Animals in Antiquity*, Oxford, Berg, 2007, 47–70.

155 Theodore A. Wertime, "The Furnace versus the Goat: The Pyrotechnologic Industries and Mediterranean Deforestation in Antiquity," *Journal of Field Archaeology* 10, 1983, 445–52.

156 Strabo, *Geography* 3.2.8, C147.

157 Clair C. Patterson, C. Boutron, and R. Flegal, "Present Status and Future of Lead Studies in Polar Snow," in C.C. Langway, Jr., H. Oeschger, and W. Dansgaard, eds, *Greenland Ice Core: Geophysics, Geochemistry, and the Environment*, Washington, DC, American Geophysical Union, 1985, 101–4.

158 J. Donald Hughes, "Environmental Impacts of the Roman Economy and Social Structure: Augustus to Diocletian," in Alf Hornborg, J.R. McNeill, and Joan Martinez-Alier, eds, *Rethinking Environmental History: World-System History and Global Environmental Change*, Lanham, MD, Altamira Press, 2007.

159 Cato, *De Agricultura* 139–40.

160 J. Donald Hughes, *The Mediterranean: An Environmental History*, Santa Barbara, CA, ABC-CLIO, 2005, 23–58.

161 Lucretius, *De Rerum Natura* 5. 247–836.

162 Frontinus, *Aqueducts* 1. 16.
163 J. Donald Hughes, "Land and Sea: Human Ecology and the Fate of Civilizations," in Michael Grant and Rachel Kitzinger, eds, *Civilization of the Ancient Mediterranean: Greece and Rome*, Vol. 1, New York, Charles Scribner's Sons, 1988, 89–133, at 125–30.

5 The Middle Ages

Gazing down over the world from above in the Middle Ages, in AD 1300 perhaps, one of the heavenly creatures which people thought existed – an angel or a dragon or the swift eagle Garuda – might have discerned changes since ancient times; swathes of forest removed; new machines being used, plowing taking place faster over longer stretches of field, trade reviving and extending further. The huge seas bore little traffic as yet, but there were daring Polynesian voyagers, Chinese junks, and Inca rafts in the Pacific, European and Arab traders on the Mediterranean Sea and Indian Ocean, Vikings venturing in the Atlantic to Iceland, Greenland, and far western Vinland, and Maya canoes in the Caribbean and Gulf of Mexico. These were widely separated pioneers on nearly empty waters. But the sky visitor would have observed more people on Earth. Built-up areas were spreading and, with them, clearance, erosion, and advancing desert. The Earth as a whole was, however, full of life in many thriving ecosystems. Parts of the continents were still covered with forests. Those places might have looked wild, but peoples had lived there for centuries or millennia, and had learned to subsist within their local ecosystems. Elsewhere, the rate at which humans were altering the face of the Earth was slow but accelerating. It was not increasing at a steady pace, but it was faster than it had ever been. Certain societies were learning skills that would in future times become more effective. They were learning to learn about the world – haltingly, with insufficient methods – but learning nonetheless. In the age to come, they would break forth upon the rest of the Earth. Preparations for rapid modern changes were made in the Middle Ages.

The Middle Ages constituted a period in which the relationship of human societies to nature varied greatly in parts of the world distant from one another. The *oikoumene*, as the Greeks called the inhabited Earth, was not what it was to become, a world united by travel and communication. While civilizations in continental regions were not completely isolated, the degree of contact was much less than it would be in later periods. Patterns of increasing economic activity and growth were sporadically interrupted by stress and decline. At times ecosystems suffered from overuse; at other times they recovered and flourished. Human societies, too, alternately burgeoned and faced disasters against which they often had no effective defenses. They worked with what they had, and demonstrated creativity in ways of dealing with the natural world. Important new discoveries occurred in technology, exploration, education, government, and agriculture. Their success or failure often depended on the degree to which they understood and were able to adapt to ecosystems. For example, during the North Atlantic warm period between 980 and 1450 settlers from relatively mild Scandinavia lived in Greenland. When the Little Ice Age arrived, climatic stress forced them to abandon their farms and hamlets, while local Inuit communities, with a cultural heritage formed in the Arctic, survived.

Technological inventions prepared the way for modern attempts to control nature, but they also enabled management of the environment to a significant extent during the Middle Ages. A new moldboard plow drawn by as many as eight oxen, and a draft harness that enabled more efficient use of horses at the same task, made possible more intensive exploitation of northern Europe's heavy, wet soils for agriculture.[1] Windmills and water wheels supplemented human and animal energy in tasks such as raising water for irrigation and the processing of grain, wood, and stone. Improved sailing vessels in Europe, China, and Polynesia made exploration and long distance trade more possible than before, along with introductions of exotic species and products valuable enough to be worth carrying. Chinese inventors devised iron plows, clocks, magnetic compasses, printing presses, and cannon long before they became known elsewhere. Mining and metallurgy improved, providing materials for tools used in agriculture, forestry, and hunting, and weapons for war, while proliferating mines and smelters increased demand for wood, depleting forests and producing pollution.

Knowledge of the natural world increased, if at a leisurely pace compared with later times. An important intellectual tool for the manipulation of nature, mathematics, received an indispensable aid with the invention of positional notation and the zero in India by the eighth century. Exploration and trade brought information about natural phenomena in distant places. Christian, Jewish, and Muslim writers of books called bestiaries recorded wonders of nature, animals both real and mythical, in words and pictures.[2] Herbals contained useful advice concerning the medical properties of plants. Chinese books on medical materials also contained descriptions of animals and plants, but were interested in them only insofar as they might benefit human health.

Biology as a science had not yet begun; Aristotle's writings had a revival in high medieval Europe, but the interest in them was mainly philosophical and theological. Cosmology received attention: European scholars viewed the Universe as a series of concentric shells leading outward from Earth through the rings of the Moon, Sun, and planets to the stars fixed in their sphere, moving in accord with the *primum mobile*. And beyond that? The greater circles of the heavens, and beyond even them, God Himself. The eyes of mystics saw that the divine was the center, and Earth, though the uttermost, worthy of contemplation as the handiwork of God. The learned doctors wanted humans to care for the world in accord with God's word and classical understanding. They thought wisdom, as embodied in scriptural revelation and classical philosophy, should precede analysis in understanding the natural world. But in that endeavor, the true order is the other way; analysis must come first. Mundane folk gathered that the immense distance that separated God and the spiritual from nature left nature open to human exploitation. Especially in urban centers, people seem to have had less sense of oneness with life, and a greater confidence that they might change the Earth for the better. In the next age, they would change it even more, but whether for the better can be argued.

Many Christian theologians thought nature could serve as a book of truth, a second "scripture" revealing the purposes of God, but the attitude of the medieval Western Church was not always so affirmative. Pope Innocent III, during the thirteenth century, saw man as equal to the beasts, but thought that this equality lowered man. He wrote:

> "The Lord God formed man from the slime of the earth,"[3] an element having less dignity than others. ... Thus a man, looking at sea life, will find himself low; looking upon the creatures of the air, he will know he is lower; and looking upon the creatures of fire he will see he is lowest of all ... for he finds himself on a level with the beasts and

knows he is like them. Therefore the death of man and beast is the same, and the condition of them both is equal, and man has nothing more than the beast. "Of earth they were made, and into earth they return together."[4]... What then is man but slime and ashes?[5]

Compared with that statement, the attitude of Francis of Assisi is like day contrasted with night. A sermon to the birds is the most widely known incident of Francis' life; he also preached to fish and flowers, and made peace with a wolf.[6] A rare spirit in the Middle Ages, he discerned the presence of God in the diversity of created beings, and desired humans to rejoice in this. He expressed a life-affirming, creation-affirming joy which did not immediately dissolve the multiplicity of created beings into oneness, but delighted in their individuality. In his poem, the *Song of Brother Sun*, he named created entities one after another and praised them with God, emphasizing their number and distinctness. The Italian word he used to describe the productions of Mother Earth is *diversi*, "diverse."[7] If God's grace is mediated to people through water, wine, bread, and oil, why cannot it also be received from any creature?[8] Francis once appeared before Innocent to ask permission to preach, and the pope demonstrated unusual good sense in not ordering the saint to keep silent.

Educational systems concentrated their efforts on the study of great classical or sacred works of the past. Older attitudes to nature were therefore preserved from generation to generation. Preparation in the Chinese examination system included study of Confucius and Mencius. Traditional attitudes prevailed in which human control of nature was assumed to be the order of things. Humans were regarded as the proper beneficiaries of human action, so it was understandable that inheritors of the Confucian tradition emphasized the use of nature, not its preservation. Indians studied the Sanskrit classics, which upheld many ideas of the gods and nature that tended to reverence for living beings. To most Indians, the world was a marvelous place, teeming with animal-shaped gods and god-bearing animals. They perceived that particular species and forests should be protected. But even there practical life made demands against the spiritual. The great Arabic schools not only closely explicated the Koran and associated Islamic traditions, but also preserved and commented on Greek works of science. Europe gave birth to the university, with all its potential for teaching and the advancement of knowledge. In the Middle Ages its leading concerns were not the natural sciences, but philosophy, theology, and the professions of law, medicine, and the Church. Still, there was much in all of this that inculcated, and reveals, attitudes to nature.

One important factor that helps determine the pattern of human effects on the environment is the degree of social management that is possible. Many countries were monarchies moderated by decentralization, as in medieval European feudalism. This made it difficult to coordinate production and use of goods and resources. Implementation of an agricultural policy, for example, would have faced local resistance. The Inca government was an exception, as a section of this chapter will explain. The raids or conquests of mobile peoples often threatened the more civilized states. States, cities, and feudal domains fought one another, and military operations inflicted damage on agriculture and the environment, sometimes deliberately.

Environmental changes caused by humans took place at different rates around the inhabited world. The expansion of agricultural land, coordinated with growth in population, was substantial in Europe before AD 1300, in China with the exception of the Mongol period, in India (where invasions episodically reversed it), in Southeast Asia, and in the Andes. Northern Europe adopted three-field crop rotation, an efficient system that raised levels of production. Irrigation and drainage works redirected water to supply agriculture and urban centers. New food plants were introduced, and others were grown over a larger area. Rice culture expanded in China, South and Southeast Asia, and Indonesia. Banana and yam cultivation spread through sub-Saharan Africa. Monks and gentry introduced rabbits to England; the first definite record is of a warren at Guildford in 1235.[9] They spread and altered the landscape by devouring vegetation.

Deforestation, already severe in areas like the Mediterranean in ancient times, took an uneven toll during the Middle Ages. The removal of Europe's forest cover was as complete by 1300 as it is today. After the Black Death disastrously reduced human population, forests spread again for a time. China lost extensive forests; India less so. In West Africa, clearings around villages extended further, sometimes joining adjacent ones, but village forests and sacred groves preserved tree cover. With a combination of selective removal of timber and planting of favored trees for fruits, nuts, and other products, the forest composition there reflected human tendance.[10] In southern Africa, the growth of Great Zimbabwe, a center of mining, metallurgy, and trade that used wood for fuel caused deforestation and erosion in its neighborhood. All the continents (except Antarctica) still possessed forests, and in areas such as the rainy tropics and the north Asian taiga and North America, they were vast, but not uninhabited. Overgrazing by the herds of pastoralists was intense in some regions, particularly the margins of the desert zone that stretches across Africa and Asia. This may have been a factor in the movements of the so-called nomadic peoples mentioned above. Both deforestation and overgrazing, by removing the vegetative cover, exposed the land to soil erosion.

The destruction of wildlife continued. Hunters killed Britain's last native brown bear in the early Middle Ages. Kings, rajahs, and emperors reserved forests for hunting, but killed thousands of animals. A single robe for Henry IV of England required eighty skins of ermine and 12,000 of squirrel. Beavers were decimated.[11] Elk, aurochs, and European bison diminished in number, as much because the expansion of agriculture restricted their habitats as from hunting.

Population increased during the Middle Ages in the areas of the world where there was intensive agriculture, and paralleled the expansion of food production. The increase, however, was far from steady, and there were episodes of depopulation. Perhaps the most severe of these episodes was the outbreak of the bubonic plague, which emerged from Yunnan in southwest China during the Mongol dynasty around 1250. China, the most populous region of the world, suffered from the plague and the disruptions that followed the Mongol conquest simultaneously, resulting in a catastrophic loss of human lives. The population dropped from perhaps 115 million in 1200 to 60 million in 1350, then recovered to 110 million in 1500 under the Ming dynasty, a rate of growth that taxed food resources.[12] Mongol soldiers spread the disease to the Crimea in 1346, from where ships carried it to Egypt and Europe. Europe lost as much as one-third of its people in a decade. While the plague may be called a natural disaster, it must be remembered that it was humans who spread its vectors – rats and fleas – across the Old World. By 1500 the European population had recovered to the level of 1300, so nature had only a brief respite.

Florence and the European scene: the barriers to growth

As I sat gazing over Florence from the Piazzale Michelangelo, I recalled the panorama landscape by Giorgio Vasari of the same view, a city set in a bowl of mountains. But they are not wild mountains. They are terraced and planted with the grapevines from which the Chianti

I was sipping is made, the fields divided by walls, dotted with villas and farmhouses. The trees are those that bear olives and other fruit, or the ornamentals that Italians love: slender cypresses, poplars, and umbrella pines that have been planted since Roman times, although today they compete with eucalyptus from Australia. Many houses are embraced by climbing bougainvilleas from Brazil with their thorns and brilliant petal-like bracts. It is an intimate, anthropogenic landscape. It has changed repeatedly since the city was named Florentia. Vasari's painting, "The Siege of Florence," dating from 1558, shows few trees; the hills look desolate even though he painted them green. Did he show the scene the way it looked in his day? That was a time of war when timber was scarce and expensive; the price of oak had tripled in the previous ninety years and would quadruple again in the next twenty,[13] and almost every available tree might have been cut. But in early medieval times the mountains above the valley of the Arno had good forests.

In the year 1054, when nine-year-old Countess Matilda became its mistress, Florence was a town of 20,000 whose trading families fortified themselves in stone-and-timber towers and talked about the need for a wall. In 1300, Florence had five or six times its former population, a wall embraced seven times the space of the earlier one, and the factious merchant oligarchy was constructing a new city hall (the Palazzo Vecchio) and a vast new cathedral dedicated to Santa Maria del Fiore. Beyond other cities Florence epitomizes the history of the European environment and economy during those centuries of growth and the disastrous decades that followed.[14]

Florence became a powerhouse of finance and trade that catalyzed economic expansion. It was the leading banking center in Europe, an industrial giant, and one of the most populous cities. Florentine merchants bustled everywhere in civilized Eurasia. Its gold coin, the florin, first minted in 1252, "was the preferred and most widely used means of payment both within Europe and beyond."[15] The ecosystem the city depended on was not limited to Tuscany, or even Italy, but covered a wide swath of Europe and the Mediterranean basin. The great Florentine companies, the Bardi and Peruzzi and their rivals, profited from the expansion.[16] They were trading firms and banking houses, and their business included buying and selling merchandise and raw materials, supervising manufacture, financing trade, exchanging currencies, lending money, and doing the accounting necessary to these transactions. Sometimes they served as tax collectors for kings and as international spies, going so far as to arrest fugitives and turn them over to their royal clients for punishment.

Florentine companies gave substantial loans to monarchs, whose position strengthened as the Holy Roman Emperor and Pope demonstrated their inability to unite Europe under secular or church leadership. Money went to the kings of France and Naples, but most importantly to the English king to secure a supply of raw material from the island, which compared with Italy was a "developing country." An agreement provided customs exemptions for wool exported to factories in Florence. The need for loans was a symptom of the fact that around 1300 the vigorous growth of the European population and economy had overshot environmental limits.

The medieval period was not the time of stagnation that the popular mind imagines. "From about the year 1000, European society embarked upon a period of sustained growth which continued until the early fourteenth century."[17] Population almost tripled, and the number of settlements increased proportionately. Large towns grew into full-fledged cities, and faced problems of waste disposal, pollution, water supply, and flooding.[18] Such rapid expansion of economy and population over a large area had never occurred before. In his definitive study of the Great Famine, William Chester Jordan presents statistics indicating an increase in England' population between 1050 and 1300 of between 330 and 385 percent, and for France between 285 and 340 percent during the same period, with similar rates for Germany and slightly smaller ones for Scandinavia. These figures cluster around 2 percent per year. While not extreme by modern standards, such an increase was then unprecedented. Jordan says, "Almost all scholars believe that these figures, however problematic any single one of them may be, reveal a population under stress, because the economic growth necessary to sustain the standard of living had slowed long before the population itself leveled off."[19] I would say that the momentum of population growth had caused it to pass the limits set by the environment and medieval technology.

A major mover of economic expansion was the wool trade; Florence sponsored it across much of the continent and became an industrial textile center. England was a major source of wool; there were more than 3 million sheep on the island by the time of the Domesday Book (1086), and they increased afterwards. Wide areas were cleared for pasture. Monasteries, particularly those of the Cistercians, deliberately sought "wasteland" for the simultaneous salvation of soil and soul.[20] They knew that the Bible portrayed forest and wild country as places of solitude, temptation, spiritual enlightenment, and penance.[21] But they also found that clearing led to profit as their herds multiplied, since wool was as good as money: the ransom of Richard the Lion-Hearted was paid in 50,000 sacks of wool.[22] Walter Map joked at the time that if monks could not find deserts to settle in, they would create them, presumably by overgrazing the pastures.[23] The appetites of sheep can be destructive, but Map's jibe was not entirely fair, because some English Cistercian abbeys had tree plantations and kept enclosures to protect the seedlings.[24]

The wool trade was an important activity of the great companies emerging in Florence.[25] It supplied raw material for an industry that produced fine woolens and added to the city's prosperity. Giovanni Villani, in the 1330s, stated that there were 200 shops belonging to the wool guild in Florence, with 30,000 employees altogether.[26] They turned out 80,000 bolts of cloth annually, selling them for 1,200,000 florins. Water from the Arno's clear tributaries was used to wash wool and provide energy for water mills used in fulling cloth. Clothmakers took every bit of wool the local sheep produced, and looked to distant sources.[27] Merchants drew on the fleeces of southern Italy, North Africa, and the Merino flocks of Spain, but the English trade had the most far-reaching effect on the Florentine economy for good and ill. The demand for wool across Europe caused a rapid augmentation of the flocks, increasing the impact on forests and grasslands. Sheep can be destructive of grass cover when there is excessive grazing, and this, especially in the highlands, contributed to soil erosion. New breeds of sheep bore more and better wool, but also tended to strip the soil of vegetation more efficiently. As greater numbers of Europeans were being clothed, the land was being unclothed.

The period of rapid growth from about 1050 to 1300 saw a transformation of the European landscape from one predominantly forested, "a sylvan sea with only isolated islands of human habitation,"[28] to one where forests had been reduced to isolated fragments. Jean Gimpel emphasizes the extent and importance of deforestation during that period, and gives many sources of evidence for the unfortunate process.[29] Landholders encouraged peasants to open lands to agriculture that had been woodlands, marshes, and moors, and to establish new villages and towns. Machiavelli opined, "There is nothing more worthy the attention of a great prince, or of a well-regulated republic, or that confers so many advantages upon a province, as the settlement of new places."[30] Settlers saw woods as a barrier and often burned them off.[31] "Everywhere, the forest receded before the logger's axe and the settler's plow."[32] The use of saws also increased.[33] The major purpose of forest removal was to expand the area under cultivation and pasture, thus increasing the wealth of the nobles and churchmen who controlled the land. Some landowners put little value on forests. Albericus Cornu, canon of Notre Dame de Paris, ordered his woods cleared, remarking that they had "for so long been so useless that they were a burden rather than a source of income."[34]

Far from being worthless wastes, forests represented an indispensable source of resources for the medieval economy. Wood and charcoal were virtually the only fuels for heating, metallurgy, and manufacture of glass, tile, bricks, and pottery. Five times the volume of wood is required to produce the same amount of heat energy at the point of use if it is first turned into charcoal rather than burned directly, but charcoal can give a more intense heat and less smoke than wood.[35] The manufacture of one ton of iron required the annual increment of 12 ha (30 acres) of productive forest.[36] Wooden torches supplied lighting. Carts and ships, weapons and musical instruments, dishes and sometimes shoes, were made of wood. Wines were stored in oaken barrels. Castles and fortifications were often constructed of timber. Even stone buildings had wooden roofs and required scaffolding during construction. This was the age when great cathedrals arose, and as Romanesque architecture turned to Gothic, spaces enlarged for multicolored windows. It took 100 square meters of forest to produce 1 square meter of stained glass.[37] All these uses hastened forest removal.

Builders noticed shortages of timber as early as the twelfth century. Abbot Suger had to search far for trees to roof his church of Saint-Denis.[38] Hansa ports such as Lübeck made a lucrative business of shipping timber and forest products including pitch, honey, and furs from Baltic lands.[39] These would not have found a market so readily had forests still been plentiful in other sections of Europe.

The lands dependent on Florence went through similar deforestation. In the eleventh century there were rich forests in the mountains of the upper Arno. "A large tract of still unexploited forest" of oak and chestnut was granted to monasteries by the bishop.[40] Collection of pig-rent indicates a sylvan landscape, since pigs found acorns and other favorite foods in forests. By the fourteenth century, sheep and cows, which prefer grassland pasture, outnumbered pigs in this area. When Florence conquered Pisa and needed to build a navy, there was no suitable timber left in its own territory, so the commune had to look elsewhere.

One indication of the loss of forests was a series of measures intended to preserve them. Monarchs in western Europe reserved large areas of forest mainly to serve as hunting reserves. A major part of the food at table in medieval courts was venison from the royal forests. An issue between king and barons in England was that Henry II, Richard I, and John had enlarged the reserved forests and subjected them to special laws. Royal forests constituted one quarter of England's territory. The famous Magna Carta wrested from King John gave permission for an inquiry into the administration of forests. The Forest Charter of 1217 allowed deforestation of lands formerly taken, another defeat for royal power. Hunting with dogs was still strictly limited, except for the king, but the penalty for taking the king's deer illegally was reduced to fine or imprisonment; formerly it was death or dismemberment. Robin Hood may be legend, but poaching of the King's deer is historical fact; in the years 1263–87, an average of eight cases of Trespass of Venison occurred annually in Sherwood Forest, and in one year there were eighteen.[41]

The king of France also appropriated forests for himself. The Ordinance of Brunoy, issued by Philip VI in 1346, ordered royal officers of *Eaux et Forêts* to supervise exploitation of forests while keeping them perpetually in "good condition."[42] The king also tried to limit deforestation on land outside royal domains, but met with opposition by nobles and parliament.

Figure 5.1 The city hall (Palazzo Vecchio) and cathedral of Santa Maria del Fiore, both of which were under construction in Florence at the height of the city's economic affluence and financial dominance near the end of the thirteenth century. Photograph taken in 1959.

The Republic of Venice saw forest protection as essential for a steady supply of ship-building timber. The doges, the elected chief magistrates, prohibited unauthorized export of timber from the neighboring Alps, and limited the glassmaking industry to the use of wood unsuited for ships.[43] Other Italian cities tried to safeguard wood supply. One commune near Siena in 1281 required every member inheriting a portion of land to plant ten trees a year.[44]

Measures such as these met with at most partial success. Dante's *Divine Comedy*, set in 1300, began in a dark forest, but there was little forest near Florence then. Stone and marble often replaced scarce timber in building. Shortages of charcoal for metallurgy appeared, and bricks, which require firing, became more expensive. Wine prices increased due to a scarcity of oak for casks. "By the end of the thirteenth century the price of wine was determined by the availability of casks rather than the quantity or quality of the vintage."[45] Loss of tree cover increased the severity of floods as water from storms poured down denuded slopes. Florence, located on the banks of the capricious Arno, was and is vulnerable to flooding. The disastrous flood of 1333 broke all four of the bridges and inundated the city center. The shrinking of forests was a pivotal cause of the environmental crisis of the fourteenth century.

The food supply was unable to keep up with the increase in population. During the twelfth and early thirteenth centuries, new land in northern Europe could be cultivated with the moldboard plow, horse power, and the three-field system of crop rotation, resulting in increased food production and population growth. These agricultural improvements, however, were directed at increasing production, not at taking care of the land. Some soil scientists believe that there was a long-term depletion of nutrients under medieval systems of cultivation and cropping.[46] Medieval agronomic writers were predominantly concerned with estate management, not sustainability.[47] Even in the south, where lighter soils still responded to older methods, more farmers meant more production until every tillable scrap of ground was utilized. But by 1300 in the European heartland, villages were everywhere and forests almost nowhere. Where could new farms be opened? The ecosystems outside those already occupied might be modified for pastoralism, but were unsuited to agriculture.

Between 1100 and 1300, food supply was adequate. While failures in distribution produced local shortages, history records no widespread famines. In the early fourteenth century, with little new land available for agricultural expansion, the increase in production failed. After that, serious famine occurred every ten years or so. Florentine grain merchant Domenicho Lenzi reported in the early fourteenth century that the surrounding territory produced only enough grain to feed the city for five months of the year.[48] The rest had to be imported, but weather, crop failures, and war made supply insecure.

From 1315 to 1317, the Great Famine ravaged northern Europe. Though heralded by an unusually wet season that was blamed for crop failures, its underlying cause lay in the uncontrolled expansion of the preceding two centuries and the disregard for the continent's ecosystems that accompanied it. Which factor weighed most heavily in causing the crises of the fourteenth century, climatic change or human activities, is a debated question. The study of climatic change in the medieval period is improving, and it seems probable that the fourteenth century heralded a period of cooler temperatures descending unevenly toward the Little Ice Age. A recent attempt to establish a global temperature record for the past few centuries using data from tree rings, ice cores, ice melt indices, and historical records of temperature and precipitation went back only to 1400.[49] The further back the data are pressed, the greater the margin of error becomes, so characterizations of climate in this period must be tentative. Some scholars maintain that the average climate between 1180 and 1299, the span of the most rapid population growth in Europe, was a warm period when agricultural production flourished, but that cooler and wetter conditions prevailed in the early fourteenth century.[50] This is the opinion of Christian Pfister and associates, who have constructed a database of climatic evidence from documentary sources.[51] "Climate" is a smooth curve showing the cumulative effect of the sharp changes we know as weather, and it is weather that directly affects the growth of crops. The wet, cold summer of 1315 and the stormy period that followed may have had more to do with the onset of the Great Famine than did a long-term variation in average climate. Either change had a disastrous effect only because the growth of the European population, and the depletion of resources, had put the Europeans in a precarious position. The weather was a sudden strain that revealed the weakness of the ecological situation.

Marginal lands had lost fertility. Although horses were a source of energy for plowing, many more were used for war, and all ate quantities of oats that might have fed the increasing numbers of poor peasants. Fewer oaks, and therefore fewer acorns for pigs, meant less pork. The medieval village was a sustainable ecosystem when it had the expansive landscape of earlier times to interact with, but in the overcrowded fourteenth century it proved unstable.

From the 1320s, crop failures struck Italy. Florence suffered; food prices were the highest in the peninsula.[52] Famine struck in 1329, and the price of wheat rose three to five times above former levels. Starvation returned ten years later at a time when it was difficult to pay for food imports because the commune had a huge war debt. This was the time of the *condottieri*, when bands of mercenary soldiers roamed the countryside and offered "protection" to cities that would hire them, as Florence had done. At that unlucky moment, another calamitous financial blow fell on the Florentines.

Edward III had squandered the money he had borrowed from bankers to prepare for what became the Hundred Years' War. By 1339, his exchequer was empty and he abrogated his debts. This disaster caused the bankruptcies of the banking houses of Bardi, Peruzzi, and seven other families.[53] The 1340s saw the lowest ebb of the Florentine economy. Hundreds of citizens went bankrupt, and hundreds starved in the famines of 1345–7. Property values plummeted and wages shrank as much as 45 percent. Wars and the need for grain from overseas raised the public debt even higher, and the Commune of Florence declared bankruptcy.

Then the Black Death arrived. Between 1347 and 1351, plague killed one-quarter to one-third of Europe's population. Three-fifths of the Florentines, about 60,000 people, died. Seven more outbreaks occurred in the following eighty years. Europe was in economic and environmental crisis already. Agricultural productivity had declined due to the mistreatment of the land during the period of unrestrained expansion. The weakened condition of the European population due to famine and lack of resources made the loss of life worse than it would otherwise have been. Some writers have suggested that the Black Death relieved the ecological crisis, reducing the population to a level that no longer pressed so hard on the carrying capacity of the land. During the following economic depression, new forests spread over depopulated land and healed wounds left by the former exploitation.

Europe recovered, as did Florence, although it took a long time. In 1850, Tuscany still had 2 million fewer people than it did in 1300.[54] But even after the disasters, Florence led the Renaissance. Assessing a suggestion made by Robert Lopez, Charles Bowlus said, "The artistic achievements of the Italian Renaissance were made possible because surplus capital, which in an earlier period would have been reinvested in commerce, agriculture, and industry, was during the fourteenth and fifteenth centuries invested in the arts due to the uncertainties of the marketplace."[55] Were Florentines reacting to bitter experience?

The environmental history of Florence in the high Middle Ages epitomizes that of Europe. Florence had taken the role of leader in the European economy, and was dependent not just on the ecosystem of Tuscany, but on all of Europe. Kings had borrowed money from the Bardi and Peruzzi and squandered it on war. Florence, and all of Europe, had borrowed environmental capital from the ecosystems of the continent, and just as surely squandered it. They might have liked to renege on their debts but, unlike money debts, environmental debts cannot be renounced. In the fourteenth century, nature sent bill collectors[56] in the shape of resource scarcities, famine, and perhaps the Black Death itself. Florence, in the context of European expansion, had come up against environmental limits.

The evidence shows that the medieval economy, at the level of technology then available, grew to the extent that the European ecosystems were no longer able to support it. While I do not impute any evil intentions in this regard to European farmers or incipient industrialists, it was human activities that caused the crisis. Jean Gimpel began his chapter on the medieval environment by saying, "The industrialization of the Middle Ages played havoc with the environment of western Europe."[57] A few paragraphs later, he underlined those words by adding a well-considered judgment that can serve as a summary: "the fact remains that medieval man brought about the destruction of Europe's natural environment. He wasted its natural resources, and very soon felt the consequences of his destructive activities."[58]

Polynesia: early impacts on island ecosystems

Headed from the open Pacific toward the island of Mo'orea, the pilot of our small craft waited for the right moment to enter the pass through the coral reef. To right and left, great waves hit the coral barrier and shot high into the air, making it vibrate. The moment came: the boat caught a wave and glided through the opening as gracefully as a surfer. A double-hull canoe that carried Polynesian voyagers, the first humans to come here, must have made a similar exciting maneuver to enter the calm lagoon surrounding the island with its tall green mountains.

Many islands of Polynesia are high, surrounded by reefs and lagoons, like Mo'orea. Others, the atolls, are coral circles around lagoons. Some, like Hawai'i, are high islands lacking reefs. Geological forces have raised still others, such as Makatea, above the sea, reefs and all.[59] All are recent, geologically speaking, except New Zealand (Aotearoa), whose two main islands are pieces of the ancient continent "Gondwanaland," and have been above water for aeons.[60]

These islands are remote and mostly small. Both these facts are ecologically important. They are small: the Big Island of Hawai'i spreads 10,450 sq km (4,035 sq mi), and Tahiti, largest of the Society Islands, is 1,043 sq km (403 sq mi). Most of the rest are very small: the atolls vary from 70 km (44 mi) across down to a mere 4 km (2.5 mi) or less. New Zealand is ten times the size of all the other islands of Polynesia combined. They are remote: islands within groups may be within sight of each other, but the groups themselves, and isolated islands like Easter Island (Rapa Nui), are often separated by hundreds of miles. The achievements of Polynesian voyagers in reaching them have dumfounded scholars, and the feats of animals and plants arriving in the millennia before human settlement are equally astonishing.

Distance acted as a filter even for winged species, but wings were an advantage. Oceanic birds had little difficulty in reaching the lands that volcanoes and coral had lifted out of the sea, and brought other organisms with them. Smaller land birds made it, too; a pigeon reached the Marquesas and a tiny flycatcher got to Hawai'i. Marine mammals thronged the shores, but among land mammals, only bats arrived. Arthropods blew across in the winds or hitched rides on birds. Biodiversity was lower in Remote Oceania than in any other lands touching the Pacific. Evolution tried to make up the deficit: in a phenomenon called adaptive radiation, one species produced variations to fill available niches in the environment. On New Zealand, one common ancestor gave rise to thirteen species of wingless birds, the moas, varying from turkey size to a giant that towered 3 m (10 ft).[61] Moas performed the functions that large grazing mammals do in continental ecosystems. In Hawai'i, a finch-like ancestor produced a remarkable group of colorful honeycreepers, their fantastic bills adapted to the shapes of long, curved flowers. The flowers also show adaptive radiation, since the arrival of seeds on an island was a relatively rare event: some could ride the winds, others could float on oceanic currents. Birds brought seeds in their digestive tracts or feathers. After millennia, most high islands were covered with forests; on the windward sides, luxuriant rainforests.[62]

Each island, generally speaking, had a unique ecosystem with its own assemblage of species. On each island, as Darwin eventually recognized, they evolved with one another into assemblages of odd, striking, and naive creatures. This evolution took place without reference to humans, because there were as yet no humans in those environments. A high proportion of species was endemic, that is, they existed there and no place else in the world. On the Hawai'ian Islands, splendidly remote in the central Pacific, gems of animals and plants evolved, the nene goose and silversword plant among others, peerlessly adapted to local conditions. In Hawai'i, 94 percent of all flowering plant species were endemic.[63] Such species are superbly adapted to local environments and the other species which share their ecosystems, but are vulnerable to competition with immigrant species brought from elsewhere, predators or other tough survivors of continental struggles for existence. Changes continued, but the changes were guided by local conditions. Species arrived from abroad, but seldom, accidentally, and only rarely successfully. The lack of browsing and grazing animals, except in the case of New Zealand with the moas, had allowed plants to flourish without evolving poisons, thorns, or other defenses.[64] On small islands, endemic species might be represented by few individuals, and a biological or meteorological disaster could quickly wipe out an entire population.

The ancestors of the Polynesians, a people called "Lapita" after the highly decorated pottery that they made, came from the arc of islands that stretches from the north side of New Guinea to New Caledonia, and first sailed across the 1,000 km (600 mile) stretch of open ocean to the closest islands of Remote Oceania, namely Fiji, Samoa, and Tonga, sometime around 1200 BC.[65] In that nucleus of islands the characteristic Polynesian culture developed, and from there the islands further out across the vast triangle extending east, north, and south were purposefully colonized. The initial phase of this new colonization probably lasted from about AD 300 to 1000, with the settlement of New Zealand later, although dates are disputed and more archaeological work is needed. But this was no chance drifting of vessels.[66] The craft were assembled with two dugout canoes 15–25 m (50–80 ft) or more long, side by side with a platform between, one or two masts and sails of plaited matting.[67] They carried, in addition to a human crew, an inventory of organisms that must have echoed a similar cargo brought by their Lapita ancestors centuries earlier, including the tubers, cuttings, and seeds of as many as two dozen plants (and, inadvertently, weeds), and pigs, dogs, and chickens, with enough food to last for a voyage of weeks. Rats came along too, as stowaways or as potential snacks. Other fellow travelers were geckos, always welcome in habitations in the tropics. What insects and microorganisms might have come along can be imagined. Not all these species reached every island.

An expedition so well-equipped, that cost so much in terms of effort and resources, must have been intended to succeed. This implies also that there were return trips, since societies will not continue to send voyages out into the unknown without knowing what they have found. The navigators knew the stars, winds, currents, clouds, flight of birds, and the light and shadows on the water and in the sky. Prevailing winds in Polynesia are from east to west, so most colonizing journeys were made against the wind, and the return would have been that much easier. Doubts of the abilities of Polynesian mariners to sail accurately over long distances out of sight of land were dispelled by the achievements of modern replicas such as the *Hokule'a*, a double-hulled canoe 19 m (62 ft) long with two sails made in Hawai'i.[68] Without instruments, guided by Mau Piailug, a traditional navigator from the Caroline Islands in Micronesia, the crew sailed the craft 5,370 km (3,340 mi) to Tahiti in thirty-two days during the summer of 1976. *Hokule'a* continued to be used for experimental archaeology, making voyages all over Polynesia, including New Zealand.[69] In 1999 it reached tiny, distant Easter Island, again without modern instruments. Similar modern re-creations of ancestral vessels have been made in Hawai'i and several other Pacific islands.

Double-hull canoes may have traveled in groups, not alone. In any case, the adventurers faced the problem of survival on uninhabited islands when they arrived. What would they eat? Few plants in pre-settlement ecosystems were edible. There were few carbohydrate-rich seeds. The voyagers had brought yams, coconuts, taro, bananas, breadfruit, and other plants in the form of seeds and seedlings, but these would take months and years to grow.[70] They would not kill the animals they had brought until they had established a viable breeding population. But they could fish, catch birds, and gather eggs, living off the protein resources of their new home until domestic plants could take root and domestic animals reproduce. Meanwhile the new residents cut trees, built houses, planted gardens, burned forests, and cleared land for agriculture. Voyagers eventually reached and colonized every inhabitable island group in the tropical Eastern Pacific except the Galápagos. They apparently visited South America as well, since they had the kumara (sweet potato) and the lagenaria gourd, both of which originated there.[71] The alternative, a South American journey to Polynesia, is considered unlikely. Genetic studies have identified remains of Polynesian-type chickens in Chile, indicating a pre-Columbian exchange in the Pacific.

From their initial settlements in the Society Islands and the Marquesas, Polynesian voyagers went onward to Hawai'i, 4,000 km (2,220 mi) to the north across the equatorial counter-current, and to New Zealand an equal distance southward, outside the warm tropical seas where almost all the other islands were found. New Zealand was a challenge; except for the kumara, no Polynesian plants could be grown outside the northernmost part of North Island. Pigs and chickens did not survive, although dogs and rats did. Settlers in the far south were forced to become hunters, fishers, and systematic collectors.[72] Fortunately for them, the huge, unwary birds and the seals and sea elephants, ungainly on land, provided a plentiful source of food. But the supply did not last. In New Zealand the new human residents hunted all the species of moas to extinction and decimated the sea mammals. On other islands, hundreds of native species, especially ground-nesting birds, fell victim to human hunters and the dogs and rats that accompanied them and ran wild.[73] In Hawai'i, for example, the Polynesians eliminated about forty of the 110 native species of land birds before the first European showed up.[74] The transformation of island landscapes had begun. Where settlement succeeded, human populations increased exponentially. There were, however, some islands such as Pitcairn where the first attempt eventually failed and any

Figure 5.2 Aerial view of Ra'iatea, a high island in the Society Islands of Polynesia. Note the
  lagoon and the pass through the barrier reef. Such islands, first settled in ancient
  times or the Middle Ages, represent on a small scale the processes of interaction
  between human populations and natural environments. Photograph taken in 2000.

survivors sailed away. The initial period, in which readily available indigenous biological
resources were exploited, lasted until those resources were severely depleted, at least along
the coasts and in lowland valleys. Evidence of change in native environments, such as charcoal from forest fires, bones of rats and other introduced animals, and the pollen of plants
that came with the Polynesians, is sometimes the earliest known sign of settlement, dating
perhaps a century or more before the oldest archaeological sites of human habitation on
some islands.[75]

A period of transition followed, during which population continued to increase. Economic activity extended inland. Primeval ecosystems, including forests, lost ground.[76]
Intensive horticulture including wet taro patches and breadfruit orchards occupied the most
fertile ground; in some places such as Hawai'i, people built terraces and stone-lined irrigation canals. In much of New Zealand, kumara was placed in insulated pits over winter.

Deforestation can be traced in pollen records deposited where dust or mud accumulated. Deforestation allowed soil erosion when the rains came, and sand and silt washed out
into lagoons, killing coral and clogging the breeding grounds of fish. Eventually all the
good arable land was occupied. Inhabitants turned to marginal resources. Competition
between groups intensified, and many societies became less egalitarian, organizing themselves under strong chiefs. This was particularly true of the increasingly militaristic Maori
in New Zealand.[77] R.C. Suggs remarked: "The cause of the intense prestige rivalry may be
seen in the relation of the population to the habitable land … As the population increased

beyond the point at which all ecological niches became filled, intergroup conflicts over land would have increased."[78] Emigration to other islands was possible, even to hitherto undiscovered ones, but this rarely relieved the pressure for long. The vessels were not large enough, and the social and resource cost of building and provisioning the craft so high that not many vessels could leave, so the resident population could not be stabilized in that way alone.

Wars of conquest were waged to gain control of fertile land, not only among the inhabitants of a single island, but to dominate other islands as well. Patrick Kirch credits Hawai'ian traditions that "speak of great navigators such as Moikeha and Pa'ao, who made round-trip voyages between the ancestral lands of 'Kahiki' [Tahiti] and the Hawai'ian group."[79] Pa'ao had contended unsuccessfully with his older brother Lonopele in Tahiti and had fled to new lands in the north. Finding an opportunity there, he sailed back to his home island, where he convinced the high-born chief Pili to accompany him to Hawai'i, bringing along a strong force of warriors. The Tahitians (actually from the religiously powerful nearby island of Ra'iatea) introduced an extreme form of chieftainship in which they became the *ali'i*, or nobles, reducing the native Hawai'ians to the status of *menehune* (commoners). Pa'ao introduced the cult of the war god Tu (Ku in Hawai'ian), along with human sacrifice. The Society Islands, smaller than Hawai'i and inhabited for centuries longer, had "filled up" to their ecological limits earlier and evolved aggressive rival chiefs. This bit of oral history could be interpreted as the story of some of these leaders venturing forth, "Vikings of the Sunrise,"[80] to acquire productive land across the water. It is quite possible that they carried new species; traditions suggest that the breadfruit tree had not reached Hawai'i before this time.[81] The new chiefs increased the pressure on native wildlife: it took the feathers of 80,000 birds to make a cloak for one of them.[82]

Aggression was not the only behavior to result from the pressure of growing population in small islands of whose ecosystems were under stress. Helen Leach has indicated that sophisticated labor-intensive horticulture may be closely linked to fine-grained ecological adaptations.[83] Polynesians also had methods of protecting aspects of nature and conserving resources for future use. Their songs and legends express great love for the land and sea, celebrating the outstanding features of nature in the islands. The Hawai'ians treated the land with the traditional ethic of *aloha 'aina*, based on love and reverence for nature, and especially the awe felt for *mana*, the spiritual energy shared by living beings. There was a sense of relatedness to other forms of life. Among the Maori, for example, "All creatures are regarded as kin, related through the *whakapapa* or genealogi[es] that trace all beings back to Papa and Rangi, Earth and Sky."[84] The concept of *tapu* (taboo), a recognition of inherent power so strong that the entities that possessed it could not be touched or approached, may have served as a force for conservation. Mountains were likely to be *tapu*, along with forests where ancestral spirits were perceived to roam.[85] Some special trees were so *tapu* that they served as shrines. The Moriori of the Chatham Islands had sacred groves.[86] Elsewhere temples (Tahitian *marae*; Hawai'ian *heiau*) were surrounded by sacred groves, as Herman Melville noted in the Marquesas.[87] The Maori could impose "a *rahui*, a form of temporary special *tapu*, for example when a species or place needed to recover from material, biological or spiritual damage, or when it had special spiritual or cultural significance."[88] On many islands *tapu* restricted the exploitation of eel weirs, shellfish beds, and certain fish in lagoons, reefs, and the outer sea; for example, Mangaunu shark-fishing enterprises were annually limited to two days.[89] In Hawai'i, certain fish species were regarded as sacred to individual gods, and catching them was forbidden during the time of year when those gods were honored. In consequence, those species could recover their numbers during critical times. There were dietary restrictions.[90] These are a few instances of widespread practices.

Measures to prevent population growth are noted, and these are often connected with social privilege; for example, on some islands such as Tahiti infants born of relations between nobles and commoners were killed. Methods of contraception and abortion were practiced. Deaths in battle, and cannibalism to ingest the vigor of brave opponents, were also an influence on population. Patrick V. Kirch, in a paper on subsistence and ecology, summed it up well:

> The prehistoric cultural sequences of Polynesia present the same scenario over and over: initial settlement by a numerically restricted group, rapid population growth, expansion into all habitable biotypes, and – frequently – intergroup conflict and degradation of the natural environment.[91]

In very small atolls, competition between chiefdoms was absent, and extremely scarce resources delimited subsistence.

In the last centuries before European contact, population on many islands reached a high but relatively stable level. Long distance voyaging declined or disappeared, which Patrick Nunn connects with the onset in AD 1270–1425 of the Little Ice Age, with changes in sea level and weather patterns.[92] Trade continued in artifacts and stone among islands in the same group. Many Polynesian islanders learned to live successfully within their limited lands, but others did not, and experienced famine from time to time. In parts of New Zealand, archaeology provides evidence of malnutrition and economic decline.

Factors of adjustment to resource scarcity were different on different islands, and the long-term results also varied. Some island societies suffered disaster. The history of Easter Island (Rapa Nui) has been presented as a cautionary tale of a human society that destroyed its renewable resources and in the process was reduced to a fragment of the population and a shadow of the culture that had marked its zenith.[93] The history of the island is obscure, since the *rongo-rongo* script carved on wooden tablets, the only indigenous writing in Oceania, has not been satisfactorily deciphered, and oral tradition was impoverished by the death of the elders entrusted with it due to slave raids and epidemics that reduced the population to 111 individuals in the late nineteenth century. Recently archaeologists have come up with evidence that makes the outlines of the cultural and environmental process clearer.

Christopher Stevenson, among others, has done this in investigation of sites there. As a volunteer, I assisted him on two of these sites.[94] At Orito, a hill that was an ancient source of obsidian, a bulldozer had laid open a series of strata in a pit. Undoubtedly the operator of the machine did not have archaeology in mind while doing it. From the base of the pit rose a layer of tightly packed roots, most of them of palm trees, incontrovertible evidence that at least this part of the island had been covered by forest for millennia before the first human inhabitants arrived. Above that was a narrower stratum containing evidence of digging, fires, and agricultural activity during the last thousand years, the period of Polynesian settlement and occupation. On top of that, the thinnest layer of the three, was the soil of grassland and the grass itself, not to mention the cows that are helping remake the present-day landscape.

Easter Island is the most isolated single island in the world. Located in the southeastern Pacific, its closest inhabited neighbor is Pitcairn Island, the refuge of the *Bounty* mutineers, 2,000 km (1,260 mi) west and itself an epitome of isolation. The coast of Chile lies 3,700

Figure 5.3 View of the lagoon and barrier reef from the high island of Bora Bora in the Society Islands. Photograph taken in 2000.

km (2,300 mi) to the east. It is a little island, 22 km (14 mi) long and 11 km (7 mi) wide, with an area of only 16, 576 ha (64 sq mi), smaller than Ni'ihau, the smallest inhabited Hawai'ian Island. So it is not surprising that it was among the last found by human beings, and one of the last noticed by European explorers. At 27° south latitude, it is just outside the tropics, comparable with Brisbane, Australia.

What was it like before the Polynesians arrived? The question is not easy to answer, since the landscape and biota were thoroughly altered by the human inhabitants. It is a volcanic island with three major calderas; all the plants and animals were progeny of a few that happened to arrive by air or sea. It was forested, and scientists have begun to understand the character of the forest ecosystem. The first Europeans to arrive found almost no trees. The native palm, the dominant species of primeval times, was by then extinct. It is known from fossil roots and tree trunks, some preserved in lava flows, and from tiny coconuts, about 25–32 mm (1–1.3 in) in diameter, found in caves. The Easter Island palm appears similar to the wine palm (*Jubaea chilensis*) that still grows in Chile. Another tree, the sophora, a legume with yellow flowers and high-quality wood, is almost extinct although efforts are being made to preserve it. As recently as 1991, evidence of few other trees was known, and the vegetation of the island before Polynesian settlement could be described as "Palm forest with Sophora and shrubs" with some grassland.[95] But soon afterwards, Catherine Orliac investigated carbonized wood at several sites and was able to identify many kinds of woody plants.[96] Some still grow as relics, but in addition Orliac identified fourteen that had never been found on Easter Island before, including some known on other Pacific islands as large trees. The picture that emerges is a complex vegetative cover including high forests containing a variety of species, with the Easter Island palm the most numerous.

There were no land mammals: no bats, no rats, and no reptiles either, although insects, spiders, and snails occurred. Marine mammals such as seals, sea lions, and dolphins were present. There were a few land birds including parrots, rails, a heron species, and an owl species, all known now only from bones. Millions of migratory sea birds found nesting sites on the rocky cliffs: terns, albatrosses, seagulls, frigate birds, tropicbirds, and others. Only a few survive today on offshore islets. There were shellfish and crustaceans such as lobsters, but fish were not as numerous as around other Pacific islands, because Easter Island's topography and climate prevented establishment of a coral reef and lagoon. Around most of the island, the cliffs fall straight into the sea and receive the force of the waves, including the occasional tsunami.

There is no doubt that Easter Islanders were Polynesians. Thor Heyerdahl[97] tried to prove they came from South America, but scholarly research demolished that theory. The Easter Islanders speak a Polynesian language, and their DNA is of Polynesian type. Still, the arrival of Polynesians on such a tiny, distant island is a wonder of human history. We do not know the date; it has been estimated from AD 300 to 1000. Radiocarbon indicates a date between AD 615 and 860.

Where did they come from? Judging from language and material culture such as stone statues, it seems the Marquesas are the most likely place: Easter Island tradition calls the ancestral land "Hiva," and the largest islands of the Marquesas are Hiva Oa and Nuku Hiva. But scholars also suggest the Society Islands and the Australs. Mangareva in the Australs, probably also settled by Marquesans, may have been the direct source of the population. In both the Marquesas and Mangareva, there was a class of men called *orogo* (pronounced "orongo") or *rogorogo* (pronounced "rongorongo") responsible for keeping history and genealogies. That the Easter Island script was called *rongorongo* strongly indicates a connection, as well as hinting that the purpose of the script was to record traditions.[98]

What did they bring with them? Domestic animals carried by Polynesian expeditions included pigs, dogs, and chickens. Of these, only chickens became established on Easter Island. Two other animals arrived, by the Polynesians' deliberate choice or as stowaways: geckos and the Polynesian rat (the latter immediately began depredations on birds and vegetation). The list of plants the voyagers brought is longer: taro, yam, sweet potatoes, sugar cane, bananas, gourds, and shrubs useful for dyes, paints, high-quality wood, and cloth, such as the paper mulberry tree.

The Polynesians first settled along the coast. They had to depend on resources they found on the island, since it would take years before the introduced crops could increase enough to feed the people. Fortunately, there were fish, birds and their eggs, and sea mammals in abundance. Unfortunately the indigenous vegetation had few edible plants, but the settlers cleared the forest by slashing and burning, and placed the familiar food plants in the soil. There was useful wood of several species. Like Polynesians elsewhere, they shaped stones and built temple platforms, called *ahu* on Easter Island.

Figure 5.4 Panorama of Ahu Tongariki, Easter Island (Rapa Nui), with its fifteen moai. These huge statues of volcanic tuff, up to 10 m (30 ft) in height, represent ancestral chiefs and undoubtedly wood from local forests was used in their transport and erection. Photograph taken in 2002.

Expanding agriculture made population increase possible in a second phase of occupation. Inhabitants increased to about 9,000 by AD 1500. This necessitated the expansion of agriculture into most of the island, including the hilly interior. Monumental architecture was erected, reflecting the development of social hierarchy. Nobles were responsible for direction of agriculture. Statues, called *moai*, figures of aristocratic ancestors as much as 10 m (30 ft) high carved from volcanic stone, were set up on the *ahu*. Progressive increase in height and weight of the *moai* indicates competition among rival communities. One, never removed from the quarry, was 21.6 m (71 ft) from head to toe and weighed 200 tons; the largest actually erected was half that size. Many of them had crowns of red scoria stone set atop their heads, and eyes of white coral with pupils of darker stone set into their faces. Meanwhile, the elite required the erection of houses whose foundations consisted of large stones called *paenga*. Moving all these masses of stone required the use of the trunks of palm trees, a major cause of forest destruction.

Trees became scarce. Along with deforestation came soil depletion and erosion,[99] water contamination, and loss of bird habitat. The resources that had supported the early expansion of the Easter Islanders began to disappear, and they depended on extension and intensification of agriculture to support their increasing numbers. The technology used to support agriculture is not as startling as the sculpture, but was even more important. With few trees, the winds had nothing to moderate their force, so farmers dug pits and surrounded them with walls to protect taro and bananas. They placed stones in the fields, forming "lithic mulch" that protected plants and conserved moisture.[100] These methods were labor intensive, and the common people had to provide the labor.

In the latter half of the seventeenth century, a convergence of crises occurred. One day the last palm tree was cut down. The statues could no longer be moved, which helps to explain today's appearance of the quarry at Rano Raraku, with sculptures in every stage of preparation, looking as if the order "tools down" had been given and all the laborers departed. The population had reached the limit of environmental support, with food shortages as a result. It was not possible for out-migration to relieve population pressure, because no materials remained for the construction of canoes large enough for inter-island voyages. Conflict increased as groups attempted to seize resources from others. The population crashed. Chickens, the major source of protein, were housed in fortress-like stone coops (*hare moa*). With starvation ever-present, the common people questioned the order of things. The direction of agriculture by the nobles had failed to provide them with ample food. The great stone statues, whose watchful presence supposedly insured safety and abundance, had also failed. There was inter-class war – a strong element in the oral tradition – and the hierarchy was overthrown. The commoners pushed down some of the statues. A few were standing when the Europeans arrived, but eventually all were toppled. What role drought, crop failure, or climatic disturbances such as El Niño played is a matter for further research, but human impact on the natural environment was the leading cause. The natural cycles of weather and climate probably added stress and exacerbated the crisis.

Environmental and social disaster made a new order necessary. This was provided by the birdman cult. Worship and labor were redirected from the veneration of the ancestors and their statues to the creator god, Make-Make. The ritual expression of this new religion was the cult of the birdman. Carvings of men with heads of the sooty tern, a migratory sea bird that by then nested only on the offshore islet of Motu Nui, cover lava exposures around the ceremonial village, Orongo. In an annual contest, young men swam out to the islet when the birds arrived. The one who brought back the first egg became birdman of the year, endowed with privileges but kept in a special house and subjected to taboos. Agricultural technology began to revive and the population survived, although within an impoverished landscape. That was the situation when Europeans arrived and began the decimation of the remaining inhabitants.

What is the lesson of Easter Island? Is it lack of foresight? Human societies organize to optimize use of natural resources, and this makes population growth possible. Consumption increases to the point where diminishing resources interfere with population growth. Faced with starvation, people devise new technologies to extract more production from the land. In times of crisis, social organizations collapse and are transformed. But there is a bottom line, and that is the ecosystem: the landscape itself with its living and non-living components. After depleting their renewable resources, the Easter Islanders used stone-based technologies to raise sweet potatoes and sugar cane on a windswept island. But they could never bring back the palm trees and the rest of the humid high forest. Birds would never nest again in great numbers on the cliffs. And without trees for building boats, the sea would no longer be a highway, but a prison. Indeed, the conviction grew through long isolation that Easter Island was the only land in the world … until the strangers came.

On other islands, population remained at a high but stable level, and resource use was sustainable after the initial period of depletion. This was true of islands such as Samoa, New Zealand, and Hawai'i. The pattern in those places was vigorous competition between strong chiefdoms combined with a deep sense of reverence for the gods of nature and the creatures and elements that shared the islands with them and on whom they depended. The motto of Hawai'i, *Ua mau ke ea o ka 'aina i ka pono*, means "The life of the land is sustained by a proper relationship."

Even Pacific islands that maintained a large population to the time of European incursion suffered great changes, including damage to landscapes and biodiversity. But one may ask what determined the difference between the trajectories of human occupation in "successful" groups like the Marquesans as compared with the "failed" Easter Islanders. The question cannot yet be answered, but I will offer a few observations. The cause cannot be ethnic: all the inhabitants were Polynesian. Nor can it be intertribal warfare, since that was rampant on all islands and archipelagos except the smallest ones. There were no important differences in basic technology between "successful" and "failed" island inhabitants. The type of island was not the sole deciding factor – there were successes and failures on both high islands and atolls – although extremely small islands did not offer much space for success. The presence or absence of resources, and differences in the list of animals and plants introduced to specific islands need further study. Contact between island groups may have played a role. Population pressure was a driving force behind environmental degradation, so that people on islands where controls on population growth were effective probably had a better chance of conserving their renewable resources. I would like to think that wise traditional leaders who knew when to place taboos on critical resources made a difference in "successful" communities.

The changes that take place on islands are local in scale, although similar events occur on worldwide scale. A more complex pattern of change occurred on larger landmasses such as pre-Columbian North America, as Flannery describes in *The Eternal Frontier*.[101] With worldwide expansion of industrial technology and the market economy, the Earth has become an island, and a pattern with trajectories of population growth, resource exploitation, depletion of biota, and inter-group conflict, observed on islands, is now occurring on a global scale. The question is: which island history will the global trajectory turn out to resemble most?

Cuzco: conservation in the empire of the Incas

Cuzco, the Inca capital, stood in a valley 3,400 m (11,000 ft) high in the Andes. Temples and residences, stairways and squares, were built of huge polished stones, ingeniously fashioned. Water flowed in rock-cut channels that paralleled the streets. Well-planned, its streets, walls, and two channeled rivers outlined the figure of a puma; the fortress was the head, the plaza the belly, and nobles had residences in the tail. Nearby were agricultural terraces built as securely and beautifully as any of the other structures. At the height of Inca power, Cuzco ruled an empire extending 4,000 km (2,500 mi) north to south. The distance is greater by road: the fine Inca ones, running straight wherever possible, had to bridge streams and switchback over passes.

The ecological variety of the Inca realm is crucial to its environmental history. There are three major regions: the dry, almost rainless hot coastal lowlands, the cold mountainous belt with its valleys and plateaus, and the steamy rainforest.[102] Each of these has its own ecological character, and the resources received by the Incas from each were different. The coast provided maize, cotton, fruit, fish, and shells; the mountains produced metals, wool, maize, potatoes and other tubers, quinoa, maguey fiber, and the wool of llamas and alpacas; and the rainforests yielded wood, feathers, fruits, and coca leaves. This ecological specialization was one factor that necessitated the Inca network of roads and the rules of production and exchange.[103]

The Incas extended their government by force over many peoples, who were adapted to various ecosystems. Cieza de León remarked in the 1550s,

> As these Incas ruled over such large provinces and such a length of territory, part of it so wild and full of mountains and snow-covered peaks and deserts devoid of trees or water, great prudence was needed to govern such a variety of peoples so different from one another in language, laws, and religion, and keep them all satisfied and in peace and friendship.[104]

The rulers maintained their hold on these peoples by reciprocal arrangements assuring a dependable supply of food, clothing, and other products for communities in every part of the empire. The conquest was swift, and the extensive empire lasted less than a century; the ruler who founded it, Pachacuti, began his reign in 1438, and the last independent Inca ruler, Atahualpa, was murdered by the Spanish adventurer Francisco Pizarro in 1533.

The central government consisted of the royal house, headed by the Sapa Inca, a monarch whose person was sacred and whose power was absolute but not arbitrary. In theory, he owned all the land and people. Under him were the governors of the Four Quarters, the divisions of the empire that lay in the cardinal directions. The priesthood, in charge of the worship of the Sun God and other deities, commanded a large segment of land and labor. The economic intent of the government was to assure adequate production of the necessities of life; to store surplus goods; and to redistribute supplies in a system of rewards, and whenever shortages occurred due to natural disasters.

When the Incas conquered a new territory, they appropriated its reproductive goods – lands, herds, forests, and waters.[105] They reserved sections of the land for the Sapa Inca and for the Sun God, and reissued some of it back to the local community, enough to meet the ordinary needs of the people. A portion was also designated for the support of widows, orphans, the handicapped, and soldiers serving in the army. In theory, the Sapa Inca received no taxes in kind (there was no currency), but only in the form of labor.

Labor was shared among the able-bodied on all the land. In the season of planting, all the men would till the ground with the foot-plow (a shovel with a footrest). No one who could work was exempt from this labor; even the Sapa Inca ceremonially turned the soil. They cultivated the lands in a set order: first those of the Sun God, second those of the poor including widows and absent soldiers, third those of ordinary peasant families, then those of the *curaca* (local chief), and only last those of the Inca.[106] The produce of the lands of the Sun God and the Sapa Inca was placed in storehouses. There were many domesticated plants, with potatoes and maize the most important. Others included quinoa and cotton. Hundreds of varieties of potatoes existed, some of which could flourish at high elevations in the Andes, where freezing temperatures were common. For storage, maize was blanched and dried; in the highlands potatoes were "freeze-dried," alternately stamped underfoot and allowed to freeze overnight, yielding a dehydrated product called *chuño*. "Both the Inca state and local societies tended to cultivate a mix of crops that permitted a sustainable rotation locally."[107]

Herders cared for the domestic llama and alpaca. Like the agricultural land, the animals and their pastures were divided between the Sun God, the Sapa Inca, and local community. The wool from the animals of the Sun God and the Sapa Inca was given to local people to weave and make into clothing, and the finished products were placed in storehouses for distribution as needed.

The state controlled surplus production, and was able to move goods on the remarkable road system from any part of the empire to any other part.[108] Labor owed by the people to the Inca was called *mita*, which included, along with service in the army, the maintenance of the infrastructure: roads, bridges, the government inns at intervals along the highways, storehouses, etc. Others served as messengers. There was a special relay system, using runners on foot, that could carry messages or objects of value. "It was said that a snail picked off a leaf at Tumi in the north of the Empire could be delivered to the Inca in Cuzco still alive."[109] The laborers gave their labor to the Inca and received food, clothing, and such amenities as coca leaves as reciprocal gifts. Laziness was detested as perhaps the worst fault.

Population policy was pro-growth. The empire had use for many laborers, and dependent peoples were required to send children to Cuzco to be educated in the Quechua language. Some were sent back to assist Inca governors in their home locales, while others stayed to serve the Inca; many girls lived in convent houses as *aclla* (virgins), weaving and playing music. Laws encouraged having a number of children, honored parents who did, allotting them more land.

Although the Incas did not have writing, they possessed the *quipu*, a sophisticated means of recording numbers by tying knots in colored cords. Record keepers kept track of production, the population of each community including births and deaths, the contents of storehouses, and every other fact necessary to run the empire efficiently.

Inca gods were nature deities, but the Inca attitude to nature was not the primal respect for other beings characteristic of hunter–gatherer societies. In the Inca universe, humans were dominant beneath the gods, and nature was managed to benefit society. Sacred places or objects associated with the gods were called *huacas*. These might be statues, mummies of former Sapa Incas, or features of the landscape. The Incas revered many temples, foremost among them Coricancha, the sun temple in Cuzco. Some temples housed oracles. Many were surrounded by walls and groves of sacred trees, such as the one at Huari-vilca.[110] Worship involved public festivals, and included sacrifice of maize, shells, and, above all, domestic and wild animals and human beings. The sacrifice of children, both boys and girls, who were buried in sacred places including mountain peaks, was considered especially effective. This was not because children lacked value – quite the opposite. The more precious the sacrifice, the more likely the gods would respond.

Inca policy conserved natural resources, and tried to assure the continued use of the environment for the benefit of human society. Their extensive terracing and water management works illustrate this. The Incas were not the first Andean society to construct terraces, but they made the most extensive systems of them, sculpturing whole mountainsides in order to expand arable land and to limit erosion. Archaeologists recognize two types of Inca terraces: production terraces and high-prestige terraces.[111] Production terraces are as elegant as those in the Mediterranean area. High-prestige terraces are even more finely worked, using large polished stones like those in palaces or temples.

Irrigation was necessary in the highlands for maize and specialized crops, and on the coastlands rivers flowing from the Andes were diverted to extend agriculture into the desert. Surviving channels, some still operating, show evidence that accomplished hydrologists designed them.[112] Garcilaso de la Vega, writing in 1590, explained how the Inca assigned irrigation engineers to direct the building of channels and terraces.[113]

Many crops, especially maize, required fertilizer. For this, human or animal manure was used. Deposits of bird droppings (guano) were taken from the offshore Chincha Islands to fertilize fields near the coast, and were valuable enough to be carried up to the mountains. Different islands were assigned to different provinces, and quantities were rationed.[114] No one was allowed to set foot on the islands during the breeding season. The penalty for anyone who killed any of the sea birds or disturbed them at the nesting period was death.[115]

Another Inca conservation measure was agroforestry. They created tree plantations and planted trees for many purposes: to surround temples, to provide amenities in towns, to shade roads and canals, and to protect the soil from erosion. The Sapa Inca himself supervised tree plantation.[116] Its success is attested by pollen studies of lake-bottom sediments. Samples from Lake Marcacocha not far from Cuzco show that in pre-Inca times trees almost disappeared, indicating overuse and deforestation. But around 1450, *aliso* (alder) pollen increased sharply, "evidence of [Inca] agroforestry using [*aliso*] on a major scale."[117] *Aliso* was used for door lintels and roof beams, and as fuel. The Incas looked on some trees as sacred, and helped to diffuse them throughout the Andes.[118] The Quechua word *malqui* means both "tree" and "ancestor."[119]

Middle elevations had humid montane forests before human impact. This is indicated by the fact that native arboreal species still exist in protected sites and demonstrate an ability to recolonize other parts of the area.[120] Pollen evidence indicates that mountain forests were diminished through clearing for agriculture by pre-Inca civilizations. There were climatic changes, which might have shifted ecological belts, including forests, to higher or lower elevations, but would not have removed the forests.

The Incas wanted to use these forests for wood and fuel and to conserve them as a resource. Wild forests were considered to belong to the Sapa Inca, and therefore were protected state property. The use of wood was regulated to prevent deforestation.[121] A special overseer, *malqui camayoc*, was appointed to enforce regulations.[122] Damaging trees was punished, according to an Inca law quoted by Guamán Poma writing shortly after 1567: "No fruit-tree, timber, woodland or straw shall be burnt or cut without proper authority on

Figure 5.5 Machu Picchu, a city of the Incas in the Andes Mountains east of Cuzco, Peru. The stone terraces held soil in which crops such as maize were grown, thereby slowing erosion. Photograph taken in 1974.

pain of death or some lesser punishment."[123] This does not mean that no trees were cut, simply that the state controlled forest use. Local communities might be allowed to gather firewood or other products. The result of Inca forest management was the maintenance of forests that existed and their expansion into additional tracts. Chroniclers of the Spanish conquest report many forests in Inca territory.[124]

The Incas carefully managed wildlife. The species considered most valuable were the llama's wild relatives, the guanaco and vicuña, which they called "llamas of the Sun." The hair of the vicuña was softer and finer than llama wool, and prized for weaving. Deer were also important as a source of meat and hides. All these animals were the sacred property of the Inca, and hunting them was forbidden except in annual ceremonial hunts. The law was: "All deer and Peruvian 'sheep' bearing wool of high quality, called guanaco and vicuña, shall be protected against hunting, capture and wanton killing so that their numbers may increase."[125] This was effective; as a result deer and guanacos came into villages; common people might chase them out of their gardens, but when the conquistadors arrived they found that the animals were without fear, and easily killed them.

The Sapa Inca himself presided over the *chacu*, a solemn royal hunt held once a year and involving thousands of people. It alternated among the four quarters of the empire, so that each section was hunted only once in four years (the wait allowed the animals to multiply and their wool to grow). Beaters formed a circle and drove the animals together. Tens of thousands of deer, guanaco, and vicuña were caught. The hunters released most of the female deer, but killed old ones past breeding, and also released the best of the males as sires. They killed the rest of the deer and divided the meat among the common people. The guanacos and vicuñas were shorn and released. Numbers were tallied on the cords of the *quipu*. "Knowing how many head had been killed and how many released alive, they could tell at what rate the game had increased at the next hunt."[126] The wool of the guanacos was distributed to be spun; that of vicuñas was reserved for the Sapa Inca.

Among other species caught by the hunters, they killed the predators, but exceptional animals might be presented to the Sapa Inca as a sign that he owned all things. The Inca view of the world of living things was not that of a community in which each species played an important role, but that of a kingdom owned and managed by the Sapa Inca for the benefit of his human subjects. To care for them, he was willing to work and to hunt.

The Inca system fell to an invasion even swifter than their own conquest of neighboring peoples had been. The Spaniards succeeded although outnumbered; Pizarro had only 167 armed men, while the Inca army consisted of 200,000. Unlike the Aztecs, the Incas commanded the loyalty of the majority in their empire. But Pizarro, luckily for him, landed during a bitter dynastic war between rivals for the Inca throne, Huascar and Atahualpa. The empire was exhausted and divided, and Spanish weaponry was superior to anything the Incas had. In addition, they brought, as unintended allies, Eurasian diseases, some of which spread to Indians who had not as yet seen a Spaniard.[127] Indeed, the Sapa Inca, Huayna Capac, whose death occasioned the struggle between Huascar and Atahualpa, had been killed by smallpox contracted during a campaign in Ecuador, and his chosen successor had died of the same disease.

A flood of looters and settlers followed the conquistadors, and the Inca state collapsed. Seizing all the gold they could (for the Incas, gold was a material for decoration, not usually a medium of exchange), the Spaniards also raided Inca storehouses, which seemed inexhaustible sources of food and clothing. They commandeered labor for the gold, silver, and iron mines they opened in the Andes. Millions of "Indios" died of unfamiliar diseases, overwork, starvation, and murder.

The ecological result was as disastrous as the political one. The forests were consumed for construction of mines and for smelting fuel. In the new settlements, the daily wood consumption of a Spaniard equaled that of an Inca peasant for a month. Between 1550 and 1650, *aliso* pollen virtually disappeared from the record.[128] The result of forest removal was erosion. Irrigation works clogged and fell apart, and terraces were abandoned, so that the area of land under cultivation shrank by at least half. Animals both domestic and wild fell to Spanish guns. The places of native animals were taken by species introduced by the Spaniards. Garcilaso observed, "[The Incas] supplied clothing for their subjects; there were no beggars." Flocks once filled the pastures so that "there was no longer any room to graze;" the Spaniards practiced "great excesses and enormous waste."[129]

The Inca system did not last long enough to judge what its ecological impact might have been if the Spanish conquest had not occurred, but it is interesting to speculate. The most likely outcome, with prevailing Inca policy, would have been a growing population pressing the limited resources of the Andean region. This had happened with pre-Inca civilizations. One study concluded, of a period around AD 1100–50, "By this time [Tiahuanaco] was sustained by a highly productive, water-dependent form of intensive cultivation on raised agricultural fields. The environmental threshold was exceeded because the raised-field system had stimulated dense human population that could not be supported during drier conditions."[130] There are few signs that this had begun to happen with the Incas, but their conservation measures could not have continued to succeed indefinitely without a population restrained to a size the limited Andean environment could support. Since their population records were the most accurate in the world at the time, they might have realized the need for limitation. If they had, it seems to me that with their organization and their control of the resources for subsistence, they had the means to succeed.

Conclusion

By the end of the Middle Ages, humankind had spread to almost every land on Earth. The few exceptions included Antarctica and some isolated oceanic islands such as the Galápagos. Most of the main inhabited lands were relatively isolated from one another; in some cases almost completely so. A series of separate worlds, culturally and ecologically, occupied the planet. The human societies in each region existed in interaction with the ecosystems characteristic of that region; while there was some trade, transfer of technology, and a few introductions of species from one part of the globe to another, wholesale translocations of biota including human populations from one region to another would reach epic proportions only in the period that followed. In the Middle Ages most ecological crises were limited to single regions. The Black Death, which spread from China to Europe, was perhaps an exception, although a very important one.

Europe, China, and the Incas experienced the pressure of population on the capacity of their ecosystems to provide food. So did the Mayas, whose classic expansion took place chronologically during the early Middle Ages, although they were considered above in Chapter 3 with other early urban societies that they resembled. So did the Polynesians, whose expansion to thousands of islands, many widely separated from one another, was driven by that pressure. The Polynesians reached New Zealand, where within a century or so they wiped out the giant birds that formerly dominated the ecosystem, and as a result suffered a food crisis. In these isolated regions, historical ecological processes can be traced whose ramifications were to be increasingly global.

Notes

1   Lynn White, *Medieval Technology and Social Change*, Oxford, Oxford University Press, 1962.
2   Florence McCulloch, *Mediaeval Latin and French bestiaries*, Chapel Hill, University of North Carolina Press, 1960; Robert Hillenbrand, "Mamluk and Ilkhanid Bestiaries: Convention and experiment," *Ars Orientalis* 20, 1990, 149–64.
3   *Genesis* 2. 7.
4   *Ecclesiastes* 3. 19–20.
5   Lothario dei Segni (Pope Innocent III), *On the Misery of the Human Condition*, ed. Donald R. Howard, tr. Margaret Mary Diaz, Indianapolis, IN, Bobbs-Merrill, 1969, 6–7.
6   Edward E. Armstrong, *Saint Francis: Nature Mystic*, Berkley and Los Angeles, University of California Press, 1973, 164 n.7; see I Celano 61; I Celano 81.
7   Francis of Assisi, "Il Cantico di Frate Sole," quoted in Roger D. Sorrell, *St. Francis of Assisi and Nature*, New York, Oxford University Press, 1988, 100.
8   I Celano 80–1.
9   H. Thompson and C. King, eds, *The European Rabbit: The History and Biology of a Successful Colonizer*, Oxford, Oxford Science Publications, 1994.
10 James Fairhead and Melissa Leach, "Culturing Trees: Socialized Knowledge in the Political Ecology of Kissia and Kuranko Forest Islands of Guinea," in Klaus Seeland, ed., *Nature is Culture: Indigenous Knowledge and Socio-Cultural Aspects of Trees and Forests in Non-European Cultures*, London, Intermediate Technology Publications, 1997, 7–18.
11 Peter Verney, *Animals in Peril* , Provo, UT, Brigham Young University Press, 1979, 40–1.
12 Colin McEvedy and Richard Jones, *Atlas of World Population History*, Harmondsworth, Penguin, 1978.
13 Carlo M. Cipolla, *Before the Industrial Revolution: European Society and Economy, 1000–1700*, 3rd edn, New York: W.W. Norton & Co., 1993, 232.
14 On medieval European environmental history, a useful collection is József Laszlovszky and Péter Szabó, eds, *People and Nature in Historical Perspective*, Budapest, Central European University Department of Medieval Studies and Archaeolingua, 2003.
15 Cipolla, *Before the Industrial Revolution*, 199; Carlo M. Cipolla, *The Monetary Policy of Fourteenth-Century Florence*, Berkeley and Los Angeles, University of California Press, 1982, xi.
16 Edwin S. Hunt, *The Medieval Super-Companies*, Cambridge, Cambridge University Press, 1994, 42.
17 David Herlihy, Robert S. Lopez, and V. Slessarev, *Economy, Society, and Government in Medieval Italy*, Kent, OH, Kent State University Press, 1969, 109–10.
18 Ronald Edward Zupko and Robert Anthony Laures, *Straws in the Wind: Medieval Urban Law in Northern Italy*, Boulder, CO, Westview Press, 1996.
19 William Chester Jordan, *The Great Famine: Northern Europe in the Early Fourteenth Century*, Princeton, NJ, Princeton University Press, 1996, 12.
20 Roland Bechmann, *Trees and Man: The Forest in the Middle Ages*, New York, Paragon House, 1990, 280.
21 Corinne J. Saunders, *The Forest of Medieval Romance: Avernus, Broceliande, Arden*, Cambridge, D.S. Brewer, 1993, 10–19.
22 M.M. Postan, *Medieval Trade and Finance*, Cambridge, Cambridge University Press, 1973, 342.
23 Christopher N.L. Brooke, *The Twelfth Century Renaissance*, New York, Harcourt, Brace & World, 1970, 134.
24 Cipolla, *Before the Industrial Revolution*, 92.
25 Hunt, *The Medieval Super-Companies*, 4.
26 Giovanni Villani, *Cronica*, Book 12, Ch. 94, quoted in Robert S. Lopez and Irving W. Raymond, eds, *Medieval Trade in the Mediterranean World: Illustrative Documents Translated with Introductions and Notes*, New York, Columbia University Press, 1955, 71–4.
27 Gene Brucker, *Florence: The Golden Age, 1138–1737*, Berkeley and Los Angeles, University of California Press, 1998, 67–8.
28 Charles R. Bowlus, "Ecological Crises in Fourteenth Century Europe," in Lester J. Bilsky, ed., *Historical Ecology: Essays on Environment and Social Change*, Port Washington, NY, Kennikat Press, 1980, 88.
29 Jean Gimpel, "Environment and Pollution," in *The Medieval Machine: The Industrial Revolution of the Middle Ages*, New York, Holt, Rinehart & Winston, 1976, 75–81.

30   Niccolo Machiavelli, *History of Florence from the Earliest Times to the Death of Lorenzo the Magnificent*, Book 2, Ch. 1.

31   Bowlus, "Ecological Crises," 95.

32   Bechmann, *Trees and Man*, 50.

33   Ibid., 68.

34   Richard Koebner, "The Settlement and Colonisation of Europe," in J.H. Clapham and Eileen Power, eds, *The Cambridge Economic History of Europe from the Decline of the Roman Empire*, Vol. 1, *The Agrarian Life of the Middle Ages*, Cambridge, Cambridge University Press, 1941, 77.

35   Bechmann, *Trees and Man*, 153

36   Based on estimates in Bechmann, *Trees and Man*, 153.

37   Bowlus, "Ecological Crises," 95.

38   Gimpel, *Medieval Machine*, 76–7.

39   Gerald A. J. Hodgett, *A Social and Economic History of Medieval Europe*, London, Methuen, 1972, 84.

40   C.J. Wickham, *The Mountains and the City: The Tuscan Appennines in the Early Middle Ages*, Oxford, Clarendon Press, 1988, 158, 162, 184–5.

41   Bechmann, *Trees and Man*, 99.

42   Ibid., 236–7.

43   John Perlin, *A Forest Journey: The Role of Wood in the Development of Civilization*, New York, W.W. Norton, 1989, 147.

44   Cipolla, *Before the Industrial Revolution*, 92.

45   Bowlus, "Ecological Crises," 89.

46   Robert S. Schiel, "Nutrient Flows in Pre-Modern Agriculture in Europe," in J.R. McNeill and Verena Winiwarter, eds, *Soils and Societies: Perspectives from Environmental History*, Isle of Harris, UK, White Horse Press, 2006, 216–43; M.M. Postan, *Essays in Medieval Agriculture and General Problems of the Medieval Economy*, Cambridge, Cambridge University Press, 1973.

47   John H. Mundy, *Europe in the High Middle Ages, 1150–1309*, New York, Basic Books, 1973, 115–16.

48   Ferdinand Schevill, *History of Florence from the Founding of the City through the Renaissance*, New York, Frederick Ungar, 1961, 235–6.

49   Michael E. Mann, R.S. Bradley, and M.K. Hughes, "Global-Scale Temperature Patterns and Climate Forcing Over the Past Six Centuries," *Nature* 392, 23 April 1998, 779–87.

50   H.H. Lamb, *Climate, History and the Modern World*, London, Routledge, 1995, 171–210.

51   C. Pfister, J. Luterbacher, G. Schwarz-Zanetti, and M. Wegmann, "Winter Air Temperature Variations in Western Europe during the Early and High Middle Ages (AD 750–1300)," *The Holocene* 8, 5, 1998, 535–52; Christian Pfister and Gabriela Schwarz-Zanetti, "The CLIMHIST Data-Base – A Tool for Reconstructing the Climate of Europe in Time and Space: The Example of the Period 1270–1525. Methodology, Coding and Software," in B. Frenzel and C. Pfister, eds, *European Climate Reconstructed from Documentary Data: Methods and Results*, Mainz, European Science Foundation, 1992, 193–210.

52   Gunnar Mickwith, "Italy," in Clapham and Power, eds, *Cambridge Economic History of Europe from the Decline of the Roman Empire*, Vol. 1, 337.

53   Cipolla, *Monetary Policy of Fourteenth-Century Florence*, 9.

54   Herlihy *et al.*, *Economy, Society, and Government in Medieval Italy*, 113.

55   Bowlus, "Ecological Crises in Fourteenth Century Europe," 87.

56   Stewart L. Udall, *The Quiet Crisis and the Next Generation*, Salt Lake City, UT, Peregrine Smith Books, 1988, 137–8. This is a paraphrase of Udall's remark on the Great Depression.

57   Gimpel, *Medieval Machine*, 75.

58   Ibid., 78.

59   Patrick V. Kirch, *On the Road of the Winds: An Archaeological History of the Pacific Islands before European Contact*, Berkeley and Los Angeles, University of California Press, 2000, 47.

60   John R. McNeill, "Of Rats and Men: A Synoptic Environmental History of the Island Pacific," *Journal of World History*, 5, 2, 1994, 299–349, at 301.

61   Atholl Anderson, *Prodigious Birds: Moas and Moa-Hunting in Prehistoric New Zealand*, Cambridge, Cambridge University Press, 1989, 2.

62   Alfred W. Crosby, *Ecological Imperialism: The Biological Expansion of Europe, 900–1900*, Cambridge, Cambridge University Press, 1986, 217.

63   Kirch, *On the Road of the Winds*, 54.

64 F.R. Fosberg, "Vegetation of the Society Islands," *Pacific Science* 46, 1992, 232–50, at 237.

65 Patrick D. Nunn, *Environmental Change in the Pacific Basin: Chronologies, Causes, Consequences*, Chichester, John Wiley & Sons, 1999, 240–1.

66 Peter Bellwood, *The Polynesians: Prehistory of an Island People*, London, Thames & Hudson, 1978, 37.

67 Malcolm McKinnon, Barry Bradley, and Russell Kirkpatrick, eds, *New Zealand Historical Atlas*, Auckland, David Bateman, 1997, 10.

68 Eric Conte, *Tereraa: Voyaging and the Colonization of the Pacific Islands*, Tahiti, Éditions Polymages-Scoop, 1992, 8–10.

69 Ben R. Finney, "Putting Voyaging Back into Polynesian Prehistory," *New Zealand Journal of Archaeology*, Special Publication, 1996, 365–76, at 372.

70 James Belich, *Making Peoples: A History of the New Zealanders from Polynesian Settlement to the End of the Nineteenth Century*, Auckland, Allen Lane, Penguin Press, 1996, 18.

71 R.C. Green, "Rapanui Origins Prior to European Contact: The View from Eastern Polynesia," in P.V. Casanova, ed., *Easter Island and East Polynesian Prehistory*, Santiago, Instituto de Estudios Isla de Pascua, 1998, 241.

72 H.M. Leach, *Subsistence Patterns in Prehistoric New Zealand: A Consideration of the Implications of Seasonal and Regional Variability of Food Resources for the Study of Prehistoric Economies*, Anthropology Department, University of Otago, Studies in Prehistoric Anthropology, Vol. 2, 1969.

73 Patrick V. Kirch and Terry L. Hunt, eds, *Historical Ecology in the Pacific Islands: Prehistoric Environmental and Landscape Change*, New Haven, CT, Yale University Press, 1997.

74 S.L. Olson, and H.F. James, "The Role of Polynesians in the Extinction of the Avifauna of the Hawaiian Islands," in P.S. Martin and R.G. Klein, eds, *Quaternary Extinctions: A Prehistoric Revolution*, Tucson, University of Arizona Press, 1984, 768–80.

75 Kirch, *On the Road of the Winds*, 60–1.

76 M.S. McGlone, "Polynesian Deforestation of New Zealand: A Preliminary Synthesis," *Archaeology in Oceania* 18, 1983, 11–25.

77 Timothy Fridtjof Flannery, *The Future Eaters: An Ecological History of the Australasian Lands and People*, New York, George Braziller, 1995, 242–53.

78 R.C. Suggs, *Archaeology of Nuku Hiva, Marquesas Islands, French Polynesia*, Anthropological Papers of the American Museum of Natural History 49, Part 1, New York, 1961, 186.

79 Kirch, *On the Road of the Winds*, 238.

80 Te Rangi Hiroa (Peter H. Buck), *Vikings of the Sunrise*, New York, Frederick Stokes, 1938.

81 Perhaps about AD 1300. Herb Kawainui Kâne, *Ancient Hawai'i*, Captain Cook, HI, Kawainui Press, 1997, 16–18.

82 Belich, *Making Peoples*, 34.

83 Helen M. Leach, "Intensification in the Pacific: A Critique of the Archaeological Criteria and Their Application," *Current Anthropology* 40, 3, June 1999, 311–39; "The Terminology of Agrricultural Origins and Food Production Systems – A Horticultural Perpective," *Antiquity* 71, 271, March 1997, 135–48.

84 John Patterson, "Respecting Nature: The Maori Way," *The Ecologist* 29, 1, January 1999, 33–8, at 33.

85 Margaret Orbell, *Natural World of the Maori*, Dobbs Ferry, NY, Sheridan House, 1985, 84–7.

86 Geoff Park, *Ngâ Uruora (The Groves of Life): Ecology and History in a New Zealand Landscape*, Wellington, Victoria University Press, 1995, 137–8, 157–9, 212, 222–3, 270.

87 Herman Melville, *Typee: A Romance of the South Seas*, New York, Heritage Press, 1935, Ch. XI, 140–1.

88 Patterson, "Respecting Nature," 44.

89 Tom Brooking, *Milestones: Turning Points in New Zealand History*, Lower Hutt, Mills, 1988, 21; Gary A. Klee, "Oceania," in *World Systems of Traditional Resource Management*, London, Edward Arnold, 1980, 245–81, at 255; Belich, *Making Peoples*, 74.

90 Michael Kioni Dudley, *Man, Gods, and Nature (A Hawai'ian Nation I)*, Honolulu, Nâ Kâne O Ka Malo Press, 1990, 102.

91 Patrick V. Kirch, "Subsistence and Ecology," in Jesse D. Jennings, ed., *The Prehistory of Polynesia*, Cambridge, MA, Harvard University Press, 1979, 286–307, at 304.

92 Patrick D. Nunn, "Human–Environment Relationships in the Pacific Islands around A.D. 1300," *Environment and History* 7, February 2001, 3–22.

93   This view is represented by, among others, Clive Ponting, *A Green History of the World*, New York, St. Martin's Press, 1991, 1–7, and Jared Diamond, *Collapse: How Societies Choose to Fail or Succeed*, New York, Viking, 2005, 79–119; see also Jared Diamond, "Easter Island's End," *Discover Magazine* 16, 8, August 1995, 36–45. The best scholarly based overview and analysis of the evidence is in Paul Bahn and John Flenley, *Easter Island, Earth Island*, London, Thames & Hudson, 1992, and a revised edition, John Flenley and Paul Bahn, *The Enigmas of Easter Island*, Oxford, Oxford University Press, 2002.

94   Christopher M. Stevenson, Thegn Ladefoged, Sonia Haoa, and Alejandra Guerra Terra, "Managed Agricultural Production in the Vaitea Region, Rapa Nui," in C.M. Stevenson, J.M. Ramírez, F.J. Morin, and N. Barbacci, eds, *The Reñaca Papers: VI International Conference on Easter Island and the Pacific*, Los Osos, CA, Easter Island Foundation and the University of Valparaíso, Chile, 2004, 125–37, is a report on the archaeological investigation in which I took part as a volunteer in November–December, 2002. I am listed among the volunteers in the article (p. 136), and made measurements for the map: "Figure 11. The elite residence (Site 161c)," p. 134. This article does not include the site at Orito, and what follows here is an informal description based on my impressions.

95   Georg Zizka, "Flowering Plants of Easter Island," *Scientific Reports PHF* 3, Frankfurt am Main, Palmengarten der Stadt Frankfurt, 16.

96   Catherine Orliac, "The Woody Vegetation of Easter Island Between the Early 14th and the Mid-17th Centuries AD," in Christopher M. Stevenson and William S. Ayres, eds, *Easter Island Archaeology: Research on Early Rapanui Culture*, Los Osos, CA, Easter Island Foundation, 2000, pp. 211–20.

97   Thor Heyerdahl, *Aku-Aku: The Secret of Easter Island*, Chicago, Rand McNally, 1958.

98   Te Rangi Hiroa (Peter H. Buck), *Ethnology of Mangareva*, Honolulu, Bernice P. Bishop Museum Bulletin 157, 1938, 304–5, 510.

99   Andreas Mieth and Hans-Rudolf Bork, "The Dynamics of Soil, Landscape and Culture on Easter Island, Chile," in McNeill and Winiwarter, eds, *Soils and Societies*, 273–321.

100 Christopher M. Stevenson and Sonia Haoa, "Diminishing Agricultural Productivity and the Collapse of Ranked Society on Easter Island," in *Archaeology, Agriculture and Identity*, Kon-Tiki Museum, Oslo, Norway, 1999.

101 Tim Flannery, *The Eternal Frontier: An Ecological History of North America and Its Peoples*, New York, Atlantic Monthly Press, 2001.

102 Luis G. Lumbreras, *The Peoples and Cultures of Ancient Peru*, Washington, Smithsonian Institution Press, 1974, 3–7.

103 Craig Morris, "Storage, Supply, and Redistribution in the Economy of the Inka State," in John V. Murra, Nathan Wachtel, and Jacques Revel, *Anthropological History of Andean Polities*, Cambridge, Cambridge University Press, 1986, 59–68.

104 Pedro Cieza de León, *The Incas (Crónica)*, [1554], tr. Harriet de Onis, ed. Victor Wolfgang von Hagen, Norman, University of Oklahoma Press, 1959, 2. 13, 185.

105 John Victor Murra, *The Economic Organization of the Inka State*, Research in Economic Anthropology, Supplement 1, Greenwich, CT, JAI Press, 1980, 94.

106 Garcilaso de la Vega, *Royal Commentaries of the Incas and General History of Peru* [1590], tr. H.V. Livermore, Austin, University of Texas Press, 1966, 1.4.7, 1.5.2, 205, 243–4.

107 Terence N. D'Altroy, "Andean Land Use at the Cusp of History," in David L. Lentz, ed., *Imperfect Balance: Landscape Transformations in the Precolumbian Americas*, New York, Columbia University Press, 2000, 357–90, at 384.

108 John Hyslop, *The Inka Road System*, Orlando, FL, Academic Press, 1984, 254–68.

109 Huamán Poma (Felipe Guamán Poma de Ayala), *Letter to a King: A Peruvian Chief's Account of Life Under the Incas and Under Spanish Rule (Nueva corónica y buen gobierno, 1567–1615)*, ed. and tr. Christopher Dilke, New York, E. Dutton, 1978, 99–100.

110 Cieza de León, *The Incas*, 115.

111 John Hyslop, *Inka Settlement Planning*, Austin, University of Texas Press, 1990, 283.

112 R.T. Zuidema, "Inka Dynasty and Irrigation: Another Look at Andean Concepts of History," in John V. Murra, Nathan Wachtel, and Jacques Revel, eds, *Anthropological History of Andean Polities*, Cambridge, Cambridge University Press, 1986, 177–200.

113 Vega, *Royal Commentaries*, 1.5.1, 241–2.

114 Ibid., 1.5.3, 246.

115 Louis Baudin, *Daily Life in Peru Under the Last Incas*, New York, Macmillan, 1962, 221.

116 Kevin Krajick, "Green Farming by the Incas?", *Science* 281, 17 July 1998, 322.

117 A.J. Chepstow-Lusty, K.D. Bennett, J. Fjelså, A. Kendall, W. Galiano, and A. Tupayachi Herrera, "Tracing 4000 Years of Environmental History in the Cuzco Area, Peru, from the Pollen Record," *Mountain Research and Development* 18, 2, 1998, 159–72, at 159.

118 León, *The Incas*, 115.

119 A.J. Chepstow-Lusty, K.D. Bennett, V.R. Switsur, and A. Kendall, "4000 Years of Human Impact and Vegetation Change in the Central Peruvian Andes – With Events Paralleling the Maya Record?" *Antiquity* 70, 270, December 1996, 824–33, at 828.

120 Chepstow-Lusty *et al.*, "Tracing 4000 Years of Environmental History," 164.

121 Louis Baudin, *A Socialist Empire: The Incas of Peru*, Princeton, NJ, Van Nostrand, 1961, 104.

122 D.R. Wickers and H. Lowdermilk, "Soil Conservation in Ancient Peru," *Soil Conservation* 4, 1938, 91–4.

123 Poma, *Letter to a King*, 50.

124 León, *The Incas*, 115.

125 Murra, *The Economic Organization of the Inka State*, 46.

126 Ibid., 326.

127 Alfred W. Crosby, Jr., *The Columbian Exchange: Biological and Cultural Consequences of 1492*, Westport, CT, Greenwood Press, 1972, 38–9, 51–5.

128 Chepstow-Lusty *et al.*, "4000 Years of Human Impact and Vegetation Change," 828.

129 Vega, *Royal Commentaries of the Incas*, 1.5.9, 257.

130 Michael W. Binford, Alan L. Kolata, Mark Brenner, John W. Janusek, Matthew T. Seddon, Mark Abbott, and Jason H. Curtis, "Climate Variation and the Rise and Fall of an Andean Civilization," *Quaternary Research* 47, 1997, 235–48, at 235.

6   The transformation of the biosphere

During the early modern period,[1] European explorers, traders, conquerors, and settlers spread through most of the rest of the world. They modified ecosystems everywhere by introducing animals and plants, extracting resources, deforesting many areas, establishing plantations, and subjugating or decimating indigenous populations that had formed their own ways of interrelating with local environments.[2]

This epoch is sometimes called the "Age of Discovery," because then European explorers sailed across the oceans, charted the coasts and islands, and led expeditions inland on the continents. Their names are familiar – Vasco da Gama, Christopher Columbus, Ferdinand Magellan – and there were many others. But discovery was not their only activity, and perhaps not even the most important one. From the moment they dropped anchor beside a new land, they began to change it. Ecosystems that had emerged in almost complete isolation for centuries or millennia, and had evolved unique biotas, suddenly began to suffer the invasion of the animals and plants that the Europeans brought with them, whether deliberately or accidentally. Other changes followed soon: fire, hunting, cutting of trees, enslaving and killing of indigenous human beings. It was "a time of dramatic and accelerating change."[3]

On many newly discovered islands, European sailors left domestic animals that could become feral and fend for themselves, such as goats and pigs. And everywhere ships moored next to a seacoast, rats made their way to land, climbing along ropes or swimming ashore. If they were lucky, and often they were, the animals found plentiful food plants, native animals that had no experience of avoiding them, and a lack of predators, so their numbers increased rapidly and they overwhelmed the local ecosystems, making many species extinct. Plants as well as animals invaded the new lands: the seeds of aggressive Eurasian weeds arrived hidden in grain and animal hair and dung.

It was a two-way exchange, although not an even one. Not many animals arrived and multiplied in Europe in the early days, although later there would be trouble with muskrats and American squirrels. But it was otherwise with plants. Tobacco, potato, maize, tomato, and sweet potato are among the domestic plants the Europeans willingly took home and soon were raising and eating (smoking in the case of tobacco). Meanwhile, European agricultural technology intruded into the rest of the world, particularly the temperate and subtropical areas, bringing machines and crops that cleared and replaced indigenous animal and plant life. Plantations of crops in demand in Europe, such as coffee and tea, replaced the biodiversity of tropical forests with monoculture.

As the early modern age went on Europeans acquired and improved technologies, including some using new sources of energy such as fossil fuels. Important inventions were made outside Europe, but it was Europeans who initially spread them around the globe. And it was in the UK and western Europe that the Industrial Revolution began, with its major impacts on human society and the community of life, including faster transport, urbanization, pollution, and scarring of the landscape. It then spread to other parts of Europe, North America, and Japan, and its effects were felt almost everywhere. Associated with the new industries were new forms of air and water pollution. Incidents of sickness and death, such as the release of toxic waste from the Müller-Pack aniline dye factory in Basel, Switzerland in 1864, led to anti-pollution legislation in several European states.[4]

The explorers had benefited from the compass and sextant, which enabled them to find direction and latitude. By the 1760s they had a ship's clock dependable enough to measure longitude. Guns, gunpowder, and the cannon gave them a military advantage over many peoples who did not yet have them. The printing press allowed the dissemination of knowledge about the discoveries. Not all Europeans were happy about technology. Soon after she took the English throne, Queen Elizabeth I decided that a new knitting machine might take jobs away from laborers, and denied it a patent.[5]

The inventions that did the most to shape the modern age were those that allowed the application of new sources of power, especially fossil fuels, to the production process. These machines became the instruments of the Industrial Revolution. Windmills turned pumps to raise water from the fields. Later pumps were powered by coal. Although crude steam engines were devised around 1700, it was a series of improvements made by James Watt, a Scottish engineer, in the latter third of the eighteenth century that enabled the mechanization of factories and eventually of transportation and agriculture. The first oil well to tap petroleum from within rock strata was drilled in Pennsylvania in 1859. The first practical internal combustion engine using gasoline, a derivative of the new fuel, was patented by German inventor Nikolaus Augustus Otto in 1876. With the manipulation of immense amounts of energy made possible by technology, the human use of resources became increasingly exploitative, with an accelerating impact of change on ecosystems around the globe. Industrial processes generated increasing levels of pollution of the air, water, and land.

The Industrial Revolution transformed agriculture in this period, marking the beginning of the end for traditional methods of farming in western Europe and North America. The products of agriculture, food and fiber, were no less important than in earlier times. In fact, they gained additional consequence because greater numbers of workers in the industrial establishments needed to be fed and clothed. New crop regimes in Europe, using fertilizers and planting nitrogen-fixing species such as clover in alternate years, did away with the need for regularly leaving the land fallow. Potatoes and turnips gave higher yields. The principles of mechanization were applied to agriculture as seed drills, harvesters, and other devices increased the area that a single worker could cultivate, while the efficiency of ever-larger agricultural businesses forced out small landholders.

The European economy came to include and in many ways to dominate most of the rest of the world, so that the world market economy came into existence. In this period, Europe drew raw materials from the rest of the world and produced manufactured goods that were sold not only at home, but also back to the countries that were the source of the raw materials. In ecological terms, this meant that Europe and other areas that came to be industrialized were using and degrading the material and energy capital of ecosystems abroad. European economic thinkers such as Adam Smith were convinced that the free market, while appearing chaotic and exploitive, is actually guided to produce the right amount and variety of goods by an "invisible hand," that is, it was the order of nature and ought not to be hindered by regulation: the so-called *laissez-faire* doctrine. His book, *The Wealth of*

Nations, expounds that while human motives are ultimately out of self-interest, the net effect in the free market would tend to benefit society as a whole.[6] Karl Marx and other socialist theoreticians pointed out that this capitalist system disregards the needs of the working class, and advocated an interim system in which governments operating in the interests of workers and peasants would own the means of production. Although Marx and Engels mentioned the basic importance of nature,[7] many Marxists emphasized the social relationships within the economy and neglected questions such as ecological relationships and the sustainability of natural resources.

During the eighteenth century, the human population of the Earth began an exponential increase that has continued to the present day. This was achieved by diverting an ever-larger percentage of the energy cycles of the biosphere into food production for humans. The mechanization of agriculture, the discovery of new sources of fertilizer such as guano and the manufacture of artificial fertilizers, and the construction of large irrigation systems contributed to this irruption of human population. Directly or indirectly, humans accelerated a process of replacing the numbers and variety of other forms of life with the sheer numbers of one species – their own – along with domestic animals and plants.

One important reason for the expansion was the spread of New World food plants such as potatoes and maize. Between 1700 and 1900, Europe's population (including Russia) more than tripled, from 122 million to 421 million, in spite of the emigration of some 40 million to the Americas and elsewhere.[8] China's people trebled in number, from 150 to 436 million, and India's almost doubled, to 290 million, in the same period. Sub-Saharan Africa's gain was also almost double – from 61 million to 110 million – in spite of the toll taken by the slave trade and the fact that the tsetse fly and sleeping sickness prevented the spread of inhabitants and agriculture to large areas. The size of the pre-Columbian population of the Americas is a disputed question, but most historical demographers agree that epidemics killed at least 90 percent of the native population during the sixteenth century.[9] By 1700 the New World had recovered to a population of 12 million, including native Americans and those of European and African descent. In 1900 there were 165 million, more than twelve times as many. The trend toward urbanization began in the latter part of this period; in 1800, just over 2 percent of the world's people lived in cities, but by 1900 it was 10 percent. The human portion of the biosphere was increasing, its demands on resources more than proportionally greater, and the other members of the community of life taxed to meet those demands.

Thomas Robert Malthus (1766–1834) lived during this period of growing population, and published his *Essay on the Principle of Population* in 1798. He observed that the mathematical principle of human reproduction is multiplication, and therefore unrestricted population growth will follow an exponential curve. To increase food production, however, depends on the incremental addition of new cultivated lands, which can be expected to follow, at best, a rising line. Even this is problematic, however, because humans usually choose to use the best soils first. The expectation for the future, therefore, is that growing population will inevitably outrun the ability of agricultural production to feed it. There have been localized famines in areas where food production could not keep pace with population growth, as occurred in Ireland in the 1840s, but technology, improved crops, fertilizers, and resultant increased yields postponed a worldwide Malthusian crisis beyond the end of this period.

Environmental thought in the early modern period began with observations by naturalists and scientists that human actions, particularly those of colonialists, were making rapid changes around the world, many of them damaging to nature and threatening to continuing subsistence. These pioneer thinkers suggested programs of conservation, forest reserves, and restoration of deteriorated landscapes. They faced apathy and opposition from others who believed that humans could not damage nature or, if they could, it was justifiable in terms of economic improvement. In the mid-nineteenth century, Darwin, Haeckel, and other scientists discovered the importance of the interaction among species and their environments in evolution, and began to conceptualize a science of ecology.[10] This was an expression of the modern rebirth of natural science. The physical sciences, in a partnership with technology, provided the means for greater impacts of humans on the rest of the natural world. The biological sciences began to supply the knowledge of how living things function and interrelate, and thus to lay the groundwork for the study of ecology.

Microscopes were made from around 1590, but it remained for Anthony van Leeuwenhoek to discover "little animals" in rainwater in 1675, and then to describe spermatozoa, yeast cells, and bacteria.[11] A vast realm of microbes, a major segment of the biosphere, had been revealed. In the course of the seventeenth century several important instruments for quantifying observations of the environment were invented, including the barometer, the thermometer, and the pendulum clock.

The founder of systematic biological taxonomy was Carolus Linnaeus of Sweden, who gave names to the genus and species of every animal and plant known to him. Without such an orderly method, the study of ecosystems would be impossible. Interestingly, his father, Nils, had created the family name after the linden tree; like most rural Swedes of his day, until then he had no last name.

Evidence that stimulated environmental ideas was gathered not only in Europe, but in remote corners of the globe where colonizing powers sent physicians, learned naturalists, and burgeoning scientists, as Richard Grove has noted.[12] Oceanic islands were particularly important in calling their attention to the relationships between deforestation, extinctions, desiccating climate, shortages of essential resources, disease, and famine. Islands were microcosms where these processes could be seen more clearly: due to their small size, limits were reached more quickly and an observer could see changes in the landscape during visits over the course of a few years or decades. The image of a lost Eden suggested itself to a number of European savants, and they gave advice on how to halt or reverse the course of destruction. Botanical gardens were established in tropical colonies, and their staffs included keen scholars who ventured beyond identifying and collecting plants to develop theories of environmental change. Since professional scientists served as advisors or even governors, their ideas were sometimes given practical trials.

One of the more telling arguments of the early scientists was that it was in the interest of colonial governments to prevent the degradation of the environment in the territories they controlled. "The state," as the economist Richard Cantillon had proposed, is "a tree with its roots in the land."[13] If the colonies were deforested, they could no longer supply timber. Deforested lands suffer erosion and decreased rainfall, so that both soil and water for food production and other crops will decline. Faced with poverty and famine, colonial peoples will become rebellious. Pierre Poivre, a French officer in Mauritius, called the treatment of the island by heedless colonists "sacrilegious," and said that deforestation had placed the "land in servitude."[14] Thomas Jefferson was attracted to many of Poivre's ideas.

George Perkins Marsh, United States ambassador to Italy, observed in the Mediterranean area and elsewhere "the character and extent of the changes produced by human action in the physical condition of the globe we inhabit," and warned in his book *Man and Nature*, published in 1864, that "the result of man's ignorant disregard of the laws of nature was deterioration of the land."[15] Differing from the prevailing economic optimism of the times, he saw "man" as the disturber of nature's harmonies. *Man and Nature* begins with an analysis of the environmental degradation of the Roman Empire and its causes. He certainly knew that the economy of the Roman Empire was organized primarily to benefit the upper strata of society, and that the balance of humankind's economy with nature's harmonies that he thought beneficial could be achieved only by considering the needs of every social stratum. He portrayed the ordinary people of the Roman Empire as forced to "struggle at once against crushing oppression and the destructive forces of inorganic nature,"[16] and judged that the fact that they had to struggle against both those opponents at once had resulted in defeat and the devastation of the natural environment.

Alexander von Humboldt had traveled to the Americas and studied the relationship of the distribution of plants to various environmental conditions. He noted that an increase in elevation is associated with a decrease in temperature. This was an initial step toward the idea that associations of animals and plants are found in zones of elevation in mountainous terrain, a concept formulated by Pyotr Petrovich Semënov-Tian-Shanskii on the basis of explorations in the Tian Shan mountains of central Asia in 1856–7.[17] C. Hart Merriam made similar studies in the San Francisco Peaks of Arizona, and published a description of "life zones" based on temperature in 1890, a step toward the concept of ecosystems.[18] Charles Darwin's contributions in this regard are discussed in a following section of this chapter.

Tenochtitlán: the European biotic invasion

> The city remained faithful to Tlaloc, the four-eyed god of rain. The colors of the forested mountains still dominated it, untouched. The swamp cypresses of Xochimilco competed for dominion over the valley with the *ahuehuetes* – the old men of the water, hung with Spanish moss – the cedars, and the bright-branched ash trees. Masses of verdure grew to the shores of the lake in which the cypresses had taken root beneath the water to anchor a vegetable city of small floating gardens called *chinampas*. As a consequence of this profuse vegetation, the rains came with a clocklike regularity.[19]

So Fernando Benítez describes the environment of Tenochtitlán as he imagines it on the day, August 13, 1521, when it finally fell to Hernán Cortez and his Spanish soldiers. Much of the Aztec capital still lies beneath Mexico City, which was built on its ruins at Cortez's order by the labor of the people the Spaniards called "Indios." Across a street north of the Spanish cathedral are the remains of the Templo Mayor, the pyramid that once elevated the twin temples of Huitzilopochtli, god of war, and Tlaloc, god of rain.[20] The National Museum of Anthropology contains sculptures reclaimed from beneath the streets, including the Piedra del Sol, the circular dial bearing the face of Tonatiuh, the Sun, surrounded by glyphs of the twenty days of the Aztec month. The past, they believed, consisted of four epochs, each of which ended in a disaster: wild animals, wind, fire, and flood. They expected the era in which they were living, the Fifth Sun, to end by earthquake, but in fact their world was deeply altered, and in many ways destroyed, by what came aboard Spanish ships.

The Aztec Empire spanned a variegated land, from tropical rainforests near the Gulf coast, past towering volcanoes to the Valley of Mexico, containing lakes with no outlet to either ocean.[21] Tenochtitlán occupied islands in Lake Texcoco. The Aztecs also controlled country westward to the Pacific, much of it covered with forests of pine and oak. The topography is dominated by complex mountain ranges, dissected by many valleys. With countless microclimates – wet and dry, low and high, hot and cool – Mexico had a vast number of local ecosystems.

When Europeans arrived in Mexico, they were amazed that the forms of life there were different from the ones they knew. Few Mexican species were initially familiar to Europeans; they had never seen hummingbirds, for example, and thought toucans fantastic. They used familiar names for unfamiliar forms, or borrowed names from the languages of the New World. They called the puma *león*, the jaguar *tigre*, and the wild canine *coyote*, from *coyotl*, its name in Nahuatl, the Aztec language. The difference was not caused simply by climate: the fact that the Americas had been nearly isolated from the Old World for thousands or millions of years meant that species had evolved separately. Mexico had an unusually large variety of native animals and plants, considering the size of the country.[22] These species were components of unique and fragile ecosystems, many of which were endangered by the onslaught of Eurasian organisms. Mexico is rich in amphibians and reptiles. The rattlesnake, a sacred symbol for Mesoamerican civilizations, does not occur in the eastern hemisphere. The vegetation was as unusual as the animal life. Mexico had 900 cacti, comprising 55 percent of all cactus species in the world. Among remarkable trees was the *ahuehuete* (tule cypress), the national tree of Mexico. It is fairly widespread in moist highland locales. The most famous specimen, near Oaxaca, measures 40.5 m (135 ft) in height, 41.7 m (139 ft) in circumference, and is over 2,000 years old.[23] The Oyamel fir, in groves at high elevation, serves as a wintering place for hundreds of millions of monarch butterflies.[24] Growing wild in the pine forest were ancestral forms of garden flowers including dahlia and zinnia.[25] The marigold was domesticated and used in quantities in Aztec fiestas.

Heirs of venerable civilizations that preceded them, the Aztecs inhabited a landscape that was already transformed. At the time of the conquest they were an urban and agricultural people. They practiced intensive farming, utilizing irrigation and terracing. Fertile valleys were covered by mosaics of productive farmland. In the lacustrine environment of the Valley of Mexico, they created richly productive *chinampas*, the famous "floating" gardens. These were platforms erected in shallow lakes, filled with mud and vegetable matter, fertilized by human excreta, and planted with food plants. Aztec agriculture utilized a wealth of native domesticated plants, including maize, beans, tomatoes, sweet potatoes, chile, *chía* (a sage), *huauzontle* (*Chenopodium*), amaranth, and squashes.[26] They had no large domestic animals, but turkeys and dogs provided meat, as did wild animals such as ducks from the lakes and deer. The use of wood as fuel and construction material had led to the deforestation of nearby mountainsides, exposing them to erosion that took soil and deposited it as silt in the lakes. Population was increasing rapidly; indeed, it may have been about to overshoot the carrying capacity of the land. An Aztec population crash might have occurred in time even if the Spaniards had not arrived. Something similar had happened to the great city of Teotihuacán. "The Spanish conquest took place at a time when the Aztecs were using all available resources, when the population of the Valley was larger than it had ever been, and before the advent of any spontaneous calamities such as those presumed to account for the downfall of Teotihuacán six hundred years earlier," says Charles Gibson.[27] That center, whose ruins were well known to the Aztecs, had expanded its population and deforested the surrounding area so that springs dried up, depleting the water supply. Eventually the city was abandoned.

Alfred W. Crosby, Jr., in his influential book, *The Columbian Exchange*, pointed out the importance of the fact that when Europeans came to the Americas, they did not come alone, but brought a number of other species in their ships: domestic animals and plants on which they had depended in their homelands.[28] Also, inadvertently, they brought along stowaways that they might have preferred to leave behind: rats, mice, aggressive weeds, and, most disastrously of all, the microbes that cause smallpox and other virulent diseases. Crosby called this assemblage of organisms from Europe (and some from Africa and Asia), a "portmanteau biota,"[29] a sailors' trunk, as it were, packed with animals and plants that would flourish in the new environment, crowding out many other species. These organisms, like the Europeans themselves, would do best on long-isolated continents and islands where climate and other conditions were most like Europe: Australia and New Zealand, Argentina and Uruguay, and much of North America, for example. Crosby applied the term "Neo-Europes"[30] to such places. They were "lands of the demographic takeover,"[31] where European populations shouldered aside the native inhabitants, in part because Eurasian crops and herds could replace native crops, and because native peoples were decimated by epidemics of Old World diseases.

Mexico was only in part a "Neo-Europe." The central plateau, including the relatively cool Valley of Mexico, came close to matching the climate of Spain. The Spaniards remarked on the resemblance, and named the province "Nueva España." But Mexico never became a "land of the demographic takeover." The Spaniards did not overwhelm the Indian population in numeric terms. Rather, from Indians who came to share the language and other aspects of Spanish culture, the Mestizos arose to become the dominant ethnic group in post-colonial Mexico. Mexicans today are proud of the heritage of this "cosmic race," and emphasize its Indian component. Mexico City has a statue to Cuauhtémoc, a leader of the Aztec resistance to the Spaniards, in a busy intersection. It has none honoring Hernán Cortez. It is impossible to deny, however, the wide-reaching transformation of the Mexican landscape by the European "portmanteau biota."

But the crucial introduction was undoubtedly smallpox. Cortez ordered an Aztec family to care for an African sick with the disease. His hosts caught it and it began to spread. Before it went far, the Aztecs rose against the outnumbered Spaniards, who were forced to flee the city. Cortez built his military strength with reinforcements, Indian allies, and ships taken overland in pieces to the lake and assembled. Meanwhile, "the great rash" raged through Tenochtitlán. "The pustules that covered people caused great desolation; very many people died of them, and many starved to death … no one took care of others any longer," an Aztec witness recorded.[32] Among those who died was Cuitláhuac, an able leader who had succeeded the discredited Moctezuma. If the epidemic had not weakened the defenders so grievously, the Aztecs might have been able to repel Cortez.

The depredations of disease did not end with the conquest. Smallpox returned and in its wake came measles, mumps, chicken pox, whooping cough, typhus, typhoid fever, bubonic plague, cholera, scarlet fever, malaria, yellow fever, diphtheria, influenza, and pneumonia.[33] All were virgin soil epidemics, that is, the populations through which they spread lacked experience of them and had acquired no immunity. Demographers estimate that of the total native population of Mexico, at least 90 percent died within the century after 1519. It was a disaster for the Spaniards as well. They had some hereditary immunity, did not catch the diseases as easily as the Indians, and their mortality rates were lower. But Spaniards in colonial New Spain depended on Indians to do all the work: in the fields, to plant and harvest the crops both peoples needed to survive; and in the mines, to bring out the ore and to smelt it so the masters could acquire wealth and buy Spanish goods.

The domestic animals brought by the Spaniards flourished in the new environment. There was plenty to eat. First to multiply and spread were pigs. It was bad enough when Spanish pigs rooted in Indian gardens; they loved maize and Indians lacked fences strong enough to keep them out. But pigs escaped and became feral. The oak forests, with plentiful acorns, were just the environment pigs loved. They devoured every reptile and mammal smaller than themselves, along with ground-nesting birds. Losing their plumpness, they reverted to a form like that of the ancestral European wild boar, since predators in the Mexican wilds took the smaller and weaker ones.

Next came cattle: by the 1540s there were so many that the price of beef plummeted. Antonio de Mendoza, the first viceroy, tried, and failed, to keep cattle ranching within limits.[34] Spanish cattlemen formed an association (*mesta*) that pressed for ever larger ranches; with the low price of meat and hides, only the biggest operations were profitable. Spanish custom declared the open range a commons, and allowed grazing in fallow fields. But the cattle hardly needed encouragement, since the vegetative cover of the land was in excellent condition. They went wild, too. Spanish cattle were tough and armed with formidable long horns. They got into Indian maize fields and forced the abandonment of villages. Observed Mendoza, "If cattle are allowed, the Indians will be destroyed."[35] The bovine population exploded when the herds reached the northern grasslands. "The sudden multiplication of cattle is one of the most astonishing biological phenomena observable in the New World."[36] Travelers described herds covering the land as far as the eye could see. Inevitably the cattle increased to the point where they consumed the range. After 1565, cattlemen noticed that the herds were no longer expanding so fast. They blamed it on a number of things: rustlers were attacking the herds, wild dogs took many calves, and the nomadic Chichimecas, who had acquired horses and a taste for beef, swooped in from the north with bows and arrows. But Martín Enríquez, writing in 1574, put his finger on the underlying reason: "Cattle are no longer increasing rapidly; previously, a cow would drop her first calf in two years, for the land was virgin and there were many fertile pastures. Now a cow does not calve before three or four years."[37] The nutritious grasses had been eaten, and the range was taken over by less palatable brush. After 1586 cattle starved by the thousands. This cycle, the explosion of the numbers of grazing animals in a new environment, the destruction of the range, and a crash from high to low numbers, is called an "ungulate irruption." The initial boom and bust was followed by smaller ones until balance was achieved between the number of herbivores and the availability of food. The final state of the vegetation, however, was severely degraded.

Horses followed the pattern described for cattle. At first there were few, but the Spaniards took them everywhere they could. They multiplied, some escaping and establishing wild herds. By the turn of the century, visitors observed that horses roaming free in Durango were beyond counting.[38] At first the Spaniards tried to keep horses out of Indian hands. But this proved impossible, since if ranchers wanted Indians to do the work of herding cattle, there would have to be Indian vaqueros. By 1580, natives were holding horse races in Tlaxcala.[39] The unsubdued tribes to the north soon had horses – they were there for the taking – and used them skillfully for attack, escape, and hunting. Horses exacerbated the impact on grassland ecosystems made by cattle.

Another equine species that adapted to Mexico was the *burro*, the Mediterranean donkey. It became a ubiquitous beast of burden, and Indians, along with poorer Spaniards, eagerly adopted it. It readily took to the wild and multiplied. Burros loved the rough terrain and had an appetite for a variety of plants including shrubs that grew in the arid northern mountains, where they were abundant by 1550.[40]

Sheep were even more destructive of the land than cattle, since they nibble the grass down to its roots, exposing the soil to erosion by rain. Mendoza imported Merino sheep, which produce superior wool but, unlike other sheep, they stay in a pasture until it is depleted. Goats, whose destruction of vegetation in the Mediterranean region was proverbial, came along with the sheep. The species together are doubly damaging: while sheep rip out the grass, goats browse on bushes and small trees. Spanish sources seldom mention

Figure 6.1 Statue of Tlaloc, the Aztec god of rain, now located at the National Museum of Anthropology in Mexico City. For a people whose agriculture depended largely on variable annual rainfall, such a deity was of prime importance. Photograph taken in 1973.

goats, but when they speak of sheep, goats may be read between the lines. Colonial drawings show goats within herds of sheep. Goats became feral and spread into mountainous terrain, but sheep lacked toughness and the ability to elude predators. Even so, with the aid of shepherds they spread across Mexico, and their numbers burgeoned astronomically. Elinor Melville, in her carefully researched book, *A Plague of Sheep: Environmental Consequences of the Conquest of Mexico*,[41] shows how the Spanish sheep-raising enterprise in the Valle del Mezquital transformed a productive mosaic of Indian intensive irrigation agriculture into the mesquite-covered desert that gave it its present name. "By the end of the 1570s the vegetation of the region was reduced in height and density. In some places it had been removed altogether and only bare soil remained. Former agricultural lands were converted to grasslands, and the hills were deforested and grazed by thousands and thousands of sheep."[42] In the process, the human population that the valley could support declined. Although Indians acquired herds of sheep, the story is for the most part one of Spanish sheep owners disregarding Indian rights.

The Mezquital is in the semi-arid highlands. In the tropical lowlands around Veracruz, according to Andrew Sluyter, "cattle and sheep were pushing at the range's ecological limit by the end of the sixteenth century,"[43] but erosion was not severe. However, the native population did not recover after disease and exploitation had decimated it, because the herds, the Spanish landowners, and African slaves had taken over the land formerly occupied by Indian farms.

Mexico had numerous native rodent species. None of them was as destructive as the rat[44] that climbed ashore from Spanish ships and dug burrows in the earth, nested in trees, devoured Indian stores of maize, crowded out or killed native species, and multiplied so rapidly that it was impossible to eradicate. In addition, it carried bubonic plague and typhus.

Nothing distinguished Spanish and Indian diets from one another as clearly as preference in cereal grains. The Spaniards brought wheat seed and insisted that Indians learn to plant and tend the crop. Within fifteen years after the conquest, New Spain was exporting wheat to the Caribbean Islands. No meal seemed complete to the Spaniards without wheat bread; they despised maize as an inferior food of natives, peasants, and animals. The Indians saw no reason to prefer wheat to maize, and even today maize is their staple. In times of hunger both peoples made exceptions, of course. Other Spanish introductions included barley and other grains, peaches, pears, oranges, lemons, chick-peas, melons, onions, radishes, and olives. They transformed the agricultural landscape. Grapes were a special problem: wine was a standard feature of the Spanish table and was necessary, along with bread, for celebration of the Catholic mass. Grapevines would grow in the highlands, but Mexico never became a leading wine region. More important in its impact on ecology and economy was sugarcane. Sugar refining became New Spain's biggest industry. Mills were brought from the West Indies and Canaries, and up to 1585 prices for sugar were so high that growers preferred sugarcane to wheat.[45] Indians did not adapt to work in cane fields and refineries, so the Spaniards brought in African slaves.

Mixed with crop seed, hiding in animal feed, spreading with dung, and stuck to clothing, came the seeds of weeds from the Old World. Among them were dandelions, nettles, clover, and tough grasses. Plowing offered them a foothold and overgrazing opened opportunities for weeds, since most of them prefer disturbed soil. Thus Mexican plant communities gained new members, but unfortunately most of them behaved badly. A major task the Spaniards required of Indian agricultural laborers was weeding, which consumed months of work time every year.[46]

From the beginnings of European exploration, colonization, and trade in the fifteenth century, the organisms they carried with them had a worldwide impact. Mexico offers an example of the way in which their onslaught altered ecosystems and reduced the abundance of native species or made them extinct. Eurasian species were not the only ones moved from place to place. Organisms both benign and troublesome were carried from Africa and Asia to other lands with similar climates aboard European ships that stopped at ports en route to the New World. For example, Spanish ships brought the banana, originally a south Asian plant, from the Canaries to the West Indies in 1516, and soon afterward to Mexico. Banana plantations replaced thousands of hectares of rainforest, and the conditions of labor on the plantations often amounted to debt slavery. *Aedes aegypti*, the African mosquito that is the vector of yellow fever, proved more than troublesome.

The result of wholesale introductions of species is homogenization of ecosystems. Like other formerly isolated lands, Mexico had biodiversity adapted to unique combinations of soils and climates over uncounted years of evolution, but the organisms brought on Spanish ships, and the social and agricultural arrangements the Spaniards imposed, gave rise to unstable situations favoring opportunistic organisms. The human species in this case proved to be, to use the phrase coined by George Perkins Marsh in 1864, "the disturber of nature's harmonies."[47]

The Netherlands: Holland against the sea

A brisk wind was blowing across the level landscape of Holland when the miller in charge of the Schermer windmill told me he would free the great sails to revolve.[48] The sight and sound were spectacular as the four arms of a circle of sails 25 m (81 ft) in diameter swung up toward the sky one after another and down almost to graze the earth. I was fortunate to be able to see it because a windmill can operate at its best only in winds of 5–10 m per second (11–22 mi per hour), and that happens only about one-quarter of the time in Holland.[49] But the most fascinating sight to me was to watch the pump when the miller pulled the lever to connect it with the gears. It was a screw pump set at an angle, with a long spiral rotor formed of wooden blades around an axle, rotating inside an open cylinder. As it turned, it continuously lifted water that churned upward and poured out the top. Part of my fascination was because I knew that the invention of this machine is often attributed to Archimedes, a Greek mathematician of the third century BC, and it was used widely in Hellenistic Egypt and the Roman Empire.[50] Its modern use in helping to keep agricultural land dry in the Netherlands began not long before the Schermer windmill was built in 1634, along with others like it – a case of hydraulic engineers looking to the ancient world for inspiration.[51]

Dutch artists, as well, used ancient imagery to express the need for pumps and other means of controlling water in their low country, much of which was and is below sea level. In a chamber of the Water Board of Rhineland in Leiden, a huge mythological painting[52] by Caesar van Everdingen and Pieter Post, measuring 218 by 212 cm (7.15 by 6.95 ft) and dating from 1655 is still visible, showing Pallas, goddess of technology, and Mercury, god of commerce, holding the sea gate against Neptune's onslaught. At the time, the Netherlands led the world in seaborne trade, which provided the financial resources for infrastructure such as dikes and the hydraulic windmills that provided protection for the country against drowning.

No one really knows who first said, "God made the world, but the Dutch made Holland."[53] It has been repeated, however, in numerous books about water management in the Netherlands, where virtually every square kilometer of the landscape bears evidence of human effort. Without constant exertion by its inhabitants, most of the Netherlands would be covered by the North Sea, but it is just as true to say that the situation of the Dutch environment is precarious in many ways resulting from the actions of people affecting it in the past. The Dutch made their country productive for human purposes, but they also exposed it to dangers against which they have had to struggle.

In the centuries before it had many human inhabitants, the major part of what is today the Netherlands was a flat coastal plain protected to some extent by a line of sand hills covered with grass, pines, and other trees, just inland from the seashore. In the south was the delta where the great rivers Rhine, Meuse, and Scheldt poured into the North Sea through shifting channels as they dropped their silty loads eroded from the continent, forming extensive fluvial deposits.

The lowland area had a high freshwater table – indeed, fens where reeds and sedges grew covered much of the land – and peat-forming ecosystems were responsible for the dominant landscape. Peat is formed by the accumulation of plant material that does not completely decay because it is soaked in water and oxygen is excluded. The preservation of organic material from decay can be illustrated by a number of human bodies that had been buried in peat; when discovered after centuries, their flesh, skin, and clothing were still relatively intact and soft. Peat moss (sphagnum) and other plants continued to grow and accumulate, so that peat bogs rose above the surrounding fens in wide, extensive pillow-like formations. Capillary action within the peat could raise the water table 4 m (13 ft) or more above the fens.[54] Sometimes trees like alder, willow, and beech grew on the peat, eventually falling and becoming part of it, making a substance called "forest peat."[55] Because peat is essentially non-oxidized organic material, rich in carbon, it makes a useful fuel when dried, and in the forest-poor Netherlands of the Middle Ages and early modern period it became the basis of an open fuel market,[56] with the result that huge volumes of the surface of the land were removed and used to generate heat. Peat was also burned to make ashes that could be leached to make salt. The result of peat consumption was a colossal human-caused alteration of the landscape.

Few descriptions of the environment of the lowlands survive from Roman times, but Pliny the Elder was there in the first century AD and wrote that the people lived on artificial mounds raised above the flood level. He also describes the use of peat in inexpert terms: "They scoop up mud in their hands and dry it by the wind more than sunshine, and with earth as fuel warm their food and so their own bodies, frozen by the north wind."[57] By the ninth century they were digging ditches to drain the peat bogs, transforming them into land for agriculture and dwellings. As peat dried, however, it decreased greatly in volume, lowering the elevation of the land surface, decreasing the efficiency of drainage, and increasing the frequency and depth of floods. The land sunk at an average rate of about 1 m (3 ft) per century, so that eventually much of the central part of the lowlands was at or below average sea level.[58] As peat miners cut deeply into the peat bogs, sections of the land became freshwater lakes that tended to become ever larger; thus the so-called "water-wolf" eating away the land was peat digging rather than natural sea flooding.[59]

The Dutch began to build dikes and dams to exclude floodwaters and to keep the farm fields dry in the Middle Ages. At first these efforts were strictly local, but it became obvious that coordination was necessary to keep one village's dikes from causing flooding in another's farmlands, so in the twelfth and thirteenth centuries water boards were organized to supervise and maintain dikes, canals, dams, locks, and sluices, and more generally the management of water in a certain region. These were among the first institutions requiring local participation that provided a base for the later Dutch republics and, eventually, the gradual development of democracy. The first to be established were the water boards of Lekdijk Bovendams (1122), Rhineland (1170), and Groote Waard (1230), and many more followed.[60] The system is still functioning today, after some reorganization and modification.

As it became evident that simple ditch drainage using gravity was no longer working, landowners began to surround units of land with dikes, creating enclosures called polders.[61] Rainwater and seepage would of course enter a polder and make drainage necessary, so sluices were installed at low points, allowing water to empty out at low tides, but were closed at other times to prevent water from flowing back in. Sluices were marginally effective, but emptying water from a polder, especially one below sea level, could be accomplished satisfactorily only with some kind of pumping, and hand-operated devices were insufficient. The land inside the polders continued to shrink, lowering the surface level, and when water was kept out by the dikes there could be no new alluvial deposits.

The answer to the question of polder drainage appeared in the early fifteenth century when the windmill, previously used to grind grain into flour, was adapted to lift water. The first record of this innovation refers to a drainage mill at Alkmaar put up by Floris van Alcmade and Jan Grietensoen in 1408. The energy from its sail arms turned a "scoop-wheel" lined with wooden blades that dipped into the water and lifted it a meter or so before discharging it at a higher level. This came to the notice of Count Willem VI of Holland, who realized its value and urged its use throughout his domain.[62] Indeed, it spread widely and hundreds were built in the following three centuries as the Netherlands moved into its Golden Age.

The major rivers of the southern Netherlands, with their interconnecting and sometimes shifting channels, presented a complicated challenge. The communities began to line their banks with low embankments and then higher dikes, and after 1300 these extended from the river mouths far upstream, but it gradually became evident that the more a river is channeled between dikes, the higher it rises and the worse floods become if the dikes are breached. Polders were built on the land between the rivers and on the islands in the delta, but they remained vulnerable to flooding from the rivers and from the sea. Careful watching of the rivers and constant labor on dikes and dams were necessities forced on the inhabitants by a continually shifting environment.

In the seventeenth and eighteenth centuries the extended lake system turned out to be of great advantage to the Rhineland water board, as compared to some neighboring water boards. The lakes became part of an enormous water reservoir in which the water board temporarily stored the water from the polders, which was pumped into the reservoir by the windmills. This was necessary because it was not always possible to get rid of the water. When strong northeastern winds blew, the water board could not open the main sluices near Amsterdam depicted on the painting by Van Everdingen and Post mentioned above, because the wind would blow the seawater into the canals and the land would flood. The neighboring Delftland water authorities did not have such large reservoirs and therefore had to invest enormously in more sluices and canals to speed up the drainage process. In Rhineland, by contrast, the number of canals and sluices remained the same until 1800. Generally speaking, the old water system primarily depended on natural rhythms and forces (wind, storms, and tides). Over time the system lost much of its dependency on the natural forces and became more technical.[63]

Constantly threatening, though, were the storm surges that swept in from the North Sea, impelled by gale winds and intensified by high tides.[64] They usually came years apart, and the Dutch had to guard against complacency in the years between, because the surges could swallow polders and villages, drowning crops, animals, and people in salt water. The natural line of high hills and sand dunes along the coast offered a degree of protection, but it was sometimes breached, and did not exist in the Zeeland delta region, where tidal streams had almost free access. Dikes were built along the North Sea coast to supplement the sand dunes and to protect them against erosion by waves. Nevertheless, a series of surges in the twelfth and thirteenth centuries broke through between north Holland and Friesland, forming and enlarging a brackish arm of the sea in the heart of the lowlands called the Zuider Zee, itself a mixed blessing because although it offered access to the world for the merchant fleet and opportunity for fishermen, it provided a dangerous portal of entry for the North Sea with its occasional surges and floods, and its constantly gnawing tides and waves. In the early modern period, the best that could be done was to line its shores with dikes to resist erosion. In those centuries, only the persistent lengthening and strengthening of dikes and dams, and the multiplication of windmills that pumped drainage water, averted the complete immersion of the lowland district.[65] As Petra van Dam noted, "The innovations in hydraulic technology of the period 1300–1600 must be understood as a response to the rapidly changing conditions in Holland's wetland environment."[66] Defenses were still inadequate against unusually strong storm surges. In 1421 one of these hit both the delta and the Zuider Zee, inundating scores of villages and killing thousands of people. Many square kilometers of polder land lost in this St. Elizabeth's Day flood were never reclaimed.

The accomplishments of the Dutch in the "Golden Age" of trade, prosperity, empire, and intellectual and artistic flowering are amazing to contemplate. It was also a time when large areas of land were reclaimed from water and hydraulic infrastructure was installed. As Simon Schama wrote, "The period between 1550 and 1650, when the political identity of an independent Netherlands nation was being established, was also a time of dramatic physical alteration of its landscape."[67] Financial resources came from the profits of the largest merchant fleet in the world, that dominated trade in the North Sea and Baltic, and the acquisition of a colonial empire that stretched from the West Indies to the Cape of Good Hope, Ceylon (Sri Lanka), the East Indies, and Formosa (Taiwan), and for a time the Dutch had exclusive access to trade with Japan. A proportion of these resources was invested at home in agriculture and land reclamation. It is astounding to remember that these developments occurred in large part while the Netherlands was fighting its war of independence, the Eighty Years' War (1568–1648) against the Spanish Empire under King Philip II, the most powerful monarch in Europe, and his successors. The war had its environmental aspects, since the Dutch at times deliberately breached dikes, flooding the land in order to impede the Spanish armies. The Treaty of Münster, which ended the war, recognized the self-government of the United Provinces (Dutch Republic).

A major reclamation activity in this period was the draining of a number of lakes, as well as small arms of the sea, beginning in north Holland. Drainage began with smaller lakes and proceeded to larger ones. This was accomplished by "empolderment," which began by encircling the lake with a dike and a drainage canal outside the dike. Windmills would be set up on the dike to pump the water out of the lake into the canal. In the case of the deeper lakes, one windmill could not raise the water high enough to reach the canal, so it became necessary to put two or more windmills in series. After a time, when the polder was empty of water, the former lakebed could be provided with drainage ditches and parceled out for farms and settlements. This system is attributed to two talented hydraulic engineers, Simon Stevin of Flanders (1548–1620) and Jan Adriaanszoon Leeghwater (1575–1650). The area of arable land added by lake reclamation was not small: between 1500 and 1800 it amounted to an increase of one-third – 100,000 ha (240,000 acres) in Holland and 250,000 ha (600,000 acres) in the entire alluvial zone of the Netherlands.[68]

As larger and deeper lakes presented a challenge, the search for more efficient pumps to drain them faster led to technological improvements. One of these was the Archimedean screw pump mentioned at the beginning of this section. Screws had been used in simple, movable pumps called *tjaskers* in Frisia since the late sixteenth century, but in 1634 a patent was granted to Simon Hulsbos of Leiden for a large Archimedean screw used in conjunction with a windmill. Several of these were installed, and gradually began to replace the older scoop-wheels, although not everywhere. These screws were angled about 37 degrees with a diameter of 1.5–1.8 m (4.5– 5.4 ft). Their advantage was that they raised the water 4–5 m (12–15 ft), much higher than scoop-wheels, so that a single windmill could replace several

Figure 6.2 Drainage windmills in the Schermermeer polder, Netherlands. These provided
energy for machines such as the Archimedes screw pump to raise water into canals
and make land available for agriculture. Photograph taken in 2007.

older ones; at 40–50 rpm a screw could lift about 40 cu m (1,404 cu ft) per minute, faster
than the scoop-wheel, and less water spilled out.[69]

The water control system as it was improved and expanded during the Golden Age
resulted from the creative energy and prosperity of that period, and provided a productive
base for the nation. Unfortunately, the eighteenth century brought relative economic
decline due to war, commercial competition with England, and depression in the agricultural sector at home. As a result, land reclamation along with the installation of new pumping
mills and the replacement of older ones slowed. There was an increase in floods from the
rivers, possibly because dike maintenance was neglected.

The flooding of the rivers was also caused by a remarkable combination of the impact of
the Little Ice Age (*c.*1400–1800) and human interference with the riverbeds.[70] In springtime, as river ice began to melt and break, huge ice mountains up to 30 m (100 ft) would
form. In that time even the large rivers got entirely blocked and all shipping stopped, and
as a result no income is noted in the toll accounts for the years when it happened. After
1600 or so, the Dutch took all sorts of measures in order to streamline and direct the flow.
One of the solutions was "groynes", small dikes built into the water to slow down the water
at the shores and prevent erosion. Exactly at the points where these obstructions occurred,
the ice mountains would form themselves, and the water would flood the dikes, undermin-
ing them and finally causing them to break. Of course the people observed this, so why did
they continue such practices? Meddling with the rivers was an ongoing process of estimating
risks. Erosion occurred everywhere and led to costs to adjoining water boards, whereas dike

breaking occurred only now and then, at certain places, hopefully in a neighbor's dike. Before 1900, no integrated management of the big rivers yet existed.

Then in 1731 a disaster struck the dikes along the North Sea coast, which were supported and faced with piles and other wooden structures. An organism called the pileworm, which is not a worm but a mollusk with awesome ability to bore into wood and rapidly turn it into a structure resembling Swiss cheese, was discovered in its millions infesting the dikes.[71] It is not known why the pileworm, also called the teredo, appeared so suddenly, although increase in salinity of the seawater in the area has been suggested; the organism prefers waters of high salinity. The teredo was already well known to sailors as the shipworm; wooden ships were often sheathed in copper sheeting to keep it from riddling their hulls and eventually sinking them. Perhaps a ship brought a new species or variety to the Dutch coast. The only way to save the dikes was to provide them with stone facing, and that was an expensive solution because the Netherlands lacks sources, and stone had to be imported from Norway and Belgium.[72]

After the French Revolution, French armies invaded the Netherlands and set up a Batavian Republic with a number of innovations including a bill of rights. In regard to the environment, one of the most important centralizing reforms was the founding of the Rijkswaterstaat, a national water administration that undertook several large water management projects and founded a professional school for hydraulic engineers.[73] It met with considerable resistance from the long-established water boards, who were jealous of their regional authority and unwilling to cede power to the center, and it took more than half a century to work out an uneasy cooperation between the national agency and the water boards. The Rijkswaterstaat proved to be a permanent institution, lasting through the Napoleonic period and the establishment of the Kingdom of the Netherlands in 1813. Weathering stormy periods, both literally and metaphorically, it exists today because historic experience has shown that the environmental survival of the Netherlands makes a national policy necessary for landscape and water management. Water boards still exist, too, with responsibility for local water control and water quality.

As the Industrial Revolution got under way in the late eighteenth century and on into the nineteenth, Dutch engineers saw the potential of steam engines to drive drainage pumps. One advantage over windmills was that steam engines could operate continuously, not just when the wind was blowing at the right speed. When driving scoop-wheels, however, steam engines proved uneconomical because they used too much expensive coal. In 1837, two steam engines turning Archimedean screw pumps were used in draining the Zuidplas polder in south Holland and proved more economical.[74] The centrifugal pump was introduced and also demonstrated superior efficiency. In the second half of the nineteenth century steam engines were commonly employed for pumping, and by 1896 they were responsible for almost three-fifths of total mechanized land drainage. In the twentieth century, experiments were made with internal combustion engines to drive several types of pump, but the source of energy that became dominant was electricity. These technological improvements rendered the windmill obsolete, and hundreds of them were demolished until it seemed that one of the most distinctive elements of the Dutch cultural landscape might disappear. A large number of them, however, were preserved as museums and other landmarks, and are highly prized by the Dutch and by tourists today.

The danger presented by the sea was not flooding alone, but also the intrusion of salt water, which could kill crops and corrupt the fresh water supply. Henrik Stevin warned about it in the seventeenth century, referring to salt as the poison of the sea.[75] Seawater moved with the tides into river mouths and seeped under the coastal dunes, and the Zuider

Zee brought it into the middle of the country. The problem grew worse as river mouths were dredged for navigation and drainage, and canals and locks were opened for ships. Saline deep ground water and a Rhine river increasingly polluted with salty industrial wastes added complications. Needs for improved barriers, testing stations, and an augmented supply of fresh water to flush out salinized areas became evident.

The twentieth century brought both human-caused and natural destruction, but also the construction of works of unprecedented size in attempts to control the sea and the rivers. The first of these was the separation of the Zuider Zee from the North Sea by a dam 32 km (19.2 mi) long between north Holland and Friesland, built 1927–32. The barrier was 3.5 m (10.5 ft) higher than the level of the highest storm surge then known. The enclosed area became a freshwater lake renamed the IJsselmeer, providing a major resource to counter salinization. Of course the saltwater fish that had inhabited it were replaced by freshwater species, and the fishing fleet had to switch from catching herring and anchovy to eel and pike-perch.[76] Several huge new polders were enclosed and drained inside the new lake, eventually creating enough arable land to add a new province to the Netherlands.

In addition to the terrible cost in human lives of the Second World War and the German occupation of the Netherlands (1940–5) – the Dutch lost 205,000 lives including 100,000 of the 140,000 Jewish citizens – there were also environmental costs. Deliberate breaching of the dikes was done by the Dutch and their allies, and by the Germans. Construction and maintenance lagged or ceased due to unavailability of materials, equipment, and fuel.

After the war, gaps in the dikes were sealed and reinforced and flooded polder lands were drained, while a series of plans was drafted to dam the Meuse River delta against salt seepage and storms. But if any doubt existed of the necessity to shore up defenses against the sea, it came to an end with the devastating storm surge of February 1953, which made scores of breaches in the dikes, inundated 200,000 ha (770 sq mi) of land, drowned 1,835 people, as well as forcing the evacuation of 72,000. Although damage was widespread, the brunt of the storm surge had hit the delta, so the Rijkswaterstaat concentrated its efforts in the decades after the flood on a large-scale delta works project to realign river dikes and to construct barriers against storm surges in the river outlets to the sea. But by 1970, when major parts of the project remained to be constructed, many of the leaders of public opinion in the Netherlands – writers, artists, and academics including scientists – raised environmental objections to some aspects of the plan and received wide support. Not only safety, they argued, but also values such as the diversity of wildlife, the beauty of the landscape, and the preservation of objects of cultural history, should receive consideration. Also, if dams blocked some of the river mouths from the sea, they observed, the resulting lakes would be dead water where pollution brought down by the rivers would stagnate.

Bureaucrats and engineers were initially outraged at these criticisms, but the design that emerged after a political struggle showed the Dutch genius for compromise and action as a community. The new plans integrated environmental concerns with technological innovation.[77] For example, the centerpiece of the project, the storm surge barrier that protects the Oosterschelde estuary, 9 km (5.4 mi) long, is provided with a 4 km (2.4 mi) section of huge vertical sluice gates with steel doors 42 m (126 ft) wide that are ordinarily open to allow the tides to pass through, but are closed when a tide or storm surge is predicted more than 3 m (9 ft) above normal high tide, an event that happened on average about once a year between 1986 (the year it was completed) and 2007. The Oosterschelde itself, with its rich variety of sea life, was thus preserved, and was designated the Netherlands' largest national park in 2002. This barrier is the largest of thirteen such barriers connecting islands and peninsulas in the Delta; in addition 300 other structures and 16,500 kilometers of dikes are part of the delta works.

At one end of the Oosterschelde barrier, an inscription proudly announces, "Here the tide is ruled by the wind, the moon, and us." Most Dutch experts would not express that much hubris, even though the project has been declared one of the world's seven engineering wonders. The sea and the rivers are ever changing, and the battle to safeguard the land is never finally won. The Netherlands is understandably one of the nations most concerned about a warming global climate, rising sea level, and a possible increase in the energy and severity of storm surges. Steps have already begun to fortify the coastal defenses. The ministry of water affairs (Rijkswaterstaat), is dredging up sand from the North Sea and dumping it in the water just in front of the beaches, so that the waves will build higher beaches. Also, it has identified the weakest low spots in the dune system, typically situated where former fishing villages border on the beach, and built artificial dunes there, planting them with marram grass intended to counter wind erosion. The effectiveness of these measures remains to be seen.

One proposal for the future includes a new, higher and wider vanguard dike along the entire North Sea coast from border to border, storage basins for river water, and new "super pumping stations" to empty excess river flow and rainfall into the sea. Such a plan would be a continuation of the trajectory of water control works in the history of the Netherlands. It would also represent an escalation of human attempts to control nature by orders of magnitude.

London: city, country, and empire in the Industrial Age

The coal burned in London during the age presided over by Queen Victoria left a dingy legacy on the city's great buildings. It was there as late as the 1950s, when I first visited England: a black deposit coated the facades, and residents and visitors alike were incredulous when reminded that St. Paul's Cathedral, Buckingham Palace, and the British Museum were not originally dark gray, but built of light-colored stone. The pollution carried in London's murky air did more than darken the architecture. It attacked the stone, creating a friable, soluble crust that accelerated erosion. Reliefs and statues took on a surrealistic look as they dissolved over the years. Victorian architects and contractors were already aware of the problem when they planned the new Houses of Parliament in 1839, and searched for suitable stone, but their choice proved unsatisfactory.[78] Travelers who knew the clear air of the nineteenth-century American West, such as Francis Parkman and James Fenimore Cooper, were offended by the pall of smoke and the besooted buildings in London, but Lord Byron saw pollution as "the magic vapour/ Of some alchymic furnace, from whence broke/ The wealth of worlds."[79] December 1879 was a month of polluted fogs, with a mortality rate that rose 220 percent. Reformers organized a Smoke Abatement Committee, which faced a problem that seemed insurmountable.[80]

The discoloration invaded the countryside, where rain falling through smoky air brought down pollution and deposited it on trees and bushes, killing lichens and other organisms. Among the darkened vegetation, blacker forms of butterflies and moths gained a protective advantage of camouflage against their predators, and outnumbered lighter-colored ones beginning in the nineteenth century, a phenomenon termed "industrial melanism."[81] First noticed in the coal and steel producing country of the English Midlands, it also occurs around London.[82]

London was the leading city of the Industrial Revolution throughout the nineteenth century, although facing competition from continental Europe and North America as the decades passed. No city had as large a number of workers engaged in manufacture, although not in iron and steel making, which were located in the north. London's manufacturers made clothing, including shoes, furniture, carriages, ships, clocks, bread, beer, liquors, leather, silk, paper, books, machinery, tools, jewelry, and musical instruments, to name a few. Many of these industries used machines powered by steam engines, and produced noxious emissions to the air and water.

The Industrial Revolution in the Victorian Age was powered by new sources of energy, led by coal and gas. In the preceding century, metallurgy and steam had depended on wood and charcoal, raising the specter of forest exhaustion, and in an attempt to prevent that, some continental governments had enacted conservation laws. But coal seemed to give the forests a reprieve. Industries and homes switched to coal for heat and gas for lighting, adding a burden to London's air. By 1880 there were 600,000 homes in the central part of the city with 3,500,000 fireplaces, virtually all burning coal. London became the most important coal-shipping center, receiving "sea coal" from Newcastle and other northern ports for its own use, but also for export.

In the 1850s, steam began to replace sail, and iron to replace wood in ships, and for a time London's shipyards continued to lead the world. Britain built the world's first great railway system, and several lines ran from the city, strengthening London's role in commerce. The economic growth of this period was phenomenal. Although the rate was not even, and spurts of growth alternated with depressions, the production of the UK more than quadrupled during Victoria's reign, an increase averaging 2.5 percent per year.[83]

The Great Exhibition of 1851 was organized by Prince Albert, Victoria's consort, as a celebration of the achievements of inventors and engineers, many of them British.[84] It was held in a huge glass and iron structure called the Crystal Palace, built in Hyde Park for the purpose, and boasted 14,000 exhibitors. Six million or more visitors viewed works of art and technology, including telegraphs, sewing machines, revolvers, reaping machines, and steam hammers. It was a defining public moment for the Industrial Revolution.

How did these far-reaching changes in methods of production, sources of energy, and means of transportation affect ecological conditions? London suffered some of the worst results, and contemporary descriptions challenge the imagination. Crowding a comparatively small area and taxing its limited facilities, the population grew to an extent never before seen in a city. Indeed, it was the world's largest city throughout the century. From about 1 million in 1801, Greater London grew to 2.3 million in 1854 and 6.6 million in 1901.[85] Most of the increase was the result of migration from rural areas and smaller towns.

Low in a river basin, London is subject to natural fogs which were greatly exacerbated by smoke and chemicals from industrial and domestic sources. Under stagnant conditions, a malodorous fog of pea-soup yellow hung over the city, reducing visibility so that people were almost blinded and were known to stumble into the river. It was suffocating, and many with lung problems died. In January 1880, during a four-day fog, there were 700–1,100 deaths in excess of the normal rate, and this is only one instance of a phenomenon that increased in frequency as the century wore on.[86] The worst incident would occur in 1952. Dr. H.A. Des Voeux of the Coal Smoke Abatement Society in London proposed in 1905 that the noxious mixture of smoke and fog be called "smog," a word that caught on.[87] Irate citizens sued polluters in court under the nuisance laws but although convictions were obtained, the fines were too small to be a deterrent. In 1891, the Public Health Act prohibited the emission of "black smoke," and shortly afterwards a generating station defeated a suit by showing that its smoke was dark brown![88] Industries were not required to halt pollution, but to show that they had used the "best practicable means" of reducing it.

Disposing of liquid and solid wastes by flushing them into a nearby river and waiting for them to be carried downstream has been a practice of cities from early times. But London's River Thames is an estuary of the North Sea, and tides sweep up it through the center of the city and above it to Teddington ("Tide-end-town"). Sewage might flow down the river during low tide, but twice a day a wall of water would carry it back upstream.[89] While the tide was at its height, the water backed up the outfalls, and precipitation of solids occurred. The mixture of organic sewage and industrial chemicals killed virtually all the fish and sea mammals. At low tides, a vast area of tainted mud was exposed to the air, producing a terrible smell. The infamous Great Stink occurred in 1858 during hot weather and a series of low tides; the odor was so intolerable that Parliament adjourned for a week.[90] Worse than the smell was the danger of disease and poisoning. Private water companies drew much of London's drinking water from the Thames or from tributaries whose condition was not much better. A cholera epidemic killed 6,800 Londoners in 1832; it was not understood that the disease is carried by fecal matter in water. Cholera recurred in 1848 and 1849 with a toll of 14,000 deaths. In 1854, Dr. John Snow provided a circumstantial demonstration that cholera is spread by drinking water. He had observed that most of the 500 people who died of it in Soho had drawn their water from one company's pump in Broad Street, but neighbors who had used local wells had escaped. When he persuaded authorities to lock the pump, the deaths ceased.[91] Medical officers took ten years to accept his findings but were not in time to prevent the last great cholera epidemic in 1865–6. With water quality much improved, London escaped the cholera epidemic of 1891 that ravaged the continent. Improvement of sewers was also needed; a Royal Commission in 1861 approved a plan to construct two main lines to intercept numerous smaller sewers and carry the effluent to an outfall below the city. A new sewer was incorporated into the design of an embankment constructed along the north side of the river.[92] Simply relocating the point at which sewage reached the estuary was not enough; treatment was later added. For a time, much of the solid matter was barged out and dumped into the North Sea, with an effect on the oceanic ecosystem that can only be guessed.

The effects on the inhabitants of London, however, could be observed. Overcrowding was extreme, especially for the poorest one-third of the city's inhabitants, who were housed in small rooms, often in basements, that contained entire families and sometimes pigs as well – that is, if they could find a roof to shelter them at all. The Rev. Andrew Mearns, who had visited many of these unfortunates, described their homes in a pamphlet, *The Bitter Cry of Outcast London*, in 1883:

> Few who will read these pages have any conception of what these pestilential human rookeries are, where tens of thousands are crowded together amidst horrors which call to mind what we have heard of the middle passage of the slave ship. To get into them you have to penetrate courts with poisonous and malodorous gases arising from accumulations of sewage and refuse scattered in all directions and often flowing beneath your feet; courts, many of them which the sun never penetrates, which are never visited by a breath of fresh air, and which rarely know the virtues of a drop of cleansing water.[93]

The environmental conditions in which the poor lived and worked contributed to a series of bread riots and other uprisings in nineteenth-century London. A bread riot in Hyde Park in 1855 involved 150,000 and led Karl Marx briefly to think that the English revolution had begun. Serious incidents followed in 1861, 1867, and 1886.[94]

There were proposals on the part of humanitarians, architects, and planners to improve conditions in the city. A number of the projects involved leveling portions of the crowded warrens for new roads, hardly a way to help the inhabitants who were displaced and had to find quarters elsewhere, usually in similar conditions. The parks movement sounded a more positive note.[95] Sometimes the amenities of open space were reserved adjuncts to the mansions of the affluent, but the more civic-minded of the elite hoped to set a moral tone for the working class, offering them a better way to spend their leisure than in pubs, gambling dens, and brothels. In 1845, the Commissioners of Woods and Forests opened Victoria Park, East London. Its trees, fountains, follies, and flowerbeds drew up to 30,000 visitors a day, offering a forum for speakers and preachers, along with habitat for birds and small mammals.[96] On the other hand, when Battersea Park replaced a marsh, nature lost a wetland and biodiversity suffered. Still, if it had not become a park, it would undoubtedly have been occupied by industries. By 1878, the City of London recognized the value of open space to the extent that it preserved 1,200 ha (3,000 acres) of Epping Forest.[97]

The city and its inhabitants, particularly the poorest ones, were increasingly cut off from the countryside by the phenomenal spread of suburbs. The area covered by Greater London grew at a rate twice that of the population.[98] William Cobbett, in his 1830 book, *Rural Rides*, called London a "Great Wen" that was disfiguring the landscape, but he could hardly have imagined the territory that the suburbs would cover in another fifty years.[99] The population of the outer ring of the urban area swelled from 414,000 in 1861 to 2,045,000 in 1901.[100] Meanwhile the resident population of central London declined. There were middle-class suburbs, and suburbs for workers who commuted daily into the city on three-horse omnibuses. Each horse dropped 8–11 kg (17–24 lb) of dung daily, which amounts to 3–4 tons annually. In 1859, work began on the first underground railway. There were electric trams after 1890. As transportation improved, the price of suburban land rose, and the financial motivation for subdivision and construction increased. With the spread of the conurbation came worsening air pollution.

The built-up area of London presented a new environment for plants and animals. Much of the wildlife disappeared and was replaced by species adapted to human constructions and human presence, such as rats, mice, flies, spiders, and cockroaches. But there were still the interactions of species that form an ecosystem. For example, house sparrows and pigeons flourished in the city, and their predators – kestrels and peregrine falcons – continued to nest in inaccessible locations on towers and high buildings.[101]

Forests suffered continuing attrition, even outside the advancing edge of urbanization. England had proportionally less woodland than any other European country at the beginning of the nineteenth century. The demand for ships' timber and fuel had taken a heavy toll from the royal forests.[102] To help assure a supply, plantations were made, such as the 2,000 ha (5,000 acres) of oak planted in the New Forest in 1819. But Sherwood Forest and less famous woods were gone. The disafforestation of Wychwood Forest, Oxfordshire, led to the creation of four farms.[103] Disafforestation meant the termination of the royal forest and usually its sale to private owners, not necessarily the removal of trees, although the latter also often happened. Along with many thousands of old trees went the wildlife, even the deer long protected by royal ownership. In 1851, pursuant to the "Deer Removal Act," Hainault Forest in Essex was disafforested and its deer removed or destroyed. There were many other such cases.[104] Wildlife also suffered from loss of habitat through the movement to remove hedgerows between fields. Many oaks from the Forest of Dean, noted as particularly fine shipbuilding timber, went to assemble the last of the great wooden warships. But if coal for fuel and iron for ships reduced the amount of wood taken from British forests or imported from abroad, timbers were needed for props in coal mines, the rapidly expanding railroad system demanded wooden sleepers beneath the rails, and the construction and furniture industries provided a growing market for wood. Between 1864 and 1899, timber imports trebled to 10 million tons.[105] Some wood came from renewable sources: one-third of English woodland was coppiced (a method in which trees were cut back, but regrew from the lower trunks and roots). Denizens of the forest were killed to meet demands of the city. Commercial hunters gathered wild birds' eggs by the thousands and sold them in London, and there was a booming market in feathers for ladies' hats. Naturalists remarked that some species such as lapwing had become hard to find. The upper classes increasingly treated forest as an amenity, an area where they could pursue foxes, shoot birds, and enjoy the scenery, which may account in part for the survival of woodland and plantations covering perhaps 10 percent of the land in Britain.

During the Victorian Age, England changed from a predominantly rural country to an industrialized, urban nation. In 1800, three-quarters of the population lived in the countryside; by 1900, a similar proportion lived in towns and cities. The percentage of the work force employed in agriculture, forestry, and fishing dropped from 25 in 1831 to 9 in 1901. In the course of the nineteenth century, Britain changed from being self-sufficient in food production, or nearly so, to importing almost half of all foodstuffs consumed.

Before the 1860s, it seemed that the mechanization of agriculture might be improving production and bringing new land under cultivation. For example, steam pumps and dredging machines made the draining of the fens possible. The wetlands north and east of Cambridge and Ely became farmland after 1820, when the first Watt engine began to drive a scoop at Bottisham Fen.[106] New machines such as seed drills, ploughs, harrows, hoes, mowers, and reapers increased speed and efficiency.[107] But agricultural laborers did not share the enthusiasm of landowners as wages fell and jobs became scarcer. Disturbances followed the passage of the Corn Law of 1815, which restricted wheat imports to protect landowner profits and resulted in higher bread prices for workers. In 1830, the Captain Swing riots destroyed 387 threshing machines.[108] Urban wages, though high only in comparison with what could be earned in agriculture, contributed to the depopulation of the countryside.

As the decades went by, more farmland was lost to factories, railroads, roads, housing, and facilities such as tips, landfills, and incinerators for the disposal of urban waste. A major depression began in 1873 and lasted into the 1890s. Within twenty years, agricultural output fell by one-half. Landowners, particularly in the western counties, found that conversion of arable land to grazing was profitable, and 911,000 ha (2,250,000 acres) was affected.[109] *The New Domesday Book*, a survey of land ownership published in 1873, showed that 363 landowners controlled 24 percent of the total land surface.[110] At the end of the century, British agriculture was at its lowest ebb. Britain was producing annually only enough grain to feed its population for eight weeks.

The difference was made up by imports from continental Europe, the United States, and the British Empire. As B.W. Clapp put it, "The acreage of land lost to houses, factories, schools, roads, and railways has been regained many times over through the use of land overseas that has supplied Britain with food, industrial crops and minerals."[111] The comment, while true, is too optimistic; there are ecological costs to such replacements. In the process of industrialization, Britain began to draw raw materials from ecosystems abroad, subjecting them to monoculture, simplification, and deterioration. This was part of the reason for the tenacity with which Britain defended and extended its empire in the Victorian Age.

London was the capital of an empire: the financial and commercial center, the administrative and military center, the nerve center of the colonial organism. This was true even at the beginning of Victoria's reign. Although the thirteen American colonies had been lost more than fifty years before, Canada remained. In addition, there were great possessions such as India, Australia, and the recently acquired Cape Colony in South Africa. Smaller dependencies existed in Guyana and on islands and outposts scattered across the Atlantic and Indian Oceans and the Caribbean and Mediterranean Seas. New Zealand was annexed in 1840, but otherwise the empire grew by little during the first thirty years of the monarch's long reign. In 1857 a part of the Indian army mutinied against its British officers, a rebellion that was crushed with much fighting and loss of life. The government in London took direct rule of India in the next year, abolishing the East India Company. The Suez Canal opened a new sea route to India in 1869, and in 1876 Victoria took the title "Empress of India." Prime Minister Benjamin Disraeli had delivered a spirited oration in defense of Britain's right to empire at the Crystal Palace in 1872; for the rest of the century fierce competition ensued among Britain and other powers for new colonies. They sliced up the African continent, Britain receiving a lion's share. British possessions expanded in south and southeast Asia and across the Pacific Ocean. At the turn of the century, the British Empire included about one-quarter of the Earth's land area and population. Just as importantly, Britannia ruled the waves. Her navy was by far the world's most powerful, and Britain had one-third of the world's merchant marine, including for a time one-half of all the steam ships. London was the busiest port, providing for those great fleets.

The reasons for the expansion are complicated, but they certainly include the desire to integrate new lands, with their natural resources and cheap labor, into the growing British

Figure 6.3 The Tower Bridge over the Thames River, in London, built in 1886–94, seems to epitomize the triumphal attitude of human conquest over nature that was characteristic of the Victorian Age and the Industrial Revolution in general. Photograph taken in 1953.

economy. Industrialization changed patterns of supply and demand and led to an uneven but inexorable aggrandizement of world markets.[112] The British government evidently decided that, with European states becoming more protectionist, it would act to secure places for trade and investment overseas.[113]

The British plan for foreign trade was to import raw materials from abroad and to sell manufactured goods in return.[114] The empire did not smooth Britain's economic course; counting all the costs, it may even have represented a net loss, and there were bank failures and depressions as the century wore on. But an empire offered the opportunity to set terms of exchange for the one-quarter to one-third of British trade that was with the colonial territories. The trade with India in cotton is a prime example. The imperial government discouraged the rise of textile manufacturing industries in India, while encouraging the planting of cotton.[115] British traders paid low prices for the raw cotton and shipped it to Britain, where it was spun and woven by machine into fabric, cut and sewn into garments. Much of the labor was done by women and children. Cotton goods were Britain's largest export, although Britain grew no cotton. And India was Britain's largest market for finished cotton goods.[116] The Civil War in America, when the Union blockade of southern ports and devastation of the South's plantation economy cut off the world's leading supply of raw cotton, encouraged the expansion of production in India and a similar scheme in Egypt.

What was the effect of the imperial economy on the ecosystems of the empire? It involved removal of forests and other native ecosystems, and their replacement by monocultures that happened to be profitable. Many of the latter were introduced species. The plantation economy expanded, especially in the tropics. Again India offers an example. Coffee and tea planters moved into the coastlands and hills of Malabar, Cochin, and Travancore; by 1866 there were hundreds of plantations.[117] So many trees of one species so close together provided the opportunity for diseases and insects to attack and grow rapidly. Coffee trees were more susceptible; in time tea plantations (and rubber in the early twentieth century) largely replaced coffee. But the original forests were almost gone, with many species greatly reduced in number or extirpated from whole districts. Since the planted trees did not protect and maintain the soil as well as the former dense vegetation, erosion was a problem.

There were animal introductions, too. Sheep spread across New Zealand. Wool exports from the Cape Colony in South Africa, produced by the voracious Merino sheep, increased from 51,000 kg (113,000 lb) in 1833 to 2,471,000 kg (5,447,000 lb) in 1851.[118] The creation of "Neo-Europes" described above was repeated in "Neo-Britains" in the nineteenth century.[119]

The establishment of centers of trade and administration gave a stimulus to urbanization. Among centers whose population and occupation of land swelled in the course of the century were Calcutta, New Delhi, Madras, Bombay, and Singapore.

The mammals and birds of plains and forests in Africa and India, and of the oceanic islands, were decimated by habitat destruction and hunting. As John MacKenzie remarked, "The exploitation of animals is everywhere in the imperial record."[120] In England, hunting had been regarded as the privilege of the elite; now, in the empire, it would become the privilege of the conquerors. The quintessential image is that of the great white hunter on safari in Africa. There were examples of this type who shot at literally every wild animal large enough for a target. Some would take trophies from the most impressive of the slaughtered beasts, and leave the rest to their African bearers or the scavengers. The predators, too, they shot as "vermin." Some of them wrote books about their exploits, while others were content just to kill thousands of antelope, elephant, giraffe, rhinoceros, and anything else they happened to see. Eventually wide swaths of land were emptied of large mammals. The great days of hunting were over in South Africa by the 1870s; there were no more buffalo in Natal and the quagga, a kind of zebra with stripes on only part of its body, was extinct.[121] The native Africans, who had managed to hunt for centuries without destroying the herds, were supplied with guns by the colonizers and induced to assist in the slaughter. Perhaps the appeal of pay and market goods won many of them away from their traditional ways. Later, British administrators tried to restrict the use of firearms to themselves.

Another stereotypical image, unfortunately all too true, is that of the lordly British hunter in India shooting tigers from high on an elephant. The hunt had been a pastime of the Indian elite, but after 1857 the Indians were disarmed and hunting tigers became a European privilege. The tiger is a keystone species in the Indian forest ecosystem, the top predator. By making tiger heads and tiger skins a mark of prestige, the great cat was eliminated from sections of India; by the twentieth century it would be on the verge of extinction. Of course the hunt was not limited to tigers. In 1875, the Prince of Wales, the future Edward VII, came to India and celebrated the installation of his mother as empress by hunting not only tigers but elephants, pigs, and other mammals and birds.[122] The lion was persecuted until only a small population remained in the Gir Forest. Cheetahs, which had often been captured and trained to hunt for the maharajahs, became extinct in India. Many other mammals such as gaur, blackbuck, and even elephant were rarer. Hunts of wildfowl were popular; when I visited the great bird sanctuary on the lakes at Bharatpur, I found names of European hunters listed on a monument with the numbers of birds killed, in the hundreds.

Hunting was also the basis of lucrative trade. The export of ivory grew as the British brought Africa into the world trade economy. By the 1880s, 12,000 elephants were killed for their ivory each year in East Africa. The trade financed the penetration of the interior by missionaries, prospectors, and entrepreneurs. Ivory was used for piano keys, billiard balls, knife handles, combs, and various ornaments and curiosities. As the substance became rarer, its price rose so that the trade continued. A similar pattern occurred with rhinoceros horns, hippopotamus teeth and hides, and ostrich eggs and feathers.

Many hunters not only wanted to display trophies of their success at gunning down unsuspecting animals, but also donated specimens to museums and wrote about the creatures they had bagged, giving the cloak of natural history to the depletion of the greatest herds surviving from the Pleistocene. The craving of collectors and museums for specimens of disappearing species sometimes led to the actual disappearance of the last few individuals, as happened to the great auk.

However, some of the scientists sent out to the colonies recognized that the activities fostered by the empire were doing environmental damage. Richard Grove, in *Green Imperialism*, has pointed out that a few individuals, some of them involved in the creation of botanical gardens and interested in research that was not necessarily encouraged by the imperial government, made observations that connected deforestation, for example, with desiccation of the climate and decline of agriculture, and subsequent increase in famine and disease.[123] They advocated reafforestation and the creation of reserves to restore the climate, water supply, and production of food, wood, and other resources. Worthy of mention is Edward Green Balfour, who was an environmentalist, a feminist who forwarded the opening of medical education to women, and eventually an anti-colonialist as well. Balfour worked in India, but many other scientists made observations on tropical islands where deforestation and other major environmental changes took place rapidly within a small area, allowing their effects to be seen clearly. But if imperial expansion provided the opportunity for environmental awareness, the empires rarely gave support to positive efforts springing from this awareness.

By the last decades of the century, the interest in natural history combined with the realization that the animals were being extirpated to produce a concern for conservation. Regulations for the protection of wild elephants were promulgated in British India in the 1870s. The Boers in South Africa enacted game laws, and created in 1898 the reserve that later became Kruger National Park. Other reserves were set aside in the face of opposition from farmers and herders who complained that they sheltered animals that raided their fields and herds, and served as reservoirs for disease-carrying tsetse flies. In 1900, the Foreign Office in London hosted the first international conference on African wildlife. The agreements reached there proved ineffective, but set a precedent for more potent efforts in the following century.

The Galápagos Islands: Darwin's vision of evolution

I thought of Darwin as I jumped out of an inflatable boat and waded up onto a short beach of greenish sand. Sea lions lounging on the shore showed minimal interest. In low trees overhanging the beach were a few small birds: "Darwin's finches," I recalled. When he was on this island, they were so unafraid that he could almost grab them by the feet. No longer so unwary after another century of human contact, they still stayed closer to me than any bird would at home. Over the crest of a lava hill lay a brackish lagoon where flamingos walked gingerly. Around the corner in Post Office Bay were rocks covered with black marine iguanas and bright red crabs. Darwin had seen all these things. He didn't snorkel among the wonderful variety of colorful little fish that I saw, but he caught ones like them, preserved them, and sent them back to England. To him, the Galápagos were a kaleidoscope of images requiring understanding. He scarcely knew where to look next, and he certainly did not suspect the power these islands would exercise upon his ideas as he reflected in the next few years on what he had seen there. He later apologized to himself more than anyone else for not realizing sooner that each island was a separate biological assemblage. He had not carefully identified which islands his specimens came from: "It never occurred to me, that the productions of islands only a few miles apart, and placed under the same physical conditions, would be dissimilar."[124]

The phenomena that were to impress Darwin existed in the Galápagos because they were islands that lacked human inhabitants, and had few visitors, from the time volcanoes built them out of the sea 3–5 million years ago until, relatively speaking, not long before his visit.[125] This meant not only that they were free from the destructive effects of settlement, but also that they were unaffected by species of animals and plants brought by humans across the seas. The organisms that reached the Galápagos got there either under their own power by flying or swimming, floated there on ocean currents, or were blown there by storms. The arrivals were few, and they survived only by adapting to the harsh local environment. Plants had to be established before land animals could survive. A species of land birds may have arrived as a single pair or a small flock. From a growing colony on one island, a few of their descendants may have made the perilous flight over water to a neighboring island. There they encountered slightly different conditions. Each island became a unique ecosystem, changing as new species arrived and others became extinct. Among those that came were land-dwelling tortoises, which may have been carried from the mainland on rafts of vegetation. They evolved into a series of species, different on each island, and eventually gave their name to the archipelago; *galápago* is a Spanish word for tortoise. Two species of iguana descended, possibly, from one that made it to the islands: a yellowish-brown land iguana and the world's only marine iguana, black and seaweed-eating. Clues endured on the

Galápagos to explain how evolution worked, because they were protected by their remoteness from changes that might have erased them.

Such changes were underway when Darwin arrived in 1835; the first settlement, a penal colony, had been established only in 1832, but before that buccaneers and whalers had often landed, looking for water and the huge tortoises that could be caught easily, dragged on board ship, and stored upside down, living for months without anything to eat or drink, as a source of meat. Rats jumped ashore and found abundant food. The seafarers marooned goats and pigs, betting that they would find numerous offspring when they returned in later years. Visitors and settlers were to bring dogs, cats, donkeys, and weedy plants, all of which did untold damage to native biota. At the time of Darwin's visit, tortoise meat was the most prevalent item of animal food in the islanders' diet. But enough continuity remained of patterns from the deep past to serve as evidence for the evolution of communities of life.

Darwin arrived in the Galápagos on the *Beagle*, a ship whose Captain Robert FitzRoy accepted him after hesitation because he thought the shape of Darwin's nose indicated lack of character. At first Darwin's father refused permission (Charles was 22), but Josiah Wedgwood talked him into changing his mind, arguing that for "'a man of enlarged curiosity,'... the voyage was a golden opportunity to see 'men and things.'"[126] It was an understatement. Later Darwin wrote, "The voyage of the *Beagle* has been by far the most important event in my life and has determined my whole career ... I have always felt that I owe to the voyage my first real training or education of my mind. I was led to attend closely to several branches of natural history, thus my powers of observation were improved, though they were already fairly developed."[127] The *Beagle* sailed on December 27, 1831. She was 27 m (90 ft) long, with two masts; one 15 m (50 ft) tall, of a class of ships called "coffin brigs" because of their propensity to sink. But FitzRoy knew how to handle a ship in a storm.

The *Beagle*'s landfall in Brazil, at Bahia, gave Darwin his first view of a moist tropical forest. "Delight ..." he wrote, "is a weak term to express the feelings of a naturalist who, for the first time, has been wandering by himself in a Brazilian forest."[128] Unfortunately, that Atlantic coastal forest is now almost gone.[129]

For the next two years the *Beagle*'s crew mapped south of Montevideo. Darwin crossed the pampas and collected fossils of extinct mammals – sloth, armadillo, and llama – apparently related to modern species. He puzzled over this "succession of types:" why was there such a parallel between extinct and extant forms? It hinted that one species could transmute into another. The Falklands, uninhabited by humans for almost all their history, were a preface to the Galápagos. The birds, and the "wolf-like fox," were unusually tame in human presence; a fox could be killed by a man with a piece of meat in one hand and a knife in the other. A Mr. Lowe assured Darwin that "all the foxes from the western island were smaller and of a redder color than those from the eastern." Darwin would remember this comment when he noted the island-specific distribution of species in the Galápagos. Darwin predicted that with settlement of the Falklands, the fox would "be classed with the dodo, as an animal which has perished from the face of the earth."[130] Lamentably, he was right; none has been seen since 1875. The comment shows that he was aware of the process of extinction; if early giant forms could die out, then the Earth's present complement of species could hardly be the same as at Creation, permanent and immutable, as Christian biologists then thought.

After sailing through the Straits of Magellan, FitzRoy decided to return to England by crossing the Pacific. On September 15, 1835, the *Beagle* arrived at the Galápagos, volcanic islands 950 km (600 mi) west of mainland Ecuador. The equatorial heat is cooled by southeast trade winds and the Humboldt Current. At first, Darwin was repelled by their aridity: "Nothing could be less inviting than the first appearance. A broken field of black basaltic lava is every where covered by a stunted brushwood which shows little signs of life."[131] Later he discovered that the higher parts of the islands catch moisture from the clouds and have luxuriant vegetation.

The *Beagle* spent five weeks among the Galápagos; Darwin went ashore to observe and collect on four of the larger islands. From the start, he speculated about the relationship of the species on the Galápagos to those on other land masses. In his diary, he wrote: "It will be very interesting to find ... to what district or 'centre of creation' the organized beings of this archipelago must be attached."[132] He soon decided that the Galápagos assemblage of animals and plants, though unique, was related to that of South America. This could not easily be explained by the idea of a separate creation on the islands, but could be the result of migration from the continent and subsequent variation.

The vegetation was remarkable. Many species and genera were new to science. High on some islands, the commonest trees (*Scalesia*) were members of the sunflower family, with hairy leaves and stems festooned with lichens giving the forest a weird aspect.[133] Darwin noted the absence of such common tropical plants as tree ferns and palms. He made a relatively complete collection of plants, only later suspecting that there was a pattern of related but distinct species on separate islands.

He was astounded to learn that the giant tortoises differed from island to island. Mr. Lawson, the governor, told Darwin "that he could at once tell from which island any one was brought."[134] Darwin was fascinated by the tortoises; he tried to ride them, confirming that they were strong enough to carry his weight.

He noticed that the birds were even more naive than those on the Falklands. He saw a boy sit by a pool of water with a stick and kill enough birds to make a pile for supper. "I pushed off a branch with the end of my gun a large Hawk," he added.[135] At the time he visited they were already becoming warier; earlier explorers had reported that they alighted on their hats and arms. Their lack of fear was not due to absence of predators – the hawk, for example, caught smaller birds – but to unfamiliarity with humans.

The first related bird species that Darwin noted were limited to certain islands were not the famous finches, but mockingbirds. One species was exclusively found in Charles Island, a second on Albemarle, and a third common to James and Chatham. As he jotted these facts in his notebook, he was reminded of the tortoises and the Falkland foxes. He was not yet sure whether these animals were species different from the others like them, or "only varieties." He continued, "If there is the slightest foundation for these remarks, the zoology of Archipelagoes will be well worth examining, for such facts would undermine the stability of Species."[136] Here is the germ of the idea that the pattern of distribution of species in the Galápagos is evidence for the way in which evolution takes place.

Darwin did not promptly observe a similar distribution among the finches because their beaks, and the birds themselves, came in such a variety of shapes and sizes that he initially thought they belonged to different genera: finches, wrens, grosbeaks, and blackbirds. It was only when he returned to England, and the ornithologist John Gould told him that they were all finches, that he realized that they might have descended from a common ancestor that came from the mainland, whose descendants developed specialized beaks for different diets. In 1839 he would declare, "It is very remarkable that a nearly perfect gradation of structure in this one group can be traced in the form of the beak, from one exceeding in dimensions that of the largest grosbeak, to another differing but little from that of a warbler."[137] Some finches use their beaks to probe flowers and bark, others crush hard seeds. Still others – the woodpecker finch and mangrove finch – use twigs and cactus spines as tools. There is even a vampire finch that wounds sea birds and drinks their blood. There are species that are found on more than one island, but no two islands have exactly the same set of species. They were such a good example of the idea he was searching for that today they are famed as "Darwin's finches." By 1845, he would venture, "Seeing this gradation and diversity of structure in one small, intimately related group of birds, one might really fancy that, from an original paucity of birds in this archipelago, one species had been taken and modified for different ends."[138]

On the return voyage, the *Beagle* called at Tahiti, New Zealand, Australia, several islands in the Indian Ocean, the Cape of Good Hope, and Bahia again before landing in England on October 2, 1836. In Australia, Darwin was amazed by a collection of animals and plants so different from any other biota he had seen that he joked that there must have been two Creators at work on Earth.[139] He already doubted biblical creation as an adequate description of the origin of the forms of life. But before he could put evolution in its place, he needed to discover how it happens, and to marshal evidence. That process would take years, but it started before the voyage was finished. In 1837, he jotted,

> In July opened first note-book on "transmutation of species". Had been greatly struck from about month of previous March on character of South American fossils, and species on Galápagos Archipelago. These facts origin (especially latter) of all my views.[140]

This is only the first of many statements by Darwin on the importance of the Galápagos organisms to his thought on evolution. Much later he wrote,

> During the voyage of the *Beagle* I had been deeply impressed by discovering in the Pampean formation great fossil animals covered with armour like that on the existing armadillos; secondly, by the manner in which closely allied animals replace one another in proceeding southwards over the Continent; and thirdly, by the South American character of most of the productions of the Galápagos archipelago, and more especially by the manner in which they differ slightly on each island of the group; none of the islands appearing to be very ancient in a geological sense. It was evident that such facts as these ... could only be explained on the supposition that species gradually became modified.[141]

Darwin's most important contribution to the explanation of evolution was the idea of natural selection. It was suggested by the work of Thomas Robert Malthus, who had pointed out that human populations tend to increase exponentially, while the amount of cultivable land, and therefore food, can be increased only in linear fashion. Thus population will grow until limited by famine or some other factor. Darwin applied this principle to all living species. If unchecked, any species will increase until it uses all the resources available to support its numbers. Then members of the species will compete against each other for resources. Darwin further maintained that as species reproduce, they give rise to variations in their offspring. Some of these variations give individual organisms an advantage in competition for resources. These individuals survive longer, and are able to pass on their favorable variations to many of their own offspring. As this process continues, a new species may gradually evolve which is better adapted to its environment.

It would be incorrect to suggest that Darwin built his system of evolution only on the observations he made in the Galápagos. He spent much of the rest of his life observing and collecting information on the ways in which breeders of domestic species produce the

Figure 6.4 A Galápagos tortoise, "Lonesome George," the last surviving member of the Pinta Island subspecies. The tortoises were decimated by sailors who caught them for food, and by introduced species such as dogs, goats, and pigs. The birds in the foreground are "Darwin's finches." Photograph taken at the Charles Darwin Research Center, Santa Cruz Island, in 1996.

amazing varieties of form one sees in pigeons, for example. But the Galápagos offered the crucial stimulus, a fact he often acknowledged. Darwin never returned to the islands, but his name has been associated with them ever since.

Natural selection is an indispensable basis for understanding how ecosystems operate over time. The species that compose an ecosystem do not evolve by themselves, but through interaction with the other species that are part of the same community. Antelope that are chased by lions experience natural selection favoring watchfulness and swiftness. Plants eaten by caterpillars experience selection for poisonous and unpalatable characteristics; and the caterpillars in turn will be selected for resistance to those characteristics. Sexual selection, in which mates are preferred because they possess certain characteristics, is an important part of natural selection. Darwin came close to discovering the concept of the ecosystem, although he never understood how variations are created and passed from one generation to another by genetic mutation and by recombination of genes through sexual reproduction. But without Darwin's ideas, there could be no science of ecology. Donald Worster, in *Nature's Economy*, names Darwin "In many ways the most important spokesman for the biocentric attitude in ecological thought."[142] Further, it is possible to find some intimations of community ecology in Darwin's thoughts on his observations in the Galápagos.

"By far the most remarkable feature in natural history of this archipelago ... is, that the different islands to a considerable extent are inhabited by a different *set of beings*," he mused, "I never dreamed that islands, about fifty or sixty miles apart, formed of precisely the same rocks, placed under a similar climate, rising to a nearly equal height, would have been differently tenanted."[143] Here Darwin is considering, not just that related species live on different islands, but that each island has a different complement of species. The communities also vary. A "*web of complex relations*" binds all of the living things in any region, Darwin writes. Adding or subtracting even a single species causes waves of change that race through the web, "onwards in ever-increasing circles of complexity."

For Darwin the whole of the Galápagos archipelago argues this fundamental lesson. The volcanoes are much more diverse in their biology than their geology. The contrast suggests that in the struggle for existence, species are shaped at least as much by the local flora and fauna as by the local soil and climate. Why else would the plants and animals differ radically among islands that have "the same geological nature, the same height, climate, &c."?[144]

Speculating on why the Galápagos organisms were so closely allied to those of South America, Darwin had recorded another thought that presaged the idea of evolution occurring within an ecosystem:

> Why, on these small points of land, which within a late geological period must have been covered by the ocean, which are formed of basaltic lava, and therefore differ in geological character from the American continent, and which are placed under a peculiar climate, – why were their aboriginal inhabitants, associated, I may add, in different proportions both in kind and number from those on the continent, and *therefore acting on each other in a different manner* – why were they created on American types of organization?[145]

That is, the species on one of the Galápagos islands interact with each other in a community in a different pattern than that found on the mainland. This is not a fully developed theory of coevolution, but it looks in that direction. Darwin's observations in the Galápagos led him not only to the theory of evolution, but also toward community ecology.

The Galápagos Islands have had an increasing measure of recorded history. After Darwin left, the raids on tortoises continued until few or none could be found, so that ships stopped coming to the islands for them. Of fourteen subspecies of tortoises in the islands, three became extinct, and another, from Pinta Island, is at least partially biologically extinct, since it was reduced to one known specimen, a male called "Lonesome George." In 2008, he apparently succeeded in fertilizing genetically similar females from the island of Isabela, who laid eleven eggs. At last report, however, none had hatched. Some of the other subspecies recovered after hunting subsided, especially on Isabela. Another onslaught of overhunting, against the fur seals, almost succeeded in making them extinct by 1900, but they are still present.

In the late nineteenth century, the human population of the Galápagos was quite small, consisting mainly of prisoners and their wardens, with a few farmers and miners. These people introduced many domestic animals and plants, added to the ones already there, which escaped, became feral, and multiplied. With goats, pigs, donkeys, horses, cattle, cats, dogs, rats, and mice searching many islands for food, the tortoises were again threatened with extinction, since few of the eggs and young survived, even though the adult tortoises could fend for themselves.

Scientific expeditions, such as that of William Beebe and the New York Zoological Society, began in the 1920s. In 1934, desiring to protect the islands, the Ecuadorian government passed laws which for a time existed only on paper. An expedition sponsored by the United Nations Educational, Scientific and Cultural Organization (UNESCO) studied the islands and their biota in preparation for the centennial anniversary of *The Origin of Species* in 1959. Finally, effective legislation was enacted to protect the surviving ecosystems of the islands and to make all of them (except the areas settled and farmed) a national park. The Charles

Darwin Foundation was established to study and restore native fauna and flora, and it received a home when the Charles Darwin Research Station was founded at Academy Bay on Santa Cruz Island in the early 1960s. The research station gathers essential data on endangered species, provides scientific information, and helps the national park with education programs, including many for Ecuadorian students. It has a project to restore the giant tortoise populations by collecting eggs, and hatching and rearing the young tortoises until they are large enough to protect themselves when they are returned to their native islands. Financial support for the station comes from, among others, the Ecuadorian government, the Smithsonian Institution, the San Diego Zoo, the Frankfurt Zoological Society, and private donors. The Galápagos have been designated a UNESCO World Heritage Site and a Biosphere Reserve under the Man and the Biosphere program.

Protection for the wildlife and natural environment of the islands, including removal of exotic species, and the opening of tourism, were the purposes of the Galápagos National Park Service, organized in 1968. Goats were eradicated from some islands.[146] Tourists came, and although permits were required and there was talk of limits, the number of visitors increased exponentially from the 1970s through the 1990s to about 60,000 per year. Tourists must be accompanied by national park guides, and are instructed not to remove anything nor to touch any wildlife. Along with tourists came an increase in residents, in large part to serve tourism. Others came to fish. The resident population rose from 1,500 in 1950 to 6,119 in 1982. In 1997, it was estimated at more than 16,000 and growing 8 percent a year. Ecuadorian attempts to limit immigration met with political stalemate. To restrict fishing and depletion of the marine ecosystem, the government established the Galápagos Marine Resources Reserve, extending 15 nautical miles (27.8 km) from the islands, in 1986. Illegal fishing continued. In the early 1990s, poachers entered Galápagos waters to take shark fins and sea cucumbers, popular delicacies in east Asia, and when park wardens tried to close down an illegal fishing camp, poachers shot and badly wounded one of them.[147]

Passed and signed in March, 1998, a new Ecuadorian law created a marine sanctuary extending 40 nautical miles (74 km) from the islands, banned industrial fishing, and directed that some revenues from tourism be designated to support conservation, including the removal of aggressive introduced species. It also established the island's first inspection and quarantine system to prevent introduction of exotic species, and granted permanent resident status only to Ecuadoreans who have lived on the islands for five years or more.[148]

Introduction of non-native plants continues to be a problem as bad as that of animals. At present there are 250 introduced plants, including aggressive weedy species such as guava and lantana. The latter, a native of Mexico, has proved to be a scourge in every part of the tropics where it has been introduced. Quinine, introduced to Santa Cruz Island in 1946, has spread through 4,000 ha (10,000 acres) of the rare Miconia vegetation zone. In 1996, a plot of kudzu was found on a Galápagos farm; in the southeastern United States, this pest plant grows into impenetrable thickets and climbs up trees that often fall under its weight. Fortunately, botanists from the Charles Darwin Research Station convinced the farmer to destroy the kudzu before it could spread.

In spite of these problems of conservation, the Galápagos continue to help in answering questions that are asked by Charles Darwin's scientific successors. One of the most interesting biological projects of recent times has been conducted since 1973 by Peter and Rosemary Grant. Peter is professor of biology at Princeton University and a graduate of the University of Cambridge (Darwin's university, where there is now a Darwin College).[149] They return each year to Daphne Major, an islet in the center of the Galápagos Archipelago, to capture, band, measure, weigh, and release every one of the 400 or so finches that live on

Figure 6.5 A cactus finch, one of "Darwin's finches," which are perhaps the most well-known examples of the evolution of species by natural selection. Photograph taken in 1996.

it, and observe their behavior.[150] They record which finches mate, how many offspring they have, and how many have survived each year. As closely as possible, they have charted the family trees of all the finches. Their observations show that the rate of evolution is much faster than had been expected. For example, the average size of the bills of a species population changes rapidly in response to stresses brought on by seasons that are wetter, such as those that come during an El Niño incident, or drier than normal. If the weather is drier, seeds of plants that withstand aridity, and are generally larger and harder, are more common, and smaller, softer seeds are in short supply. Among seed-eating finches, this situation favors the survival of larger birds with larger, heavier bills. Males survive better than females, and the surviving females choose to mate with the largest males. The average beak size increases by a millimeter or two, and this is observable over a period of years, not of centuries or millennia. But it is something like the stock market; a series of wet years favors the survival of birds with smaller bills. Still, the Grants have observed evolution occurring in Darwin's finches, something that Darwin would have been happy to know. It would have surprised him that sometimes evolution can move with more than glacial speed.

Conclusion

Early modern times saw a greater transformation of the biosphere, and of the face of the Earth, than any previous period. One reason for this was the acceleration in the growth of the human population that began in the later part of this period and would continue in the twentieth century. Another cause was the explosive dispersion of European explorers, traders, conquerors, and colonists into virtually every other part of the world, along with the other forms of life they brought with them, intentionally or not. Not only did they spread European organisms, but they also transferred animals, crops, and diseases from various parts of the tropics to others. The result was a tendency to homogenize ecosystems and to drive unique native species into decline or extinction. Introduced diseases decimated populations unfamiliar with them. Settlers and exploiters removed forests, killed vast numbers of animals, and established plantations of products valuable in world trade.

These were centuries of technological invention. Europeans were often the inventors, but they also realized the potential of innovations made elsewhere and exploited them. They did this in their homelands, often regarding their achievements as human triumphs over nature. The story of "Holland against the sea" is an example. It was the Europeans who first embraced the Industrial Revolution, and it gave them a margin of military and economic lead for a time, indeed a long time, over other peoples. It also produced unprecedented amounts of pollution of the air, waters, and soil. Transportation, especially by sea, colonization, and the industrialization of increasingly urbanized countries created the first world trade economy worthy of the name. It operated for the benefit of the metropolitan states, and made possible the exploitation of resources located in far distant parts of the globe.

Science began its modern odyssey in this period, although like Odysseus it was not always aware of where its quest would take it. At times it seemed to offer humankind the ability to understand and, in association with technology, to control the processes of nature. But it also began to learn how the various forms of life are interconnected, and the dangers of the incipient destruction it had helped to make possible. Ideas of preserving and restoring nature appeared, even among the colonizers, and were to gain greater currency as the decades passed.

Notes

1   For the purposes of this chapter, "early modern" is taken to mean the period from Columbus to the reign of Queen Victoria, or roughly 1492 to 1890. The latter date is chosen to indicate the beginning of the period of conservation, with the creation of wildlife reserves and national parks.

2   John F. Richards, *The Unending Frontier: The Environmental History of the Early Modern World*, Berkeley and Los Angeles, University of California Press, 2003.

3   Neil Roberts, *The Holocene: An Environmental History*, Oxford, Blackwell, 1989, 155.

4   Anthony S. Travis, "Poisoned Groundwater and Contaminated Soil: The Tribulations and Trial of the First Major Manufacturer of Aniline Dyes in Basel," *Environmental History* 2, 3, July 1997, 343–65.

5   Victor Ferkiss, *Nature, Technology, and Society: Cultural Roots of the Current Environmental Crisis*, New York, New York University Press, 1993, 48.

6   Adam Smith, *An Inquiry into the Nature and Causes of the Wealth of Nations* [1776], Cambridge, MA, Harvard Classics, 1909.

7   John Bellamy Foster, *Marx's Ecology: Materialism and Nature*, New York, Monthly Review Press, 2000; Paul Burkett, *Marx and Nature: A Red and Green Perspective*, New York, St. Martin's, 1999; Reiner Grundmann, *Marx and Ecology*, New York, Oxford University Press, 1991.

8   Paul Demeny, "Population," in B.L. Turner II, William C. Clark, Robert W. Kates, John F. Richards, Jessica T. Mathews, and William B. Meyer, eds, *The Earth as Transformed by Human Action: Global and Regional Changes in the Biosphere over the Past 300 Years*, Cambridge, Cambridge University Press, 1990, 41–54.

9   William H. McNeill, *A World History*, New York, Oxford University Press, 1999, 306; McNeill, after reviewing the literature, concluded that the population of the New World fell from 50 million in 1492 to 4 million in 1650.

10 Donald Worster, *Nature's Economy: A History of Ecological Ideas*, Cambridge, Cambridge University Press, 1977.

11 Anton van Leeuwenhoek, letter to the Royal Society of London, in Forest Ray Moulton and Justus J. Schifferes, eds, *The Autobiography of Science*, Garden City, NY, Doubleday, 1945, 158.

12 Richard H. Grove, *Green Imperialism: Colonial Expansion, Tropical Island Edens and the Origins of Environmentalism, 1600–1860*, Cambridge, Cambridge University Press, 1995.

13 Ibid., 221.

14 Ibid., 203, 206.

15 George Perkins Marsh, *Man and Nature: The Earth as Modified by Human Action*, New York, 1864 (reprint, Cambridge, MA, Belknap Press of Harvard University Press, 1965), 10–11.

16 Ibid., 11–12.

17 W. Bruce Lincoln, *The Life of a Russian Geographer*, Newtonville, MA, Oriental Research Partners, 1980, 30–3.

18 Keir Brooks Sterling, *Last of the Naturalists: The Career of C. Hart Merriam*, New York, Arno Press, 1974, 270.

19 Fernando Benítez, *The Century After Cortés*, Chicago, University of Chicago Press, 1965, 1–2.

20 Eduardo Matos Moctezuma, *The Great Temple of the Aztecs: Treasures of Tenochtitlan*, London, Thames & Hudson, 1988, 25.

21 Victor E. Shelford, *The Ecology of North America*, Urbana, University of Illinois Press, 1963, 469.

22 David Rockwell, *The Nature of North America: A Handbook to the Continent*, New York, Berkley Books, 1998, 290–1.

23 M. Walter Pesman, *Flora Mexicana*, Globe, AZ, Dale S. King, 1962, 115–16.

24 Eric G. Bolen, *Ecology of North America*, New York, John Wiley, 1998, 325–6.

25 Pesman, *Flora Mexicana*, 118–19.

26 Charles Gibson, *The Aztecs Under Spanish Rule: A History of the Indians of the Valley of Mexico, 1519–1810*, Stanford, CA, Stanford University Press, 1964, 319.

27 Ibid., 5.

28 Alfred W. Crosby, Jr., *The Columbian Exchange: Biological and Cultural Consequences of 1492*, Westport, CT, Greenwood Press, 1972.

29 Alfred W. Crosby, Jr., *Ecological Imperialism: The Biological Expansion of Europe, 900–1900*, Cambridge, Cambridge University Press, 1986, 89–90, 162, 270, etc.

30 Ibid., 2.

31 Alfred W. Crosby, Jr., *Germs, Seeds, and Animals: Studies in Ecological History*, Armonk, NY, M.E. Sharpe, 1994, 29.

32 Florentine Codex 12. 29, in James Lockhart, ed., *We People Here: Nahuatl Accounts of the Conquest of America*, Berkeley and Los Angeles, University of California Press, 1993, 182.

33 Ibid., 12. 57.

34 François Chevalier, *Land and Society in Colonial Mexico*, tr. Alvin Eustis, Berkeley and Los Angeles, University of California Press, 1963, 59.

35 Ibid., 94.

36 Ibid., 93.

37 Ibid., 103.

38 Crosby, *Germs, Seeds, and Animals*, 55.

39 Benítez, *Century After Cortés*, 89.

40 Crosby, *Columbian Exchange*, 82.

41 Elinor G.K. Melville, *A Plague of Sheep: Environmental Consequences of the Conquest of Mexico*, Cambridge, Cambridge University Press, 1994.

42 Ibid., 53.

43 Andrew Sluyter, "From Archive to Map to Pastoral Landscape: A Spatial Perspective on the Livestock Ecology of Sixteenth-Century New Spain," *Environmental History* 3, 4, October 1998, 508–28, at 519, 522.

44 These were probably mostly black rats, skilled climbers that are common aboard ships. Crosby, *Columbian Exchange*, 97.

45 Chevalier, *Land and Society in Colonial Mexico*, 75.

46 Gibson, *Aztecs Under Spanish Rule*, 231–5.

47 Marsh, *Man and Nature*, 1864.

48 I am grateful to Petra J.E.M. van Dam, who took me to the Schermer Windmill Museum in September 1998, and who made the request to the miller-curator to start up the mill and pump. She also showed me the Rijnlandshuis, headquarters of the Rhineland water board, including the allegorical painting mentioned in the following paragraph.

49 Richard L. Hills, *Power from Wind: A History of Windmill Technology*, Cambridge, Cambridge University Press, 1994, 128–9.

50 John G. Landels, "Engineering," in Michael Grant and Rachel Kitzinger, *Civilization of the Ancient Mediterranean: Greece and Rome*, Vol. 1, New York, Charles Scribner's Sons, 1988, 323–52, 345–7.

51 Jan de Vries and Ad van der Woude, *The First Modern Economy: Success, Failure, and Perseverance of the Dutch Economy, 1500–1815*, Cambridge, Cambridge University Press, 1997, 28, 344.

52 Its title is: "Count William II of Holland gives the founding charter to the water board of Rhineland." Christopher Wright, *Paintings in Dutch Museums: An Index of Oil Paintings in Public Collections in the Netherlands by Artists Born before 1870*, London, Sotheby Parke Bernet, 1980, 119; Albert Blankert *et al.*, *Gods, Saints and Heroes: Dutch Painting in the Age of Rembrandt*, Washington, DC, National Gallery of Art, 1980.

53 Paul F. State, *A Brief History of the Netherlands*, New York, Facts on File, 2008, 1.

54 William H. TeBrake, "Taming the Waterwolf: Hydraulic Engineering and Water Management in the Netherlands during the Middle Ages," *Technology and Culture* 43, 3, July 2002, 475–99, at 479–80.

55 G.P. van de Ven, ed., *Man-Made Lowlands: History of Water Management and Land Reclamation in the Netherlands*, The Hague, Uitgiverij Matrijs, 1993, 17–21.

56 Charles Cornelisse, *Energiemarkten en energiehandel in Holland in de late Middeleeuwen*, Hilversum, Verloren, 2008, English summary, 287–97.

57 Pliny the Elder, *Natural History* 16.4, tr. H. Rackham, Cambridge, MA, Harvard University Press, Vol. 4, 1968, 389.

58 Petra J.E.M. van Dam, "Ecological Challenges, Technological Innovations: The Modernization of Sluice Building in Holland, 1300–1600," *Technology and Culture* 43, 3, July 2002, 500–20, at 505.

59 Petra J.E.M. van Dam, "Sinking Peat Bogs: Environmental Change in Holland, 1350–1550," *Environmental History* 5, 4, 2000, 32–45; TeBrake, "Taming the Waterwolf," 476.

60 Van de Ven, *Man-Made Lowlands*, 67–70.

61 Charles Singer, E.J. Holmyard, A.R. Hall, and Trevor I Williams, eds, *A History of Technology*, Vol. 2, Oxford, Clarendon Press, 1956, 682–3.

62 Arne Kaijser, "System Building from Below: Institutional Change in Dutch Water Control Systems, *Technology and Culture* 43, 3, July 2002, 521–48, at 531.

63 Milja van Tielhof and P.J.E.M. van Dam, "Losing Land, Gaining Water: Ecological and Financial Aspects of Regional Water Management in Rijnland, 1200–1800," *Water Management, Communities, and Environment: The Low Countries in Comparative Perspective c. 1000–c. 1800, Jaarboek voor Ecologische Geschiedenis (Yearbook for Ecological History) 2005/2006*, 10, Gent, Academia Press, 2006, 63–94.

64 Van de Ven, *Man-Made Lowlands*, 27.

65 TeBrake, "Taming the Waterwolf," 497.

66 Petra J.E.M. Van Dam, "Ecological Challenges, Technological Innovations: The Modernization of Sluice Building in Holland, 1300–1600," *Technology and Culture* 43, 3, July 2002, 500–20, at 506.

67 Simon Schama, *The Embarrassment of Riches: An Interpretation of Dutch Culture in the Golden Age*, New York, Alfred A. Knopf, 1987.

68 De Vries and van der Woude, *First Modern Economy*, 31.

69 Hills, *Power from Wind*, 120; van de Ven, *Man-Made Lowlands*, 148–9.

70 Van de Ven, *Man-Made Lowlands*, 168.

71 De Vries and van der Woude, *First Modern Economy*, 22.

72   Van de Ven, *Man-Made Lowlands*, 289.
73   Harry Lintsen, "Two Centuries of Central Water Management in the Netherlands," *Technology and Culture* 43, 3, July 2002, 549–68.
74   Hills, *Power from Wind*, 130.
75   Van de Ven, *Man-Made Lowlands*, 204.
76   Zuiderzeepolders Development and Colonization Authority, *The Enclosure of the Zuiderzee and the Reclamation of Polders in the Yssel-Lake*, The Hague, Zwolle, 1957, 29.
77   Lintsen, "Two Centuries of Central Water Management," 567.
78   M.H. Port, ed., *The Houses of Parliament*, New Haven, CT, Yale University Press, 1976, 97.
79   Peter Brimblecombe, *The Big Smoke: A History of Air Pollution in London since Medieval Times*, London, Methuen, 1987, 90–1; the Byron quotation is from *Don Juan*, 1819.
80   David Stradling and Peter Thorsheim, "The Smoke of Great Cities: British and American Efforts to Control Air Pollution, 1860–1914," Environmental History 4, 1, January 1999, 6–31, at 10–11.
81   Brimblecombe, *Big Smoke*, 156.
82   H.B.D. Kettlewell, *The Evolution of Melanism*, Oxford, Clarendon Press, 1973; A.W. Mera, "Increase in Melanism in the Last Half-Century," *London Naturalist*, 1926, 3–9.
83   François Crouzet, *The Victorian Economy*, New York, Columbia University Press, 1982, 33.
84   Asa Briggs, *Victorian People: A Reassessment of Persons and Themes, 1851–67*, rev. edn, Chicago, University of Chicago Press, 1970, 15–51.
85   Stephen Inwood, *A History of London*, New York, Carroll & Graff, 1998, 411.
86   Brimblecombe, *Big Smoke*, 124.
87   Ibid., 146, 165.
88   Ibid., 163.
89   Dale H. Porter, *The Thames Embankment: Environment, Technology, and Society in Victorian London*, Akron, OH, University of Akron Press, 1998, 56.
90   B.W. Clapp, *An Environmental History of Britain since the Industrial Revolution*, London, Longman, 1994, 76.
91   Anthony S. Wohl, *Endangered Lives: Public Health in Victorian Britain*, Cambridge, MA, Harvard University Press, 1983, 124–5.
92   Porter, *Thames Embankment*, 120–1.
93   Andrew Mearns, *The Bitter Cry of Outcast London*, ed. Anthony S. Wohl, New York, Humanities Press, 1970.
94   Inwood, *History of London*, 503, 512, 602.
95   Hazel Conway, *People's Parks: The Design and Development of Victorian Parks in Britain*, Cambridge, Cambridge University Press, 1991.
96   Inwood, *History of London*, 666–8.
97   Clapp, *Environmental History of Britain*, 134.
98   Inwood, *History of London*, 569.
99   William Cobbett, *Rural Rides*, 1830 (1967, ed. G. Woodcock), 165, 216, 229.
100   Asa Briggs, *Victorian Cities*, New York, Harper & Row, 1965.
101   R.S.R. Fitter, *London's Natural History*, London, Collins, 1945, 101, 118.
102   John Harold Clapham, *An Economic History of Modern Britain*, 3 vols, Cambridge, Cambridge University Press, 1930–8, Vol. 1, 9–10.
103   J.T. Coppock, "Farming in an Industrial Age," in Alan R.H. Baker and J.B. Harley, eds, *Man Made the Land: Essays in English Historical Geography*, Totowa, NJ, Rowman & Littlefield, 1973, 181–92, at 186.
104   Clapham, *Economic History of Modern Britain*, Vol. 2, 501.
105   Clapp, *Environmental History of Britain*, 107.
106   Clapham, *Economic History of Modern Britain*, Vol. 1, 18.
107   Coppock, "Farming in an Industrial Age," 184.
108   Crouzet, *Victorian Economy*, 153.
109   Coppock, "Farming in an Industrial Age," 191.
110   Crouzet, *Victorian Economy*, 176.
111   Clapp, *Environmental History of Britain*, 119.
112   B.R. Tomlinson, "Empire of the Dandelion: Ecological Imperialism and Economic Expansion, 1860–1914," in Peter Burroughs and A.J. Stockwell, eds, *Managing the Business of Empire: Essays in Honour of David Fieldhouse*, London, Frank Cass, 1998, 84–99, at 90–1.

113 C.C. Eldridge, *Victorian Imperialism*, London, Hodder & Stoughton, 1978, 143.

114 D.K. Fieldhouse, *Economics and Empire, 1830–1914*, London, Weidenfeld & Nicolson, 1973, 10–14.

115 Ibid., 58–63.

116 Crouzet, *Victorian Economy*, 193.

117 Richard P. Tucker, "The Depletion of India's Forests under British Imperialism: Planters, Foresters, and Peasants in Assam and Kerala," in Donald Worster, ed., *The Ends of the Earth: Perspectives on Modern Environmental History*, Cambridge, Cambridge University Press, 1988, 118–40, at 133.

118 John M. MacKenzie, *The Empire of Nature: Hunting, Conservation, and British Imperialism*, Manchester, Manchester University Press, 1988, 92.

119 Thomas R. Dunlap, *Nature and the English Diaspora: Environment and History in the United States, Canada, Australia and New Zealand*, Cambridge, Cambridge University Press, 1999; Alfred W. Crosby, "The British Empire as a Product of Continental Drift," in *Germs, Seeds, and Animals*, 62–81, at 66.

120 MacKenzie, *Empire of Nature*, 7.

121 Ibid., 149–56.

122 Ibid., 171, 179, 187, 193–4.

123 Grove, *Green Imperialism*, 441–56.

124 Charles Darwin, *The Works of Charles Darwin*, ed. Paul H. Barrett and R. B. Freeman, *Journal of Researches*, 2 vols, New York, New York University Press [1839], 1986, 474, 629.

125 Allan Cox, "Ages of the Galápagos Islands," in Robert I. Bowman, Margaret Berson, and Alan E. Leviton, eds, *Patterns of Evolution in Galápagos Organisms*, San Francisco, CA, American Association for the Advancement of Science, 1983, 11–24.

126 Adrian Desmond and James Moore, *Darwin*, New York, Time Warner, 1991, 102.

127 Charles Darwin, *Charles Darwin's Autobiography*, ed. Francis Darwin, New York, Henry Schuman [1876], 1950, 38.

128 Darwin, *Journal of Researches*, 11.

129 Warren Dean, *With Broadax and Firebrand: The Destruction of the Brazilian Atlantic Forest*, Berkeley and Los Angeles, University of California Press, 1995.

130 Darwin, *Journal of Researches*, 250–1, 476–7.

131 Ibid., 454.

132 Charles Darwin, *The Works of Charles Darwin*, ed. Paul H. Barrett and R. B. Freeman, *Diary of the Voyage of H.M.S. Beagle*, New York, New York University Press [1836], 1986, 337.

133 Michael H. Jackson, *Galápagos: A Natural History*, Calgary, AB, Canada, University of Calgary Press, 1993, 86–9.

134 Darwin, *Journal of Researches*, 465.

135 Darwin, *Diary*, 334.

136 Nora Barlow, ed., *Charles Darwin and the Voyage of the Beagle*, New York, Philosophical Library, 1946, 246–7; Nora Barlow, "Darwin's Ornithological Notes," *Bulletin of the British Museum (Natural History)*, Historical Series 2, 1963, 201–78, at 262.

137 Darwin, *Journal of Researches*, 1839, 462.

138 Darwin, *Journal of Researches*, 1845, 379–80.

139 F.W. Nicholas and J.M. Nicholas, *Charles Darwin in Australia*, Cambridge, Cambridge University Press, 1989.

140 Darwin, *Journal (Pocket Book)*, 1837.

141 Darwin, *Autobiography*, 52–3.

142 Worster, *Nature's Economy*, 180.

143 Darwin, *Journal of Researches*, 1845, 393–4.

144 Jonathan Weiner, *The Beak of the Finch*, New York, Vintage Books, 1995, 225–6.

145 Darwin, *Journal of Researches*, 1845, 393.

146 Ole Hamann, "Changes and Threats to the Vegetation," in R. Perry, ed., *Key Environments: Galápagos*, Oxford, IUCN and Pergamon Press, 1984, 115–31.

147 Eliecer Cruz, "News from Academy Bay," *Noticias de Galápagos* 58, May 1997, 2–3; "Galápagos: Too Many People," *The Economist*, May 10, 1997, 44.

148 "New Law," *La Carta* (Charles Darwin Foundation), Spring 1998, 3.

149 Peter R. Grant, *Ecology and Evolution of Darwin's Finches*, Princeton, NJ, Princeton University Press, 1986.

150 Jonathan Weiner, *The Beak of the Finch: A Story of Evolution in Our Time*, New York, Alfred A. Knopf, 1994.

7 Exploitation and conservation

Human exploitation of the natural world increased on an unprecedented scale in the period between the last decade of the nineteenth century and the 1960s. Within one human life-time of "threescore and ten," humankind experienced both escalating economic activity and a widespread depression. Viewed on a world scale, the two great wars were the most destructive of life, both of humans and of the biosphere, in history. The ecosystems of the Earth were damaged in ways unknown before, although few of the writers who commented on the fact expressed it in those terms. Rather, they talked about the depletion of natural resources. A few, like Fairfield Osborn, wondered if the cornucopia was about to run out of riches.[1]

Among the forces driving exploitation was the continuing growth of human population, again unprecedented in history. From 1890 to 1960 human numbers about doubled, from 1.57 billion to 3.02 billion. The numerical increase was greatest in Asia and Europe, but these were already the most populous continents, so they grew by just under 80 percent. Population in the Americas and Oceania tripled, and in Africa (where there are few reliable statistics) probably doubled. Population increase acts as a multiplier of human impacts on other parts of the Earth, but more than simply increasing effects, it may carry them beyond critical thresholds. Renewable resources can absorb use up to a certain level, but beyond that level there are diminishing returns, and eventually exhaustion. Non-renewable resources may be exhausted.

Urbanization was a major process of change. The size and number of metropolitan areas increased, along with density of occupation. In 1890, there were nine cities with over 1 million inhabitants; by 1960, sixteen cities had over 4 million each, and cities of over 1 million numbered over 80. Such large urban concentrations occupied ever-larger expanses of land, replacing natural ecosystems and agricultural acreage and reaching outward over greater distances for food and other resources. The spread of metropolitan populations and urban land uses reshaped natural landscapes and environments, altering ecosystems. Cities also affected the climate in their neighborhoods, increasing average temperature, clouds and precipitation; decreasing humidity, winds, and hours of sunshine; and adding pollution to the air and waters.

Another factor adding to human exploitation of the planet was technology. Generation of power from fossil fuels expanded in quantity and in kinds. Coal production, which had increased during the nineteenth century, passed 500 million metric tons[2] by 1890. In 1960, it reached over 2,600 million, five times as much, but was surpassed by oil, refined into petrol and diesel, and natural gas.[3] Much of the new energy was generated and transmitted in the form of electricity. In the same period, steel manufacture multiplied by twenty-eight times. An improved internal combustion engine using petrol, light enough to be used in vehicles, was invented by Gottfried Daimler in the late 1880s, revolutionizing transport and mechanized agriculture. Gasoline-powered tractors and bulldozers were followed by chain saws that transformed forestry. Andreas Stihl invented a gas-powered chainsaw in 1929, but it was hand-held, lightweight models, which gained widespread use after the Second World War, that had a key impact. Before the chainsaw, loggers with a hand-drawn crosscut saw might have taken two hours to fell a tree that a chainsaw could take down in two minutes. It is one of the forces accelerating the deforestation of the Earth's landmasses. This invention may epitomize the technological developments of the century. All these mobile and efficient engines vastly increased the speed and energy at human command in clearing, plowing, and sculpturing land. Agricultural businesses were able to afford the tools of the new technology, and used them to cover large areas with monocultures of such products as sugar, tobacco, bananas, tea, coffee, and rice. Insects and fungi found opportunities to spread in these plantations, so the entrepreneurs had to seek effective chemicals to use against them, often using airplanes to dust the crops. Small farmers, less able to afford machines and chemicals, gradually lost ground to larger corporations.

Trucks and airplanes, along with refrigeration, enabled the increasing spatial separation of production and consumption,[4] while buses and automobiles allowed workers to live at greater distances from their jobs, so that suburbs spread further outward from core cities. Generally their density of residence was less than nearer the centers, so that they occupied proportionately more land. Roads, parking lots, and fuel supply facilities began to use more space in the cities than residences, other businesses, and green space.

Airplanes also presented the danger of the rapid spread of organisms between distant land masses. This often happened with microorganisms that produce disease in humans; earlier, when ships were the only means of overseas travel, there was a greater chance that the incubation period would pass and the carriers could be quarantined. Exotic agricultural pests, as well as animals and plants, that could invade and harm native ecosystems often stowed away on airplanes successfully, in spite of measures taken to prevent it.

The prevailing attitude toward technological innovation in the early twentieth century was optimistic, although some perceptive writers such as Aldous Huxley and H.G. Wells observed the tendency of technology to grind down the individual and debase social life, and extrapolated future horrors along those lines.[5] On the other hand, most people had an easy confidence in the capacities of technology to increase human power, improve economy, and to solve the very problems it created.[6] This confidence was shaken by the weaponry of two world wars, including the invention and use of nuclear bombs near the end of the Second World War, and it was hardly improved by the arms race between Communist and Western nations during the Cold War. But technology is an aspect of human creativity, which can also turn it to positive purposes. The feared annihilation did not occur, and technology continued to offer both achievements and new challenges.

A third force that impelled the human exploitation of the natural world was economic growth, which had its ups and downs, but proceeded at a rate exceeding the increase in population if the entire period 1890–1960 is considered. The emergence of the world market was a conspicuous element of the burgeoning economy. The accumulation of capital in the industrialized nations, primarily western Europe and the United States, and its investment in foreign countries, is noted in economic histories, but such a process would have been impossible without tapping natural capital. The greater part of the real wealth of nations, as S.R. Eyre and a few other economists have reminded us, lies in the renewable organic productivity of the Earth and in its reserves of non-renewable resources.[7] Economic growth cannot continue steadily or indefinitely if business enterprises overdraw on the natural production of renewable resources, using them at a rate exceeding their replacement in natural systems. To do so is obviously to liquidate living capital; to kill the goose that lays the golden egg. Nor can it continue if they squander non-renewable resources. Yet both of those modes of operation were the rule in the early twentieth century, and continued afterwards. The Great Depression, "an economic bankruptcy ... closely related to a bankruptcy of land stewardship," as Stewart Udall put it,[8] served as a warning that the world market economy might not operate well on the *laissez-faire* principles of the past. After the Second World War intervened with its artificial stimulus to national economies, and its temporary destruction of some of them, the financial experts of the capitalist nations began to put together a structure that would encourage free trade and open the resources of the world, renewable and non-renewable, to exploitation by private enterprise now receiving an ever-higher level of public support, and by multinational corporations. Two elements of this structure, the International Monetary Fund (IMF) and the General Agreement on Trade and Tariffs (GATT), grew out of discussions at a conference in Bretton Woods, New Hampshire, in 1944.[9]

Russia after the 1917 Bolshevik Revolution, and eventually other communist countries, opted out of the capitalist world economy. It might have been expected that environmental conditions there would have been better, since socialist theory held that nature should be managed for the benefit of society. Karl Marx wrote, "Nature is man's inorganic body ... That man's physical and spiritual life is linked to nature means simply that nature is linked to itself, for man is part of nature ... Estranged labor ... estranges man's own body from him as it does external nature and his spiritual essence, his human being."[10] In other words, he held "that human beings and the environment are parts of a dialectically interactive whole."[11] Friedrich Engels turned to ancient history for cautionary conservationist precedents. Citing the devastation visited by human misuse on Mesopotamia, Asia Minor, and Greece, he concluded, "Let us not, however, be very hopeful about our human conquest over nature. For every such victory, nature manages to take her revenge."[12] V.I. Lenin provided the initiative for more than 200 regulations for the conservation of nature in the period 1917–22, including a decree "On Nature Preservation." In reality, however, economic and political priorities took the upper hand, since Soviet planners insisted that socialism should outproduce capitalism. Well-intended environmental programs and laws were not funded or enforced. As Marxist thought coalesced, its emphasis on the conditions of human labor eclipsed any serious consideration of the natural environment. Madhav Gadgil and Ramachandra Guha perceptively observed,

> the mode of production concept is not adequately materialistic in the first place. This may seem an ironic accusation against a doctrine as supposedly materialist as Marxism, yet a little reflection bears it out. Marxist analyses usually begin with the economic "infrastructure" – the so-called relations of production and productive forces – without investigating the ecological context, i.e. the soil, water, animal, mineral and vegetative bases of society in which the infrastructure is embedded.[13]

In practice, Russia and other communist states devastated the environment in a race for higher levels of production. Their record was at least as bad as that of the capitalist countries, if not as far-reaching. The contrast between communist East Germany and capitalist West Germany after the Second World War illustrates this. Prior to 1970, both societies devoted their energies to economic growth with little concern for environmental health; pollution was, on the whole, worse in the West because the expansion was more rapid. After

1970 the affluent, democratic West was to initiate environmental improvements while the East continued on its damaging course.[14]

Of the major impacts on the ecosystems of the Earth caused by human exploitation in the late nineteenth and early twentieth centuries, perhaps the most visible was deforestation. It occurred to varying degrees on each of the six continents that had forests, and on many islands. Asia, mainly because of its vast size, lost the largest extent of forests, but the highest percentage losses were in the Americas and Australia. Europe, where forest removal had been going on since medieval times, and where forestry was practiced in several of the large states, had the smallest decrease in both absolute and percentage terms.

Around 1910, the United States was clearing forest land at a rate of 96,000 sq km (37,000 sq mi) annually, an area larger than the state of Indiana. By 1920, "stumpland" covered a swath of the South as large as Virginia, Kentucky, and North Carolina combined, and in the Great Lakes region, stumpland exceeded the state of Ohio in total area.[15] During the 1930s, the combined extent of forest fires in an average year was 16,000 sq km (61,000 sq mi), an area larger than Michigan, some of it on land already cut over. While timber became scarcer in the regions just mentioned, the raid on the forests of the Pacific Northwest accelerated.

The most serious deforestation in Latin America occurred in the Atlantic coastal forests of Brazil, where railroads gave access to formerly isolated land, and the area in coffee plantations increased more than seventeen times between 1920 and 1931.[16] In south and southeast Asia, forests decreased by one-quarter between 1880 and 1950. That was the proportion in Thailand, while rice paddies there tripled in acreage, and rubber plantations were augmented from 1,600 ha (4,000 acres) in 1913 to 531,000 ha (1,312,000 acres) in 1953.[17] In India, the forest exploitation in the south described in this chapter was more than matched by the inexorable march of deforestation into the Himalayas and the northeastern region of Assam.[18]

The disappearance of habitat for plants and animals of forest ecosystems was one of the primary causes for the decline of biodiversity, more serious in this period than during any preceding time. Biodiversity is reduced by restriction of the range of species, decline in species populations, and most seriously and finally by extinction of species. The rate of extinctions has increased with each century, and in the twentieth century it began to increase noticeably with each decade. Some famous disappearances occurred in the period considered here. The last known passenger pigeon in the wild was shot in 1900, and Martha, the last captive of that species, died in Cincinnati Zoo fourteen years later. The last definitive sighting of the Carolina parakeet in the wild was in 1904, although unconfirmed reports continued for decades, and a captive specimen survived until 1918, again in Cincinnati Zoo.[19] The thylacine (Tasmanian marsupial tiger) made its last appearance in 1934. The ivory-billed woodpecker disappeared by 1944 after a remaining fragment of its old-growth forest habitat, the Singer Tract in Louisana, was cut down by the Chicago Mill and Lumber Company (more recent sightings have not been confirmed).[20] A series of colorful Hawaiian birds and butterflies, and snails as well, made their final exits, mostly due to predators introduced to the islands. But these noted extinctions are the tip of the iceberg. Scores of others were recorded, and the majority of actual extinctions were probably of insects that had never been observed and given names. Many plants perished, too.

The seas, 70 percent of the planet's surface, were depleted almost as critically as the land. Some species became extinct, such as the Caribbean monk seal. Whalers with factory ships almost destroyed the great whales. In 1952, the killing of 49,752 whales was recorded.[21] The remaining number of humpbacks fell to a low of 500, and gray whales were thought to be extinct before, under protection, they restored their population. The International Whaling Commission was established by the Washington Convention of 1946 to oversee the taking of whales by the nations with whale-hunting fleets, rationing the catch among member states and thereby hoping to preserve the whaling industry. Catches of commercially preferred fish, such as halibut, cod, salmon, and herring, decreased, especially in the traditional fishing grounds in places like the North Atlantic. More fishing boats had to travel further, use more sophisticated technology, and take species that used to be rejected, to stay in business. California sardines seemed to be gone by the 1950s, and Cannery Row, made famous by John Steinbeck, had to shut down.[22] Pollution of waters just offshore on the continental shelves poisoned the spawning grounds of many fish species.

Water pollution from human wastes continued to increase. China and India customarily used human excrement as agricultural fertilizer. In the US, the sewage of about two-thirds of the population was collected in sewage systems in 1954, and two-thirds of that was treated.[23] In many parts of the world the situation was worse. In East Pakistan (now Bangladesh), a low-lying country subject to chronic flooding, cholera, typhoid, dysentery, and diarrhea periodically spread among the people. Industrial wastes such as detergents, acids, lead, mercury, and various chemicals, caused another set of problems. In 1960, releases of oil and grease into the Hudson and Raritan Rivers emptying into New York Harbor reached 368,000 tons annually.[24]

Air pollution escalated through the late nineteenth and early twentieth centuries with the spread of coal-burning industry to many parts of the world. Later in the period, automobiles and other internal combustion engines proliferated, using petroleum fuels and producing a new class of pollutants. Serious concern escalated after the Second World War, when cities such as Los Angeles often lay under a layer of photochemical smog generated by the operation of sunlight on effluents from motor vehicles and industries. Severe episodes of air pollution made hundreds of people ill, and killed many, as happened in Donora, Pennsylvania in 1948 and in London in 1952. The effects of air pollution on other forms of life were also harsh; pine trees died in the mountains above Los Angeles, for example, and damage to agricultural crops was observed. Public interest groups and scientists advocated anti-pollution measures, but adoption and enforcement took time, and increases in population and the number of automobiles made progress difficult.

Soil erosion was not a new problem,[25] but it reached unprecedented severity in the twentieth century. Its basic causes are removal of vegetative cover, overgrazing, and plowing in arid regions or steep terrain, particularly when furrows run down the slope. The first three causes were present in the high plains of the western United States when a period of drought in the 1930s followed a time of optimistic agricultural expansion that had broken vast stretches of short-grass prairie. The crops failed, and the strong winds typical of the region carried the topsoil high into the atmosphere. Most of it settled out in choking clouds, forming dunes on the plains, but some was carried thousands of kilometers to the east, darkening the sun over New York and Washington and dusting the decks of ships on the Atlantic Ocean. This, the Dust Bowl, was the most visible and newsworthy example of soil erosion, but it was hardly the only one. Similar incidents occurred in central Asia, the loess zone of China, and in the Sahel, the southern margin of the Sahara in Africa. Drought and water erosion have always been a problem in regions of cultivation. The Mississippi annually carried more than 200 million tons of sediment into the Gulf of Mexico, and the Ganges-Brahmaputra system delivered 1.7 billion tons to the Bay of Bengal.[26] Soil erosion in South Africa caused great concern in the 1940s and 1950s, manifested in the work of soil conservation associations among both native peoples and settlers.[27] Paul B. Sears commented on many of these phenomena, and compared them to similar processes of damage to the land in ancient times in his *Deserts on the March*, published in 1935.[28]

There were those, the conservationists, who looked at the ways in which humankind was wasting the Earth's riches and argued that humans should exercise restraint and save something for future generations. They did not always agree with each other as to how, or why, that should be done. John Muir, a leading advocate of untrammeled nature, offered two reasons for wilderness preservation. First, because wild things exist in their own right, and man is a fellow creature, not the lord of creation. Second, contact with wild nature freshens, cures, and expands the human spirit. It is wrong to destroy wilderness, because it is a constant source of inspiration and creativity for human beings. To those who maintained that love of nature is merely a preference of the few, Muir replied that there is a love of wild nature in everybody, whether recognized or not, and that even children from the slums discover it when given the opportunity. Muir urged that more national parks should be established to protect wild lands and to enable people to experience them. With like-minded people, he organized the Sierra Club in 1892.

To Muir, the idea of "managing" nature seemed an unwarranted presumption. It was quite another matter for the foresters of continental Europe, Germany and France in particular, who were making silviculture a science and formulating practical principles of sustained yield, the doctrine that forests can be managed so as to produce timber and other products forever, if the average annual "harvest" did not exceed the yearly increment minus losses to factors such as diseases, insects, and fire. There was little wilderness left in Europe, but some forests had survived centuries of exploitation or could be restored on appropriate lands. Germany even had some well-managed forest plantations with trees standing in neat rows. When the British government took control of immense forests in the Indian subcontinent, it appointed German forester Dietrich Brandis to supervise them according to the best silvicultural standards. Bernhard Fernow, another German professional forester, introduced forest science to the US as the first chief forester of the Department of Agriculture and founded a forestry school at Cornell University in 1898. After Brandis' return to Europe, he had an American student protégé, Gifford Pinchot, who became Fernow's successor and headed the US Forest Service in its formative years.[29] Pinchot's philosophy of natural resource use was summarized in his famous dictum, "Conservation means the greatest good to the greatest number for the longest time."[30]

Conservation received the enthusiastic support of the American presidency with Theodore Roosevelt's entry into the White House in 1901. T.R. was an outdoorsman and a friend of both Pinchot and Muir. He appointed Pinchot chief forester and often asked his advice; the policy of his administration was the one Pinchot enunciated, often called Progressive Conservation. Roosevelt believed that the federal government needed to take the lead in assuring the thrifty use of natural resources in ways that would assure their availability in the future. He convened a White House conference in 1908 to signal the adoption of conservation as a national policy, followed by an international conference on conservation in the next year.[31] He created many national forests, parks, monuments, and wildlife reserves. During his administration, public land policy changed from one favoring disposal at little or no cost into private hands to one of federal resource management.

Conservation also received official recognition in many colonies in the still-growing empires of European powers around the world. When they had first been seized, their resources had often been thrown open to exploitation, but by the end of the nineteenth century, governments had abolished chartered companies and instituted regulations of their own. These, like the 1878 Forest Act in India, were intended to take control of resources and to assure their continued use for the benefit of the occupying nations. Forests and wild-life sanctuaries were reserved and declared off limits to indigenous peoples who often had traditional uses that had lasted for centuries. In many colonies, Cyprus for example, local people who could no longer use nearby forests set them on fire instead. In African game reserves, poaching became a means of political protest by native people.[32] In the context of predatory colonialism, perhaps the best apology that can be made for the administrators is that "Colonial forest conservation, as an early form of 'sustainable management,' prevented what might have been an even more disastrous transition under an unbridled capitalist regime of resource extraction."[33]

Advances in technology made vast water reclamation projects possible, and the movement to build ever-larger dams for power generation, irrigation, and flood control swept around the world. One of the achievements of the Theodore Roosevelt administration was the passage of the Reclamation Act of 1902 and the creation of the Bureau of Reclamation. A federal program for water use, particularly one that recognized its scarcity in the arid lands of the American West, had been one of the ideas most cogently argued by John Wesley Powell (explorer of the Grand Canyon and head of the US Geological Survey) a generation earlier; he died in the same year.[34] Soon promoters were representing dams as a means of water conservation; others observed that they flooded forests, fields, habitats, and towns. Whether the distribution of water and power was equitable could also be questioned.

The alliance of advocates of wilderness preservation with progressive conservationists was a fragile one, and its breakup was symbolized by a political controversy over the application of the City of San Francisco to build a dam for water supply in Hetch Hetchy, a beautiful valley in Yosemite National Park. Muir fought against the proposed dam as an invasion of wilderness and as a precedent-setting invasion of a national park, while Pinchot and Roosevelt joined those who favored the use of the site for domestic water. Muir lost when the US Congress and President Wilson authorized the project in 1913.

Franklin Delano Roosevelt's twelve-year presidency marked the second great advance for federally directed conservation in the US. New national parks and monuments were designated. As a measure to counter unemployment in the Great Depression, FDR launched the Civilian Conservation Corps, putting young men to work in national forests and parks and elsewhere on the land planting trees, fighting fires, and building structures, roads, and trails. The Soil Conservation Service, organized in 1935 under the energetic leadership of Hugh H. Bennett, forwarded terracing, contour plowing, planting, and other means to stabilize the soil and restore ground cover in the Dust Bowl area and elsewhere. Walter C. Lowdermilk, Bennett's assistant, had experience in soil conservation work in China, and on the eve of the Second World War undertook a survey of historic land use practices and their results in Europe, the Mediterranean basin, and the Near East, and published a report that is a landmark in the writing of environmental history of the ancient world.[35] During the FDR administration, giant dams were constructed at Grand Coulee in Washington State and above the Central Valley of California, and the Tennessee Valley Authority directed a massive experiment in planning for a whole river basin.

Major dams were continually part of the Soviet Union's plans, and workers constructed many, with Herculean efforts, from the Dnepr Dam in the Ukraine, begun in 1927, to Bratsk Dam on the Angara River in Siberia, in 1957, and beyond. Dams, especially gigantic ones, came to be a symbol of "development" for nations around the world. Many dams and irrigation projects had been built under British rule in India, but with independence in 1947, India began an ambitious program of dam construction. Others were built in Africa, flooding stretches of forest and savanna. When Kariba Dam was closed in 1959 on the

Zambezi River, and a lake began to rise that would cover 5,700 sq km (2,200 sq mi), the world's newspapers carried stories of an "Operation Noah." A journalist, three game rangers, and a few helpers in a slow, clumsy boat rescued "baboons, monkeys, civet cats, ant-bears, porcupines, and all kinds of buck that had been marooned on temporary islands by the rising waters."[36] Readers possibly did not realize that without the submerged habitat, few of the displaced animals could survive. A lacustrine ecosystem replaced forests, but it was not as productive of fish as had been hoped. An account of Egypt's experience with the dams at Aswan is given later in this chapter.

A final word must be said about one of the intellectual bases for conservation in this period. From its origin in the late nineteenth century to the 1960s, ecology was the chosen subject of a growing number of scientists, but had not yet received much attention from managers of natural resources, nor wide recognition in the public sphere.[37] Ecologists were struggling to define the methods of their science, and were arguing over what should be the conceptual foundations of their subject. One of the leading ideas whose meaning they debated was that of community ecology. Victor Shelford, an animal ecologist, defined ecology as "the science of communities," and joined with the plant ecologist F.E. Clements to describe a broader "biotic community" that included both animals and plants, indeed all living organisms in a described territory. There was general agreement among ecologists that such a community exists, but sharp disagreement between those who, like Clements and Shelford, used the metaphor of an organism to describe it, and those such as Charles Elton who preferred the images of sociology and economics, or who, with A.G. Tansley, preferred the metaphor of a machine. This controversy has been well described by Donald Worster in *Nature's Economy.*[38]

The relationship of humans to the ecosystem, or to the biosphere, was also a matter for discussion. When the agrarian philosopher Liberty Hyde Bailey said, "The living creation is not exclusively man-centered; it is biocentric,"[39] he was attacking the generally held presumption that human cultures, particularly the advanced ones, have transcended the limitations that circumscribe the world of nature. In other words, he rejected the common misconception that somewhere along the evolutionary track humankind had adopted culture, left nature behind, and resigned from the community of life.

Aldo Leopold, a founder of the discipline of wildlife management, agreed with Bailey on that point. Indeed, Leopold took the principles of ecology as he understood them and gave them practical application in his field. For him, the community of life was the foundation of ethics. "A thing is right," he wrote, "when it tends to preserve the integrity, stability, and beauty of the biotic community. It is wrong when it tends otherwise."[40] This ethical standard places humans firmly within the community of life and subject to its laws. Commenting on those who, like Leopold, translated community ecology into "land ethics," Frank Benjamin Golley observed,

> There was, it seems, some connection between their interest in the ecosystem concept and their environmental concern ... It is not clear to me where ecology ends and the study of the ethics of nature begins, nor is it clear to me where biological ecology ends and human ecology begins. These divisions become less and less useful. Clearly, the ecosystem, for some at least, has provided a basis for moving beyond strictly scientific questions to deeper questions of how humans should live with each other and the environment. In that sense, the ecosystem concept continues to grow and develop as it serves a larger purpose.[41]

In the latter part of the twentieth century, the question of how ecology should inform human attitudes toward and treatment of the natural environment was to move into the arena of public debate on a scale that could scarcely have been imagined by Aldo Leopold and the scientists of his day.

The Western Ghats: tradition and change

The practice of protection of patches of woods as sacred is ancient. The Roman poet Ovid said, "Here stands a silent grove black with the shade of oaks; at the sight of it, anyone could say, 'There is a god in here!'"[42] One might think that such a grove, and such an idea, are things that passed away with ancient times. But scores of sacred groves persist in many parts of the world. Traditionally, they were owned and protected by small communities, and where that is still true they represent one of the last vestiges of local power over local resources

I visited several village groves in the Western Ghats, mountains by the western seacoast of India, with Dr. M.D. Subash Chandran, an ecologist who has published excellent studies of them. One of these villages was Mattigar, a community of hunters and gatherers who engaged in shifting cultivation and have recently turned to settled agriculture. Surrounded by an area cleared for crops stood the majestic grove, Devaravattikan, a fragment of the original evergreen forest, tall, cool, dark in color, a few acres in size. We entered with due respect; offerings had been placed, but we could see no temple or carved stone. As we left, we met an old man who explained, "There is no image. The gods there live among the trees."[43]

Groves like this one operated as refuges for biodiversity and helped maintain a balance between human groups and the ecosystems of which they are part.[44] Dedication of sacred groves is one among many traditional Indian practices of nature conservation. Veneration of sacred groves, however, must be distinguished from worship of individual trees. Species such as fig are revered everywhere in India, but a grove is sacred independently of the species it contains. The case with sacred animals is similar. A grove may shelter animals that are considered sacred, such as monkeys, cobras, etc., but in a grove all species are protected.[45]

Sacred groves are segments of landscape, containing trees and other forms of life, that are protected by human societies because it is believed that to keep them undisturbed is an expression of relationship to the divine or to nature. Wherever they occur, sacred groves are of ecological and cultural interest.

Ecologically, sacred groves are fragments of the original ecosystem in a region. They are refugia that often shelter plant and animal species that have disappeared elsewhere in the region.[46] Thus they are possible centers of restoration. They are not immune to ecological change, however. As a rule, groves are small, from a fraction of a hectare to a few square kilometers in size, so are island-like, or like individual pieces in a landscape mosaic.[47] Such fragments may suffer extinctions, invasions of weedy introduced species, and natural or human-caused disasters.

Culturally, sacred groves are of interest because they exemplify phases of social interaction with the local ecosystem. Practices permitted or forbidden in them reveal attitudes to nature. Also, as examples of local autonomy, they may serve as rallying points when it is threatened.[48]

In sacred groves some or all ordinary activities are prohibited, such as tree felling, gathering of wood, plants, and leaves, hunting, fishing, grazing of domestic animals, plowing, planting or harvesting crops, and building ordinary dwellings.[49] Worship, including offerings and sacrifice, may take place inside groves. *The Golden Bough*, James George Frazer's *magnum opus* on cults and myths, gives evidence of groves in almost every major area of the world.[50] Recently, it has become clear that sacred groves are still a feature of village life in India, ub-Saharan Africa, and parts of the former Soviet Union, east Asia, and Oceania. Evidence indicates that they were once common in Europe, the Mediterranean lands, and pre-Columbian America.

Uttara Kannada in the central Western Ghats is a mountainous region with a monsoonal climate; from June to September, the highlands receive 5,000 mm (200 in) of rainfall. *Kans*, as they are called there, are sacred groves, patches of tropical evergreen forests protected and used as places of worship by local communities. There are hundreds of them. Some villages have several groves; Arendur has thirteen.[51] The characteristic sacred grove includes a source of water: a spring, stream, or pool. Dense forests regulate the runoff, preventing floods and releasing a year-round flow.[52]

Though it is problematic to talk about ownership of groves, since they are considered to belong to the gods, the responsibility for protecting them and enforcing rules was assumed by the local community.[53] The grove was and is an integral part of village life. Ceremonial events were held in or beside them. The land of the community was delineated to some extent by the location of its groves. Trenches or fences bordering the *kans* excluded grazing animals. Violations of them were adjudicated by village councils. However, protection of the groves was enforced not by human authority alone, but by the gods as well. When woodcutters were killed by a falling tree, it was thought to be a punishment by the deity, not an accident.[54] A man who entered a grove to hunt might fall ill.[55]

Madhav Gadgil, a leading writer on ecology and environmental history, observed that sacred groves belong to a variety of cultural practices that from early times helped Indian society maintain an ecological balance with wild living resources.[56] *Kans* are centers for the conservation of plant diversity.[57] They represent surviving fragments of climax evergreen forests; enormous changes have taken place outside sacred groves. When the original evergreen forest is cut, it is replaced by deciduous forest of low tree density. Viewed from a distance, the majestic dark *kans* stand out from surrounding forest not only by color and density, but also by the height of canopy trees. Within the *kans*, from thirty to sixty species of trees can be found in one hectare, a diversity characteristic of rainforests.[58] Scores of species, including trees and lianas, are found only in the groves. Some are rare or endangered. For example, *gurjan*, a mighty evergreen tree, has the isolated northerly end of its range in Karikan, a mountain grove dedicated to the "Mother of the Dark Forest," and in the fine wetland grove, Katlekan, also the locale for a notable species of wild nutmeg and a rare palm.

Sacred groves are the only surviving habitat for a number of species of small animals. The larger mammalian wildlife of Uttara Kannada has suffered attrition due to mass slaughter in the British period as well as curtailment of habitat: tigers, leopards, elephants, and gaur are seldom seen, and the groves are too small to protect them. But the endangered lion-tailed macaque has been observed in Katlekan.[59] Many birds, resident and migratory, frequent the groves and nest in them; about half the bird species occur in the sacred groves.[60] Without that refuge, a number would likely become extinct. Birds that prefer thick forest survive mainly in sacred groves. Among these are crested goshawk, lesser serpent eagle, grey jungle fowl, blossom-headed parakeet, blue-eared kingfisher, brown-headed stork-billed kingfisher, crimson-throated barbet, and Nilgiri flowerpecker. It must be noted, however, that surviving *kans* are reduced in number and area, so that changes in temperature and humidity,

Figure 7.1 Karivokkaliga peasant praying in the grove near votive offerings, metal tridents symbolic of a primordial local god now identified with Shiva. Photograph taken in Devaravattikan sacred grove in Mattigar village in the Western Ghats of south India, by Dr M.D. Subash Chandran in 1995.

windfalls, and encroachment of sun-loving introduced plants occur and lessen the qualities that make the groves refuges.

Cultural change threatens the sacred groves. The local deities of the groves are part of a religion that pre-dates the written texts of the Sanskrit tradition. They are not the characteristic gods of widespread Hindu devotion worshipped in temples throughout India in ceremonies led by Brahmin priests. They are nature deities that invisibly permeate the entire grove, but may be represented by vacant spots, stones, or termite mounds. Often they are called "mother" or "father." There are also animal deities such as serpent and tiger. The ability of Brahminic Hinduism to rationalize local deities as forms of the great gods and goddesses, and to provide icons in the form of reliefs and sculptures, is a danger to the continuation of the old religion and to the preservation of the groves. It has encouraged some communities that hitherto kept the groves to identify the local with the universal, and to replace the devotion once accorded to the groves by that symbolized by images and temples. Wood for temples may be cut from the sacred grove. In Waghoba Deorai (Tiger Grove) near Pune in 1991, settlers felled three big trees to construct a new temple.[61] Rules protecting the groves relax as the center of ritual moves away from the trees and toward the temple and the images it contains.

Every stage of the evolution of grove into temple can be seen in Uttara Kannada: a grove with no icon, but perhaps with a sacred spring and a termite mound; then a carved relief standing under the trees; then a small temple enclosing the spring or mound; then a larger and more ornate temple, as the grove decays; finally a temple with a sacred tree or two beside it, the grove forgotten. This process recapitulates the history of Hinduism in much of India. The Brahmin priesthood undoubtedly intended the process not to destroy the groves, but to improve their own position, convinced of the superiority of literary Hinduism to village animism. Some Brahmins regard the primal deities as "demons" and inculcate a more spiritual belief. But Hinduism generally is tolerant and willing to absorb local spirits.

Local pride and the desire to outdo neighbor communities motivate village leaders to build temples and deemphasize groves. Trees are sold to get money to build temples. But in order to place this minor entrepreneurship in perspective, it is necessary to say something more about the process of change since the onset of British rule.

The sacred groves of the Western Ghats are in danger of destruction. A recent survey in one section of Uttara Kannada indicates that 95 percent of the groves existing at the beginning of the colonial period have disappeared.[62] For this, imperial exploitation and the demands of a world market economy must be blamed. Appropriation of resources by British entrepreneurs and bureaucrats, and by proponents of economic growth since Independence, has damaged or annihilated groves throughout India. They are seen not as sacred reserves but as sources of materials such as timber, firewood, leaf manure, bamboo, and pepper.[63]

When the British occupied Uttara Kannada after 1799, they found it well stocked with forest resources, and opened the doors to a period of unbridled exploitation.[64] But the colonial power's interests lay in taking over as much of the country's resources as possible, and the British later began to control forest use. In 1864 the Forest Department was created to supervise management, and Dietrich Brandis, educated in scientific forestry in Germany, was appointed inspector-general. The Forest Act, amended in 1878, asserted British ownership of forest resources. "Reserved" forests were closed to all uses except those planned by the state, principally timber production. By the turn of the century, working plans were initiated for controlled exploitation, while lip service was given to principles of sustainable harvest. "Protected" forests could be used for timber, fuel, grazing, leaf manure, and other needs by the rural population, but without formal rights. Communities were deprived of power to keep others out of protected forests, including sacred groves, or to regulate harvests by their own members. Formerly communal property became open-access resources liable to exhaustive usage, a classic case of "the tragedy of the commons."[65]

Brandis noticed the widespread occurrence of the groves, and called them "the traditional form of forest preservation:" "Very little has been published regarding sacred groves in India, but they are, or rather were, very numerous. I have found them in nearly all provinces ... These sacred forests, as a rule, are never touched by the axe, except when wood is wanted for the repair of religious buildings."[66] Francis Buchanan, a British traveler, wrote in 1870, "The forests are the property of the gods in the villages in which they are situated, and the trees ought not to be cut without having leave from the Gauda or headman of the village, whose office is hereditary, and who here also is priest (*pujari*) to the temple of the village god."[67] But Buchanan added that sacred groves are a "contrivance" designed to prevent the government from claiming its rightful property. The state refused to recognize the sacred character of the groves. *Kans* were often included in reserved forests, and their takeover was followed by the introduction of a destructive contract system for exploitation of resources which replaced village community management.

A proper demarcation of the *kans* was not conducted by the government, so that many of them merged with ordinary forests and lost their identity. When restrictions on biomass removal from state forests were imposed on the population, pressure increased on village *kans*. Encroachment on the *kans* by land-hungry farmers reduced their area.

Since the *kans* contained softwoods, unmarketable at the time, little state timber exploitation was carried out in them almost to the end of the British period. The emergencies of the Second World War, however, were made excuse for "war fellings" throughout the forests, including the groves. Dipterocarps, fine straight trees typical of south Asian moist evergreen forests, which survived in this area mainly in the *kans*, were cut for railroad ties and plywood.

When India achieved independence in 1947, the Forest Department continued the methods of professional forestry with central state management that it had inherited. It was a disastrous model. Madhav Gadgil stated the case incisively:

> This whole system of resource management initiated under the British rule and further elaborated after independence is based on alienating local people from control of and access to resources. Its primary objective has been to make resources available as cheaply as possible to the elite, be it the British ruling classes or subsequently the industrial complex of Indian society. The elite that benefits from resource mobilization is shielded from the ill effects of the degradation of the resource base, since it can shift to the use of other resources or resources from other regions as the occasion demands.[68]

The new Indian government launched a drive for industrialization that included leasing reserve forests to companies producing plywood, paper, matches, and packing cases.[69] Hydroelectric reservoirs submerged other forests. These leases and projects superseded forest sanctuaries that had been created, and many *kans* as well. For example, a forest working plan for one district in 1966 included over 4,000 ha (9,900 acres) of *kans* for timber exploitation by forest industries. Such working plans included "improvement felling" in *kans* because the magnificent old trees were regarded by foresters as "overmature" specimens impeding the rapid growth of younger trees. In Menasi village, a *kan* was clearcut and converted into a eucalyptus plantation. During 1976, despite protests from the local community, the *kans* of Muroor-Kallabbe village, which had been in an excellent state of preservation by the people, were leased to a plywood company which extracted hundreds of logs, with attendant damage. To add insult to injury, sacred ponds were poisoned to kill native fish, and restocked with carp.

Local residents saw their exclusion from reserved forests as attacks on the community-based system of use, and the first case of forest resistance in the district occurred in 1886. Agitation on behalf of ancestral rights continued in the 1920s, and was coopted into the Gandhian *Satyagraha* of the 1930s. Groves became rallying points. A demonstration against deforestation and commercialization of forests called *Appiko* began at Salkani, "goddess forest," which was cut for plywood:[70]

> Inspired by Chipko [the "tree-hugging" peasant movement in the Himalayas], in August 1983 the villagers of the Sirsi area requested the forest department not to go

ahead with selection felling operations ... When their requests were unheeded, villagers marched into the forest and physically prevented the felling from continuing. They also extracted an oath from the loggers (on the local forest deity) to the effect that they would not destroy trees in that forest.[71]

The practice of honoring sacred groves was part of a pattern that helped to make possible a sustainable way of life within forest ecosystems. This positive function has not disappeared; it is more important today. Sacred groves, wherever they still exist, serve as historical evidence for the relationship of human beings to nature. The people of rural and tribal communities in Uttara Kannada once protected groves as fragments of living ecosystems, and can do so again if they are respected as partners in the conservation effort. To quote Madhav Gadgil,

Figure 7.2 The moist evergreen forest canopy of Devaravattikan sacred grove in Mattigar village in the Western Ghats of south India, showing "crown avoidance" between trees. Such sacred groves have been protected by villagers since time immemorial, and many species of trees, as well as birds and other animals, survive within them after disappearing elsewhere. Photograph taken in 1994.

For local people, degradation of natural resources is a genuine hardship, and of all the people and groups who compose the Indian society they are the most likely to be motivated to take good care of the landscape and ecosystems on which they depend. The many traditions of nature conservation that are still practiced could form a basis for a viable strategy of biodiversity conservation.[72]

To maintain sacred groves and even to promote their expansion would assure the health of the environment in and around the villages, the survival of many species, and a continuing supply of important biotic resources. A human community, the village, could again take responsibility for a biotic community, the sacred grove, as an indispensable part of its own ecosystem.

Grand Canyon: preservation or enjoyment?

Lipan Point rises above the southeastern rim of the Grand Canyon, commanding a wide view of vertical walls 14 km (9 mi) apart. Almost a mile (1,500 m) below, the Colorado River flows between narrow margins of green. To the west are huge masses of rock, remnants of erosion, whose shapes suggested ancient temples to modern mapmakers: Vishnu Temple, Wotan's Throne. This is near the place where members of Coronado's expedition saw the canyon in 1540 and, like many current visitors, greatly underestimated its size and that of the river. In my eleven summers as a seasonal ranger-naturalist at Grand Canyon, this point became one of my favorites. Every time I return, I find that although I know the Canyon's dimensions, something in my mind is slow to appreciate the canyon's true vastness, but quickly recognizes its color, monumentality, and unusual beauty.[73]

The impact of the canyon's form is so overwhelming that only after a while do visitors begin to notice the life that is all around them. The air in the canyon is full of birds. White-throated swifts zoom after insects, red-tailed hawks soar near the cliffs looking for small mammals, and ravens play, squawking and doing midair somersaults. Even California condors, recently reintroduced after recovery from near-extinction, are frequently seen today. In the rocks at the edge of the chasm, begging ground squirrels make themselves known. The rim is framed by a low forest of twisted pines and junipers. One can often see deer, or more seldom be surprised by a bobcat crossing the road with a rabbit in its jaws, or even, in the dusk, spot a mountain lion.

Topography on a grand scale and wildlife are two aspects of the national park experience, and among the reasons why national parks were created in the United States. The purpose of national parks, according to an act of Congress in 1916, "is to conserve the scenery and the natural and historic objects and the wild life therein and to provide for the enjoyment of the same in such manner and by such means as will leave them unimpaired for the enjoyment of future generations."[74] Many historians and commentators on the national parks have noted that this sentence contains two purposes – "to conserve" and "to provide for enjoyment" by the public – which were likely to come into conflict with each other, and indeed have done so.[75] Hotels, even if well-designed using natural materials as architect Mary Jane Colter did with Phantom Ranch at canyon bottom, would intrude on the natural scene, as would trails like the Bright Angel, not to mention roads and automobiles, largely unforeseen in 1916 but soon afterward to invade almost every national park along with gas stations, parking lots, scenic pullouts, and signal lights. The rims of Grand Canyon would be inundated with visitors' cars. The act did not envision the parks remaining pristine, since it allowed the secretary of the interior to sell timber "to control disease or conserve scenery," to grant leases to concessionaires, and to allow other uses. These actions were intended to be exceptions, however, and in that fact lies the difference between national parks and national forests. In national forests, timber sales, location of mines, hunting, grazing, and other uses of the land would be allowed or encouraged. In national parks, they would be discouraged or eliminated. Recreation would be the leading use to be encouraged in the parks.

The first part of the statement of purpose directs that four things characteristic of national parks be conserved: scenery, natural objects, historic objects, and wildlife. In the early days of the designation of parks, the first two received the greatest emphasis. There were already thirteen national parks in 1916. All of them were places primarily noted for monumental scenery except Mesa Verde, where "historic objects," the ancient cliff dwellings, were the main interest. It is notable that the first proposal for a national park advocated the preservation of animals, vegetation, and native people. In 1832, the artist George Catlin envisioned that an area on the Great Plains be preserved as "a nation's park, containing man and beast, in all the wild and freshness of their nature's beauty ... What a beautiful and thrilling specimen for America to hold up to the view of her refined citizens and the world, in future ages!"[76] Catlin's suggestion was resisted as far as the Great Plains were concerned, and steps to create a national park there came a century later, after the land had been plowed and the bison had almost disappeared. It is striking that Catlin, many of whose paintings were of Indians, considered the Native Americans as appropriate dwellers in a national park. Grand Canyon was part of the homelands of several tribes, and among them the Havasupai actually lived, and still live, within the canyon. But relations between the Havasu people and the administrators of the national park would often be painful for both. Tribal ancestors traditionally hunted and gathered plants and minerals in a large area of the canyon, but officials sometimes treated the Havasupai as interlopers and tried to move them out of places like Grand Canyon Village and Indian Gardens and to limit them to their tiny reservation of 210 ha (519 acres) tucked away in a western tributary canyon.[77]

The first national Park, Yellowstone, had its own Grand Canyon, along with waterfalls and geysers. Its herds of megafauna were also something to see, but by themselves could not have generated the railroad tourism desired by its promoters and the congressional designation that came in 1872. The primary purpose of parks then was to save the crown jewels of America's natural scenery. Yosemite was designated a park for its waterfalls and granite domes, and Sequoia and General Grant for giant redwoods, biological phenomena indeed, but so large and old that they were awe-inspiring features of the landscape. These early parks were created before the science of ecology, with its concepts such as the ecosystem, had received wide recognition, so backers of the parks in those days had only a general sense of protecting nature, along with a desire to encourage people to visit the areas. Mount Rainier, Crater Lake, Rocky Mountain, Mount Lassen: the theme was evident, and no feature of the American earth fit it better than the Grand Canyon.

John Wesley Powell, who led expeditions by boat down the Colorado River through the canyon in 1869 and 1871–2, urged that the Grand Canyon be made a national park because of its grandeur and geological interest. When John Muir saw the Grand Canyon in 1896, he repeated the call for park status, because of its superlative scenery.[78] President Theodore Roosevelt first visited seven years later, and voiced similar thoughts:

> Leave it as it is. You cannot improve on it. The ages have been at work on it, and man can only mar it. What you can do is to keep it for your children, your children's

children, and for all who come after you, as the one great sight which every American … should see.[79]

Roosevelt gave the canyon all the protection he could. Congress then was receptive to mining companies who sought bonanzas, ranchers who feared curtailment of grazing rights, and timber concerns who wanted access to forests. Since chances of passage for a national park bill seemed slight, he took an unprecedented action, invoking the Antiquities Act to create Grand Canyon National Monument in 1908. The area included was an eastern section regarded as most scenic, with narrow strips of land along the rims that avoided impinging too far on commercial timber and grazing interests. There was opposition, however, and a suit challenging the proclamation on the grounds that the Antiquities Act did not authorize making national monuments of large natural features went to the Supreme Court, which eventually ruled in the president's favor.[80]

Up to this time, national parks had been administered, whether effectively or not, by agencies including the General Land Office, the US Army, and the Forest Service. When Congress passed the 1916 act creating the National Park Service (NPS), a more consistent management of the parks could be envisioned.

Arizona became a state in 1912, and local pride and hope for a bigger tourist industry strengthened the movement to create a national park. The first director of the NPS, Stephen T. Mather, supported making the Grand Canyon a park, and his close associate and eventual successor, Horace M. Albright, worked with Representative Carl Hayden and Senator Henry F. Ashurst, both of Arizona, to get a bill through Congress which was signed by President Wilson in 1919.[81] The area included was almost the same as the monument; the intent was clearly to protect the scenic and geological features of the canyon itself, and only a small slice of neighboring forests and wildlife.

Another purpose of national parks, however – the protection of wildlife, and what would come to be recognized as assuring that ecosystems would continue to function as whole systems – was beginning to be recognized. John C. Merriam, head of the Carnegie Institution, urged that national parks be regarded as laboratories where natural processes could be observed and studied.[82] The Grand Canyon has served as a treasury of evidence about evolution of ecosystems and species in the Earth's past. The geological sequence in the rocks of the canyon and its vicinity include fossil records of forms of life in many major eras in the history of the planet. Scientists, in seeing the Grand Canyon as a treasure trove of evidence for the evolution of life on Earth as well as its present ecological interactions offered a reason for the preservation of the Grand Canyon in that it contributes to understanding the origin and nature of the living community.

Vernon Bailey, chief naturalist of the US Biological Survey, in agreement with Merriam, argued that the boundaries of most national parks, including Grand Canyon, had been located without sufficient attention to the need to provide wildlife with habitat during all seasons of the year.[83] The area included, he maintained, should be large enough to sustain a viable population of animals under the most natural conditions possible. In 1929 he recommended an expansion of Grand Canyon National Park. This suggestion became lost in inter-agency disputes.

The idea of an expanded park did not die, but the argument in favor of it that would prevail was the old idea of monumental scenery: the existing national park embraced only 169 km (105 mi) of the canyon's total length of 446 km (277 mi). The omission was partly repaired by President Herbert Hoover's proclamation of a new national monument in 1932 adjoining the national park on the west and extending 64 km (40 mi) down the Colorado

River. Ironically, in the same year Congress authorized Boulder Dam, later called Hoover Dam, which created Lake Mead, a reservoir extending into the lower Grand Canyon and drowning some of the famous river rapids there.

Glen Canyon Dam, a short distance above Lees Ferry, authorized as a storage and power generation facility, was completed in 1964. When the gates closed, the color of the Colorado River below the dam turned from red to green during most of the year, and its temperature and rate of flow dropped. Plans to build two additional dams in the Grand Canyon itself, Marble Canyon Dam and Bridge Canyon (Hualpai) Dam, caused acrimony between conservationists and developers, between the Upper and Lower Basin states along the Colorado River, and between California and Arizona, from the time of the completion of Hoover Dam to 1968, when Congress authorized the Central Arizona Project and placed a moratorium on dams within the Grand Canyon. The decision against the dams was mostly the result of a complicated political compromise,[84] but also of public opposition aroused by environmentalist organizations such as the Sierra Club under its activist executive director, David Brower, which placed ads in the nation's largest newspapers with slogans such as, "Now only you can save Grand Canyon from being flooded ... for profit."[85]

The idea of a national park embracing the entire length of the Grand Canyon, except for the portions within Indian reservations, gained the support of Senator Barry Goldwater of Arizona, the NPS, and environmentalists. A bill to enlarge the national park, and also to guarantee Havasupai access and expand their reservation, was signed into law in 1975. It almost doubled the size of the park, to 4,900 sq km (1,892 sq mi). But the idea that the national park was intended to protect scenery was implicit in the fact that the new boundaries mostly ran along the rims, putting the interior of the canyon within the national park and leaving the areas above the rims, with their wildlife habitats, in other jurisdictions. A new national monument to include some of these areas was proposed in 1999 by Secretary of the Interior Bruce Babbitt, former governor of Arizona, and was proclaimed by President Bill Clinton in 2000.

Grand Canyon historically has provided a great amount of evidence for the understanding of living communities. Even without the scenic monumentality of the canyon, there would be enough biological and paleontological interest to justify its designation as a national park. In 1889, C. Hart Merriam, chief of the US Bureau of Biological Survey, studied the distribution of plants and animals in the Grand Canyon region. Within a range of 3,000 m (10,000 ft) elevation from the Colorado River at canyon bottom to the top of the San Francisco Peaks he distinguished seven "life zones," that is, "areas inhabited by definite assemblages of animals and plants."[86] Merriam's ideas represented a step toward the concept of the ecosystem.[87] His life zone theory based on temperature and humidity had great influence.[88] When he wrote, "The Grand Canyon of the Colorado is a world in itself, and a great fund of knowledge is in store for the philosophic biologist whose privilege it is to study exhaustively the problems there presented,"[89] he aptly described himself.

The purpose of national parks was questioned and to some extent redefined as a result of a crisis of wildlife management that occurred in the Kaibab Forest north of the Grand Canyon in the mid-1920s. The theory of game management then was that "good" species such as deer should be protected, but that predators including wolves and mountain lions should be "controlled," that is, exterminated. James T. "Uncle Jim" Owens was appointed warden by the Forest Service. In the twelve years preceding the establishment of the national park, he killed 532 mountain lions. Among those who used his services as a lion-hunting guide were the writer Zane Grey, Buffalo Jones, and Theodore Roosevelt (who came to hunt in the game reserve he, as president, had created) with his sons Archie and Quentin and nephew Nicholas. The policy of destroying predators continued in both national park and national forest until 1931.[90] As a result, lions and bobcats were greatly reduced in number and wolves were extirpated, but coyotes continued to flourish. The Kaibab herd of mule deer, spared from most predation, increased from 4,000 in 1906 to about 100,000 in 1924. The swollen numbers of deer ate every green thing they could reach, and the forest took on the appearance of a clipped city park. The Forest Service inaugurated limited hunting, fawns were captured and transplanted, and there was a disastrous attempt to drive deer across the canyon by trail to the South Rim, all to little avail.[91] During the severe winter of 1924–5, thousands of deer died of starvation. In defiance of logic, predator control continued for several years. Game managers such as Aldo Leopold, who had worked at Grand Canyon, were convinced by the tragedy in the Kaibab Forest that "predators are members of the community,"[92] and that overpopulation was more dangerous to deer, and to the land, than any predator could be. Subsequently, the Park Service policy came to be the restoration of a functioning ecosystem by protection of all native species including predators, herbivores, and plants and allowing their natural interactions. Some parks later created, such as Everglades and the rainforest sections of Olympic National Park, were designated because of their biological interest and are ecosystem preserves. Unfortunately, most parks, even including the expanded Grand Canyon National Park after 1975, are too small to protect all important members of the ecosystem, especially the larger animals.

The NPS adopted the "Leopold Report"[93] in 1963, changing its wildlife management policies to protect interactive complexes of species. The plan advocated that large national parks be managed as biotic wholes or "original ecosystems."[94] Where the parks were not large enough to encompass entire ecosystems, the surrounding areas would be managed as peripheral zones with the parks as core areas, similar to a plan for biosphere reserves then being discussed by United Nations agencies. The latter aspect of the plan was not implemented, but the idea remains an option.

Certain species in the Grand Canyon area have received study and protection. The Kaibab squirrel is limited to the ponderosa pine forest on the Kaibab Plateau north of the Canyon. It is a tassel-eared squirrel related to the widespread Abert squirrel. The Abert is gray, with white underparts. The Kaibab is dark gray or black, including the underparts, and has a striking white tail. Both species depend on pines for food and shelter, and could not exist outside an ecosystem dominated by ponderosa pine. The two populations do not overlap; the Kaibab squirrel has been isolated by the Grand Canyon and surrounding deserts for thousands of years, and its evolution has taken a separate path; the white tail may help it hide from predators during snowy winters at 2,500 m (8,000 ft) above sea level. Due to its narrow range and small population, it has been listed as an endangered species.[95]

Restoration of species formerly present in the Grand Canyon region has been advocated, and tried with varying degrees of success. The California condor, the largest living land bird, once flourished in the canyon, but the last known individual there was shot in 1881. The condor was included on the first list of endangered species in 1967. By 1985, only nine condors existed in the wild, in California. All were captured and placed in zoos along with birds previously captured, for breeding purposes. The program raised the captive population to 71. Some were released in the Coast Range of California, but encountered dangers from power lines, chemicals, and shooters. In 1996 six were released in the remote Vermillion Cliffs 50 km (30 mi) north of the Grand Canyon, and now are often observed soaring above the canyon.[96]

Some aggressive non-native species have damaged natural ecosystems in the Grand Canyon. Prospectors' burros escaped in the 1880s; adapted to aridity and lovers of desert

Figure 7.3 View of the Grand Canyon from the North Rim near Bright Angel Point. Despite its formidable topography and climate, the Grand Canyon is home to many species in life zones occupying bands of elevation, as well as microclimates in locations with unusual conditions. Photograph taken in 1955.

vegetation, they proliferated in every corner of the canyon, destroying native plants, fouling springs, and competing with wildlife such as desert bighorn sheep. The NPS tried to eliminate the invaders, at one stage using helicopters to locate them and to bring in rangers with rifles. But a children's book, *Brighty of the Grand Canyon*,[97] romanticized the burros, helping to gain popular support for including a prohibition against shooting feral burros in congressional legislation to protect wild horses on federal lands. So the Park Service arranged for the Fund for Animals, Inc., to round up humanely and airlift all the burros in 1981. Many were adopted by private citizens. At the moment they are absent from the national park except for a few that wander in from the Lake Mead area from time to time. Another introduced species is the tamarisk, a bushy tree which, like the burro, is native to the Mediterranean. It has spread along the river margins along with other introduced plants, replacing native vegetation and degrading habitat for native birds and animals. There appears to be no effective means of eradication.

The Colorado River and the life dependent on it form a key ecological component of the Grand Canyon that has changed radically since the completion of Glen Canyon Dam in 1964.[98] The water that enters the canyon at its upper end comes from the lower part of the reservoir. It is now at a constant 9°C (48°F) instead of the fluctuating temperature of the free-flowing river, which varied from 1°C to 29°C (33°F to 85°F).[99] Since the silt carried by the river now settles in Lake Powell, the Colorado, named for the red color of the turbid sediments it once carried (as much as half a million tons a day), is clear for a varying distance below the dam, so that sunlight enters it and green algae can grow. Many native fish species have become extinct or nearly so, and the dominant fish is now the introduced rainbow trout.

The clear stream carries away sand instead of depositing it, so beaches have eroded. From a wild river that had powerful floods each spring, the Colorado has become a controlled river whose flow depends on fluctuations in power generation. It took floods to move large rocks, so rapids have become worse as flash floods coming down side canyons dump boulders into the main stream. Vegetation along the river margins increased without floods to wash it away. A major exception was the great flood of 1983, when managers at the dam, underestimating the amount of runoff from snow and rain in the upper basin, allowed Lake

Figure 7.4 A Havasupai Indian harvesting alfalfa beneath the Wigleeva rocks, near the village located deep in a tributary chasm within the Grand Canyon. The Havasupai, or "Blue-green Water People," have subsisted as farmers and hunter-gatherers in this place for many centuries, since before the arrival of the Europeans. Photograph taken in 1961.

Powell to remain at too high a level until late in the spring, hoping to maximize revenues from power generation. Suddenly, they had to open the concrete-lined spillways that tunnel through solid rock on either side of the dam. The spillways on one side began to disintegrate, spitting rock from the walls. Only narrowly, with stopgap efforts, did engineers keep the lake from overtopping the dam. The flood greatly accelerated erosion and removal of vegetation.

Ecological interpretation and management of national parks has gained in recent years in the US. There is popular recognition of the value of places like the Grand Canyon and the need to preserve them. The parks have been spared from the more extreme pressures of development that would surely have overwhelmed them before now if they had not been set aside, in view of the terrifying numbers of visitors that besiege them every year – some 5 million annually come to the rim of the Grand Canyon. But it is unclear whether the ecosystem can maintain integrity in the face of an increase in human activity that seems certain to continue. Plans to reduce impact are being implemented: for example, the West Rim Drive is closed to private cars during high visitation and free buses powered by natural gas (and commercial tour buses as well) transport people along the route. A plan approved in 1999, but not implemented, might have reduced automobile congestion along the South Rim by locating new facilities outside the national park, and building a light rail system for transit to viewpoints and trails. But these attempts at amelioration of human impact are technological fixes that could yield a profit to concessionaires. In the decades-long competition between two purposes of the national parks, to conserve and to enjoy, enjoyment seems the clear winner, now and in the immediate future.

Aswan: the dams and their effects

The revolutionary government of Egypt under Gamal Abdel Nasser announced in 1952 its decision to build a high dam on the Nile River at Aswan. Since the project's completion, it has been acclaimed a national treasure and criticized as an ecological catastrophe, and has aroused more controversy than any other resource development project. It illustrates the principle that dams "are not just engineering works but also constitute social institutions."[100] Assessment of the dam's social and environmental effects shows mixed results and indicates a missing element in many large development projects: a careful examination of perspectives that environmental history could provide.

The Aswan High Dam represents a massive break with the past and, as in all such cases, it had results beyond those intended by its planners. Eclipsing an earlier dam built by the British, it was a gigantic step in transforming Egypt from traditional agriculture to an adjunct of the world market economy. It is 111 m (364 ft) high, its length across the valley is more than 3.75 km (2.3 mi), and its reservoir, Lake Nasser/Nubia, can hold two years' average flow of the Nile. The dam ended the annual flood and converted the river below it into an open aqueduct.

There were several purposes of Nasser's decision to impose such a traumatic alteration on the Nile and the people who depend on it. Egypt is unique in that all its cropland is irrigated. The dam would make perennial irrigation, and a second or even third crop, possible on all cultivated land. The additional production would be in export crops, especially cotton, sugar, and rice. There would be water to expand cropland by as much as 810,000 ha (2 million acres).[101] Maintaining food production to feed Egypt's growing population was undoubtedly one goal, but subsidiary: witness the fact that Egypt, largely self-sufficient before the High Dam, now imports 70 percent of its food. A second function of the dam would be generation of electric power for industrialization.[102] A third purpose was that the planned reservoir allowed for containment of large floods. Any hydraulic engineer knew these three purposes would interfere with each other. It is impossible to maximize two independent variables over time, much less three. Irrigation would require releases of water at times not optimum for power generation, and vice versa. Power generation is most efficient with a full reservoir, and flood control requires a lower level to receive surges from upstream.

There were also political purposes for building. "Egypt is the epitome of the downstream state."[103] The Nile, the world's longest river, flows 6,400 km (4,000 mi) from its sources to the Mediterranean. Only the last 1,500 km (950 mi) are within Egypt. A former development plan adopted in 1948, termed the "Century Storage Scheme" because it would allow for extremes of high and low water in a hundred-year period, would have kept the Old Aswan Dam, and added a series of dams, reservoirs, and canals in upstream states. At the time, almost all the headwaters of the Nile were in British hands. The later division into independent nations, several chronically unstable, presents obstacles to a basin-wide agreement. Nine upstream states supply the Nile, and it is noteworthy that Ethiopia provides 86 percent of the water that reaches Aswan. The High Dam allows Egypt to control storage, granted that part of the reservoir is in the Sudan and makes cooperation between the two states essential.

Another political purpose of the High Dam was fulfilled by the grandeur of the project. It is seventeen times the volume of the Great Pyramid. It would be a lasting monument to Nasser and Egypt's independence.[104]

It was no wonder that Nasser quickly announced that it would be built. Within four years, when the West reneged on promises of aid, Nasser seized the Suez Canal with the intent of using its revenues for the dam, fought a war, and turned to the Soviet Union for assistance. In that context, discussions of possible negative effects had to be circumspect. One technician expressed the situation, quoting the Persian poet Omar Khayyam: "When the King says it is midnight at noon, the wise man says, 'Behold the moon.'"[105] Engineers asked questions concerning effects on stream flow, removal of the silt load and erosion of the channel, evaporation from the reservoir, sedimentation, degradation of the Delta coastline, induced seismic activity, and seepage, during the early days of planning. But since the government was committed to the dam, it became less receptive to these questions even though answers to them might have improved project design, and discussion was "suppressed."[106] Dr. Abd al-Aziz Ahmad, chair of the Hydroelectric Power Commission, criticized the project in British journals, citing the danger of excessive evaporation. Egyptian leaders thought he was offering ammunition to adversaries abroad, and he suffered professional ostracism. Ali Fathy, a professor of irrigation, said, "It became clear that competent technicians in government circles were collectively determined to overlook any signs of the deterioration of soil fertility as a side effect of the High Dam, even as a hypothesis. This was the result of what might be called the 'High Dam Covenant,' a psychological state born of political and other circumstances which ... cloaked the project from its very inception."[107]

If those who raised technical questions risked careers, those who warned of negative social or environmental consequences had more to fear. As Hussein Fahim put it,

> Government policy was not to be debated publicly before being formally adopted. Policies were to proceed from the top downward. The ... public channels of technical and political dialogue were blocked ... Anything that was described less than superlatively became potentially treasonous. As a result, the reasonably balanced combination of the political and the technical in the execution of big development schemes, designed

to avoid the waste of scarce resources, was undermined ... [T]he late 1950s and the ... following decade witnessed a total blackout of any [discussion of] mistakes or malfeasance connected with the Aswan High Dam.[108]

Foreign consultants also muted their criticisms. An International Bank for Reconstruction and Development "review of its own involvement in the scheme revealed that the ecological ramifications of the dam ... did not figure prominently in its own positive evaluation of the project."[109]

Gilbert White termed the construction of the Aswan Dam "a massive, unique intervention in physical, biological, and human systems."[110] Such an intervention always has unintended consequences, foreseen or not. The reservoir loses 20 percent of annual flow by evaporation and seepage.[111] An effect of evaporation is to increase the salinity of the stored water; the salt content of the river entering Lake Nasser is 200 ppm; that of the water leaving it is 220 ppm.

The most serious ecological factor is that the sediment and nutrients carried by the river settle out in the reservoir at 130 million tons each year, and will fill it, perhaps in five centuries. Windblown sand dunes spill into the reservoir, adding to fill and altering chemical composition.[112] Water leaving the dam is virtually free of suspended solids. Without silt, the erosive power of flowing water is greater. The river below the dam scours and lowers its bed, making it harder to get water into canals, and caving of banks sweeps away soil.

With elimination of sedimentation, the Delta loses land to coastal erosion.[113] The shoreline retreats 30 m (90 ft) a year. Eighty percent of Egypt's agricultural land is in the Delta. What ultimate good is a dam if Egypt loses a major part of the Delta? Sea water invades, further encroaching on farmland and wetlands inhabited by birds and other wildlife. The fish catch, once supported by Nile nutrients, has declined and many species have disappeared, although a growing catch from Lake Nasser is partial compensation. With silt gone, brickmakers began to strip topsoil.[114]

A shift to perennial irrigation was a purpose of the High Dam. Basin irrigation was used throughout Egypt before the nineteenth century.[115] Earth banks divided the land into basins of 400–18,000 ha (1,000–40,000 acres). Farmers allowed flood water into these and held it there forty to sixty days, during which it dropped its silt, forming a flat surface.

Perennial irrigation, universal today, runs water through canals onto the land every two or three weeks throughout the year. Without adequate drainage, it inevitably waterlogs the soil. Unfortunately, planners abandoned a 1958 project to install field drains because they thought a lower Nile would improve drainage. The opposite happened. The water table rose from 15 m (50 ft) below surface to 3 m (10 ft). In Cairo, the water table is only 81 cm (32 in) below the surface.[116] Ninety percent of cultivated land in Egypt is waterlogged, and 35 percent is salinized. When water evaporates from soil without adequate drainage, salts accumulate at as much as a ton per hectare per year.[117]

Since perennial irrigation provides no nutrients, fertilizer is applied, and the rate has increased exponentially. A fertilizer factory uses much of the power from the dam. Fertilizers and pesticides pollute drainage, yet that is what Egypt plans to pump to new croplands.

Fertilizers in water cause growth of algae. Stimulated by sunlight penetrating the clearer water, they multiply enormously and have clogged Cairo's purification system. Invasive water hyacinths cover 82 percent of water courses,[118] and their transpiration increases annual evaporation perilously,[119] but massive herbicide applications intended to control them also destroy non-target plants and animals.

An increase in schistosomiasis was predicted, caused by parasitic worms that pass into water in urine and feces and infect snails as alternate hosts. The debilitating disease can lead to death. Since the dam, urinary schistosomiasis, previously common, has decreased due to public health measures. But the intestinal form has spread.[120] Perennial irrigation keeps farmers in contact with water through the year and gives snails a permanent habitat.[121] Malaria has become more of a problem in Nubia and the Sudan with the increase in slack water.[122]

The reservoir displaced more than 110,000 Nubians, whose villages and the land itself were sacrificed for the prosperity of more numerous people downstream. The Egyptian government resettled many in "New Nubia" north of Aswan, providing education, health care, and land, but requiring them to raise sugar cane, a crop unfamiliar to them. For Nubians, it was an alien environment too far from the Nile.[123] Sudan moved its Nubians to settlements near Khashm el Girba.[124] They received community services and leased land, and were directed to raise cotton and peanuts. Herding folk who already used the land resented the intrusion. Many Nubians avoided resettlement areas, seeking city jobs. Others refused to leave Nubia, or returned, where some farm or provide tourist services.[125]

A purpose of the High Dam was to open new areas for cultivation.[126] By 1982, work had begun on 400,000 ha (almost 1 million acres), but irrigation reached less than 20 percent. Total acreage declined due to urbanization, brickmaking, waterlogging, and salinization, but productivity increased due to the shift of 365,000 ha (900,000 acres) to multiple cropping.

Predictions for reclaimed land are optimistic because they assume fertility equal to old lands, but soil is a living ecological community, not just a substrate. Heavy desert soils will not produce without expense of energy, materials, and time. Soils in new lands are poor and unsuited to export crops.[127] Still Egypt proceeds with projects to reclaim the desert without evidently recognizing that water may prove inadequate in quantity and/or quality.

Could the planners who considered building the High Dam have avoided some of the worst mistakes in this situation of running up against inexorable limits? The modern environmental history of Egypt, including the first Aswan Dam and its heightenings, could have provided warnings that might have helped prevent some of the damaging effects of the High Dam, or led to a decision not to build it.

The transformation of Egypt from a society dependent on traditional agriculture to an adjunct of the world market economy began with Mohammed Ali,[128] who ruled Egypt in the early nineteenth century. To enrich himself through commerce, he began cotton cultivation in the Delta. Primarily for export, cotton could be grown as a second crop but required irrigation when the Nile was low. He envisioned a double barrage where the Rosetta and Damietta branches of the Nile diverge north of Cairo to raise the level and divert the flow into canals.[129] He was barely persuaded not to use stones from the pyramids to build it.[130] Construction was finished in 1861. The venture prospered when the American Civil War deprived world markets of the largest source of cotton. Exports rose sharply and the cropped area expanded by over 50 percent. Most important, the transition from basin to perennial irrigation began in earnest. It would transform the landscape and the ecology of the soil, and create a demand for "timely water," that is, water for cash crops in the former fallow season.

The British, who seized control in 1882, pushed the agricultural revolution for their own benefit. Lord Cromer brought in engineers to repair the irrigation infrastructure. They strengthened barrages, dredged canals, and tried to disentangle the lines of irrigation from those of drainage. They noted the problems of waterlogging and salinization.

In 1894, Sir William Willcocks, born and educated in India, proposed a dam at Aswan to create a reservoir holding 2.4 billion cu m (84 billion cu ft) of "timely water."[131] He remarked, "It will be an evil day for Egypt if she forgets that ... the lessons which basin irrigation has taught for 7,000 years cannot be unlearned with impunity. The rich muddy water of the Nile flood has been the mainstay of Egypt ... and it can no more be dispensed with today than it could in the past."[132] He knew silt would clog his reservoir unless he could let through the first part of the annual flood, containing almost all the sediment. The dam could hold the last portion of the flood because the water then was relatively clear. Using Periyar Dam in South India as model, he included in the plan for Aswan 180 sluices with gates designed to let the flood pass.

The project waited four years while Cromer looked for funding. Sir Ernest Cassel, multi-millionaire friend of Edward VII, offered a loan. He was not disinterested, since his company had Egyptian land investments. Meanwhile, another issue agitated the world of arts and letters: the dam would periodically flood the temples of Philae, graceful buildings surviving from antiquity. With money for the dam at a premium, the cost of moving the temples was prohibitive.[133] Philae should be sacrificed "to the welfare of the world," said a young officer on a Sudan expedition, Winston Churchill.[134] Little concern was voiced for Nubian villagers whose lands and homes would also be flooded. Sir Benjamin Baker, consulting engineer, scaled down Willcocks's dam by 6 m (20 ft) to protect Philae, though water entered the temples when the reservoir filled. This lowering reduced volume to 1 billion cu m (35 cu ft). Construction lasted from 1898 to 1902.[135] The dam was 1,950 m (6,400 ft) long and 20 m (66 ft) high. It held the end of the flood and spread it out over a longer time, but reduced its height. In order to make water available at a higher level, additional barrages arose downstream.

Soon the managers discovered that the low dam did not retain enough water to supply its backers, particularly Cassel and his desert irrigation project at Kom Ombo. Archaeologists' opposition could not hold back the compelling arguments of commerce, so by 1912 the dam was raised 7 m (23 ft), to 27 m (89 ft), providing the capacity Willcocks originally wanted.[136]

A commission in 1920 recommended two dams in the Sudan to supplement flood control at Aswan and provide water for a cotton scheme. Sennar Dam on the Blue Nile was completed in 1926, followed by the Jebel Auliya reservoir in 1937. The latter, on the White Nile, was a white elephant; while it stores 3.6 billion cu m (124 billion cu ft), it evaporates 2.8 billion (75.6 billion cu ft) per year.

Demand for irrigation continued to grow, and since the Nile must be shared with Sudan, a second heightening was proposed to store more of Egypt's water in Egypt. By 1933 the Aswan Dam was raised again, this time by 9 m (30 ft), to 36 m (118 ft).[137] Capacity more than doubled, to over 5 billion cu m (179 billion cu ft).

The Old Aswan Dam, with its two heightenings, produced many, but not all, of the side effects that later appeared with the High Dam. As noted, the engineers chose a design that would allow the annual flood, with its deposition of silt, to continue. Even so, there was some downstream scouring and lowering of the riverbed.

The dam flooded part of Lower Nubia every year, displacing an increasing number of Nubians with each raising of the structure. These people relocated themselves to villages north of Aswan, or to Cairo.[138] They received small compensation, but no aid in relocation.

The worst environmental problems, such as inadequate drainage, waterlogging, and salinity, appeared as early as 1890 and worsened after 1902. In some areas, the schistosomiasis rate rose from 21 percent to 75 percent.[139] These were results of the shift to

Figure 7.5 Aerial view of the Aswan High Dam on the Nile River in Egypt; Lake Nasser to the right. This dam, which ended the annual flooding of the Nile below it, impounds two years' flow of the river. It has been hailed as one of the greatest projects of the twentieth century and denounced as a major ecological catastrophe. Photograph taken in 1981.

perennial irrigation accelerated by the dam. The need to address the drainage problem was clear. Retreat of the coastline at the outlets of the Nile and invasion of the Delta by seawater were observed after the 1933 heightening.[140] Works to counter coastal erosion began, but proved ineffective.

The environmental effects of the High Dam were not entirely hypothetical at the time of the decision to build it. But in authorizing the High Dam, the historic negative effects of the older dam were not given serious consideration, since those who could have commissioned studies were already committed to the project. Yet historical precedents were available from a dam on almost the same site. As Waterbury observed, "The history of this project is testimony to the primacy of political considerations determining virtually all technological choices with the predictable result that a host of unanticipated technological and ecological crises have emerged that now entail more political decisions."[141] He terms Egypt's policies leading up to the dam "short-sighted" and "non-integrated."[142]

The antidote for shortsightedness is careful consideration both of environmental history and the need for sustainability in the future. The antidote for a non-integrated approach is consideration of the many facets of the ecosystem, including the fact that humans cannot control every aspect of it, since massive actions always have massive unintended effects, nor can humans exceed the limits of the ecosystem without catastrophic results for themselves.

At least two problems lessen the possibility that Egypt can arrive at a sustainable level of production within the limits set by water, land, and the Nile Valley ecosystem. The first is population. At the time the first Dam was under construction, Egypt had 10 million people. With the High Dam rising, the population passed 30 million. In 1995 it was 63 million, heading toward 97 million in 2025, in spite of one of the lowest growth rates in Africa. This pattern indicates expanding demand for water in the future. Where will it come from?

Second is urbanization. Every year a larger percentage of Egyptians live in cities, particularly Cairo, which had 7.5 million people in 1976, and reached 17.3 million in 2000, 25 percent of the population. Industrial, commercial, and residential building, with infrastructure, uses space and water, in spite of a 1984 law prohibiting urban development on agricultural land. Estimates indicate a water deficit for Egypt of 14 billion cu m (491 billion cu ft) by 2025.

The Nile will not grow, but upstream projects might send more water to Egypt. Most ambitious is the partially constructed Jonglei canal in southern Sudan, intended to carry water past the Sudd swamps and end the evaporative loss of half the flow of the White Nile, but now halted by war. By drying up a huge wetland, Jonglei would damage a unique ecosystem and decimate wildlife. Sudan has treaty rights to half the additional Jonglei water. What of the other upstream states? Ethiopia's population, growing at twice the rate, will soon surpass Egypt's, and could reach 127 million in 2025. Ethiopia wants irrigation projects using the headwaters. Nasser, Sadat, and Mubarrak each threatened war with any state that takes "Egypt's water."[143] Sooner or later a plan for the watershed must be negotiated. But no plan can meet the desires of every nation concerned to support its growing population and to achieve economic growth by producing more for world markets.

As far as the decision to build the Aswan High Dam is concerned, if the experience of the past had lessons to teach, they seem not to have been learned. The present ecological situation of Egypt is precarious. It is difficult to imagine what the path to sustainability might be, since the constraints of politics convince planners, and they rarely consider the limits of the ecosystem. But planning will be misleading until it takes account of the ecological–historical perspective.

Conclusion

In his environmental history of the twentieth-century world, *Something New Under the Sun*, John R. McNeill observes that the last century was unique in the extent and intensity of changes in the natural environment, and centrality of human agency in causing those changes.[144] He asks why this was the case, and finds a threefold answer: accelerating use of resources, especially through conversion to a fossil fuel-based energy system; very rapid population growth; and the ideological commitment of nations and corporations to military power and economic growth swelled by mass production. His analysis is convincing.

During the period before 1960, all of these factors were evident, and for virtually all of the leaders of society in the industrialized world and beyond were causes for pride and self-congratulation. People whose ways of life were swept away by the trend objected, but lacked power, and their traditions were discredited as old-fashioned and unscientific. The very rapidity of change, however, allowed its destructive aspects to be noticed more clearly than in any previous era, and advocacy of conservation of resources and the preservation of natural areas and wildlife gained momentum. This happened most noticeably in North America and Europe, but was not limited to them. In some places conservation measures were adopted; forest and wildlife management, soil conservation, and the creation of parks and reserves appeared, usually under governmental direction. But the extent to which human efforts were affecting the environmental systems of the Earth had as yet hardly been realized. A few prophetic voices such as Paul Sears and Fairfield Osborn warned of the consequences of soil erosion and unbridled exploitation of resources, but even they hardly suspected the range and scale of environmental damage that would be detected in the following decades.

Notes

1   Fairfield Osborn, *The Limits of the Earth*, Boston, Little, Brown & Co., 1953.
2   Again, 1 metric ton equals 1.1 English/US tons. Since the two units are relatively close in scale, only the weights in metric tons will be given.
3   *Environmental History* 3, 2, April 1998, is a special issue containing several articles (by Paul Sabin, Myrna Santiago, Nancy Quam-Wickham, and Brian Black) on the environmental effects of the oil industry in the twentieth century.
4   Michael Chisholm, "The Increasing Separation of Production and Consumption," in B.L. Turner II, William C. Clark, Robert W. Kates, John F. Richards, Jessica T. Mathews, and William B. Meyer, eds, *The Earth as Transformed by Human Action*, Cambridge, Cambridge University Press, 1990, 87–101.
5   E.g. H.G. Wells, *The Time Machine*, New York, Random House, 1931; Aldous Huxley, *Brave New World*, London, Chatto & Windus, 1932.
6   Carl O. Sauer, "The Agency of Man on the Earth," in William L. Thomas, Jr., ed., *Man's Role in Changing the Face of the Earth*, Chicago, University of Chicago Press, 1956, 49–69, at 66.
7   S.R. Eyre, *The Real Wealth of Nations*, New York, St. Martin's Press, 1978.
8   Stewart L. Udall, *The Quiet Crisis and the Next Generation*, Salt Lake City, UT, Peregrine Smith Books, 1988, 137–8.
9   A.G. Kenwood and A.L. Lougheed, *The Growth of the International Economy, 1820–1960*, London, George Allen & Unwin, 1971, 239, 273.
10 Karl Marx, *Economic and Philosophical Manuscripts of 1844*, Moscow, Foreign Languages Publishing House, 1961, 74–6; quoted in Charles A. Ziegler, *Environmental Policy in the USSR*, Amherst, University of Massachusetts Press, 1987, 12.
11 Douglas R. Weiner, *A Little Corner of Freedom: Russian Nature Protection from Stalin to Gorbachev*, Berkeley and Los Angeles, University of California Press, 1999, 400.
12 Friedrich Engels, *Dialectics of Nature*, New York, International Publishers, 1940, 292.
13 Madhav Gadgil and Ramachandra Guha, *This Fissured Land*, Berkeley and Los Angeles, University of California Press, 1992, 12.
14 Raymond Dominick, "Capitalism, Communism, and Environmental Protection: Lessons from the German Experience," *Environmental History* 3, 3, July 1998, 311–32.
15 Michael Williams, "Forests," in Turner II et al., eds, *Earth as Transformed by Human Action*, 179–201, at 185–7.
16 Warren Dean, *With Broadax and Firebrand: The Destruction of the Brazilian Atlantic Forest*, Berkeley and Los Angeles, University of California Press, 1995.
17 Williams, "Forests," 132–5.
18 Richard P. Tucker, "The Depletion of India's Forests under British Imperialism: Planters, Foresters and Peasants in Assam and Kerala," in Donald Worster, ed., *The Ends of the Earth: Perspectives on Modern Environmental History*, Cambridge, Cambridge University Press, 1988, 118–40.
19 Mikko Saikku, "The Extinction of the Carolina Parakeet," *Environmental History Review* 14, 3, Fall 1990, 1–18.
20 Jerome A. Jackson, 2002, "Ivory-Billed Woodpecker (*Campephilus principalis*)," *The Birds of North America Online*, ed. A. Poole, Ithaca, NY, Cornell Lab of Ornithology, http://bna.birds.cornell.edu /bna/species/711.
21 Michael Graham, "Harvests of the Seas," in Thomas, Jr, ed., *Man's Role in Changing the Face of the Earth*, 487–503, at 493.
22 James O'Connor, *Three Ways to Think about the Ecological History of the Monterey Bay Region*, Santa Cruz, CA, CNS, Center for Political Ecology, 1995.
23 Martin V. Melosi, *Effluent America: Cities, Industry, Energy, and the Environment*, Pittsburgh, PA, University of Pittsburgh Press, 2001.
24 Joel A. Tarr and Robert U. Ayres, "The Hudson–Raritan Basin," in Turner II *et al.*, *Earth As Transformed by Human Action*, 623–39, at 638.
25 J.R. McNeill and Verena Winiwarter, eds, *Soils and Societies: Perspectives from Environmental History*, Isle of Harris, UK, White Horse Press, 2006.
26 Ian Douglas, "Sediment Transfer and Siltation," in ibid., 215–34, at 217–18.
27 Farieda Khan, "Soil Wars: The Role of the African Soil Conservation Association in South Africa, 1953–1959," *Environmental History* 2, 4, October 1997, 439–59.
28 Paul Bigelow Sears, *Deserts on the March*, Norman, University of Oklahoma Press, 1935.

29   Harold T. Pinkett, *Gifford Pinchot: Private and Public Forester*, Chicago, University of Illinois Press, 1970; Char Miller, *Gifford Pinchot and the Making of Modern Environmentalism*, Washington, DC, Island Press, 2001.

30   Gifford Pinchot, *The Fight for Conservation*, New York, Doubleday, 1910, 48–9.

31   Samuel P. Hays, *Conservation and the Gospel of Efficiency: The Progressive Conservation Movement, 1890–1920*, New York, Atheneum, 1980, 176–88; Paul Russell Cutright, *Theodore Roosevelt: The Making of a Conservationist*, Urbana, University of Illinois Press, 1985, 228–31.

32   Richard H. Grove, *Ecology, Climate and Empire: Colonialism and Global Environmental History, 1400–1940*, Cambridge, White Horse Press, 1997, 206–16.

33   Ibid., 222.

34   Donald Worster, *Rivers of Empire: Water, Aridity, and the Growth of the American West*, New York, Pantheon Books, 1985, 132–42.

35   Walter Clay Lowdermilk, "Lessons from the Old World to the Americas on Land Use," *Annual Report of the Board of Regents of the Smithsonian Institution, 1943*, Washington, DC, Government Printing Office, 1944, 413–27; revised and reprinted as *Conquest of the Land Through 7,000 Years*, Agriculture Information Bulletin 99, Washington, DC, Government Printing Office, 1953.

36   Frank Clements, *Kariba: The Struggle with the River God*, New York, G. Putnam's Sons, 1959, 183.

37   Robert McIntosh, *The Background of Ecology: Concept and Theory*, Cambridge, Cambridge University Press, 1985, 28–33.

38   Donald Worster, *Nature's Economy: A History of Ecological Ideas*, 2nd edn, Cambridge, Cambridge University Press, 1977, 208–20, 291–315.

39   Liberty Hyde Bailey, *The Holy Earth*, New York, 1915, reprint, Ithaca, New York State College of Agriculture, 1980, 30–1.

40   Aldo Leopold, *A Sand County Almanac and Sketches Here and There*, New York, Oxford University Press, 1949, 224–5.

41   Frank Benjamin Golley, *A History of the Ecosystem Concept in Ecology: More Than the Sum of the Parts*, New Haven, CT, Yale University Press, 1993, 205.

42   Ovid, *Fasti* 3. 295.

43   Dr. Subash Chandran also noticed that this magnificent grove, Devaravattikan, was being invaded by domestic water buffaloes in spite of the fact that the Karivokkaliga villagers had erected a wooden fence in an attempt to keep them out. I found financial support among friends, and he arranged for the villagers to buy materials and paid them for the labor to build a more substantial fence that effectively protects the grove and the rare plants of its understory. We were guided by the advice of the village elders, who ascertained that the gods inherent in the sacred grove would not be angered by the fencing, but accepted it as an offering.

44   R.T.T. Forman and M. Godron, *Landscape Ecology*, New York, John Wiley & Sons, 1986.

45   M.D. Subash Chandran, "Peasant Perception of Bhutas: Uttara Kannada," in Kapila Vatsyayan and Baidyanath Saraswati, eds, *Prakrti: The Integral Vision*, Vol. 1, *Primal Elements: The Oral Tradition*, New Delhi, Indira Gandhi National Centre for the Arts, 1995, 151–66, at 161.

46   N.V. Joshi and Madhav Gadgil, "On the Role of Refugia in Promoting Prudent Use of Biological Resources," *Theoretical Population Biology*, 40, 1991, 211–29.

47   Larry D. Harris, *The Fragmented Forest: Island Biogeography Theory and the Preservation of Biotic Diversity*, Chicago, University of Chicago Press, 1984.

48   J. J. Roy Burman, "The Institution of Sacred Grove," *Journal of the Indian Anthropological Society*, 27, 1992, 219–38, at 232.

49   M.D. Subash Chandran and Madhav Gadgil, "Sacred Groves and Sacred Trees of Uttara Kannada," in Baidyanath Saraswati, ed., *Life-Style and Ecology*, New Delhi, Indira Gandhi National Centre for the Arts, 1998, 85–138, at 98.

50   James George Frazer, *The Golden Bough: The Magic Art and the Evolution of Kings*, 3rd edn, 12 vols, New York, Macmillan, 1935. The work was first published in 1890.

51   Chandran and Gadgil, "Sacred Groves and Sacred Trees of Uttara Kannada", appendix.

52   Ibid., 29.

53   Madhav Gadgil and M.D. Subash Chandran, "Sacred Groves," *India International Quarterly*. 19, 1–2, spring–summer 1992, 183–7.

54   Madhav Gadgil and V.D. Vartak, "The Sacred Uses of Nature," in Ramachandra Guha, ed., *Social Ecology*, Delhi, Oxford University Press, 1994, 82–9, at 87.

55   Gadgil and Vartak, "Sacred Uses of Nature," 87.

56 Chandran and Gadgil, "Sacred Groves and Sacred Trees of Uttara Kannada," 85.

57 Ibid., 110, 115.

58 Ibid., 111.

59 Gadgil and Chandran, "Sacred Groves," 183–7.

60 R.J.R. Daniels, "A Conservation Strategy for the Birds of Uttara Kannada District," Ph.D. Thesis, Centre for Ecological Sciences, Indian Institute of Science, Bangalore, 1989.

61 Burman, "Institution of Sacred Grove," 223–4.

62 Chandran and Gadgil, "Sacred Groves and Sacred Trees of Uttara Kannada," 109.

63 Ibid., 100–4.

64 Madhav Gadgil, "Conserving Biodiversity as if People Matter: A Case Study from India," *Ambio* 21, 3, May 1992, 266–70, at 267.

65 See above, Chapter 1, section on "Ecological Process."

66 Dietrich Brandis, *Indian Forestry*, Woking, Oriental Institute, 1897.

67 Francis D. Buchanan, *A Journey from Madras through the Countries of Mysore, Canara and Malabar*, Madras, Higginbothams & Co., 1970, vol. 2.

68 Gadgil, "Conserving Biodiversity as if People Matter," 268.

69 Madhav Gadgil and M. D. Subash Chandran, "On the History of Uttara Kannada Forests," in John Dargavel, Kay Dixon, and Noel Semple, eds, *Changing Tropical Forests: Historical Perspectives on Today's Challenges in Asia, Australasia and Oceania*, Canberra, Centre for Resource and Environmental Studies, 1988, 47–58, at 51.

70 Jeremy Seabrook, "Uttara Kannada," in *Notes from Another India*, London, Pluto Press, 1995, 65–83, at 65–6.

71 Madhav Gadgil and Ramachandra Guha, *This Fissured Land*, University of California Press, 1992, 224.

72 Gadgil, "Conserving Biodiversity as if People Matter," 267.

73 Stephen J. Pyne, *How the Canyon Became Grand: A Short History*, New York, Viking, 1998, explores some of the meanings that visitors of many kinds have given to the Grand Canyon.

74 US National Park Service Act, 1916, *The Statutes at Large of the United States of America from December 1915 to March 1917*, Washington, DC, Government Printing Office, 1917, Vol. 39, Part 1, 535, quoted in Carolyn Merchant, ed., *Major Problems in American Environmental History*, Lexington, MA, D.C. Heath, 1993, 394.

75 Alfred Runte, "National Parks and National Park Service," in Richard C. Davis, ed., *Encyclopedia of American Forest and Conservation History*, 2 vols, New York, Macmillan, 1983, 464–7.

76 George Catlin, *Letters and Notes on the Manners, Customs, and Condition of the American Indians*, 2 vols, reprint, Minneapolis, Ross & Haines, 1965, Vol. 1, 261–2.

77 Stephen Hirst, *Life in a Narrow Place*, New York, David McKay, 1976; Mark David Spence, *Dispossessing the Wilderness: Indian Removal and the Making of the National Parks*, New York, Oxford University Press, 2000.

78 John Muir, "The Wild Parks and Forest Reservations of the West," *Atlantic Monthly* 81, January 1898, 28.

79 New York *Sun*, May 7, 1903.

80 Douglas Hillman Strong, "The Man Who 'Owned' Grand Canyon," *American West* 6, September 1969, 36. In its 1920 decision, the Supreme Court ruled that as one of the greatest examples of erosion in the world, the Grand Canyon was clearly an object of unusual scientific interest and therefore could be set aside by proclamation under the Antiquities Act of 1906.

81 C. Gregory Crampton, *Land of Living Rock: The Grand Canyon and the High Plateaus of Arizona, Utah, and Nevada*, New York, Alfred A. Knopf, 1972, 206.

82 Barbara J. Morehouse, *A Place Called Grand Canyon: Contested Geographies*, Tucson, University of Arizona Press, 1996, 66.

83 Ibid., 55–62.

84 Byron Eugene Pearson, "People Above Scenery: The Struggle Over the Grand Canyon Dams, 1963–1968," Ph.D. Dissertation, Department of History, University of Arizona, 1998.

85 Roderick Nash, *Grand Canyon of the Living Colorado*, New York, Ballantine Books, 1970, 132–3.

86 Keir Brooks Sterling, *Last of the Naturalists: The Career of C. Hart Merriam*, New York, Arno Press, 1974, 294.

87 Clinton Hart Merriam and Leohard Stejneger, "Results of a Biological Survey of the San Francisco Mountain Region and Desert of the Little Colorado, Arizona," *North American Fauna*, 3, US Depart-

ment of Agriculture, Division of Ornithology and Mammalogy, Washington, DC, Government Printing Office, 1890.

88 Sterling, *Last of the Naturalists*, xiii.

89 Joseph Wood Krutch, *Grand Canyon: Today and All Its Yesterdays*, New York, Doubleday and the American Museum of Natural History, 1962, 12.

90 Robert Wallace, *The Grand Canyon*, New York, Time-Life Books, 1972, 56.

91 J. Donald Hughes, *In the House of Stone and Light: A Human History of the Grand Canyon*, Grand Canyon, AZ, Grand Canyon Natural History Association, 1978, 90.

92 Aldo Leopold, *A Sand County Almanac and Sketches Here and There*, London, Oxford University Press [1949] 1970, 211.

93 Named for A. Starker Leopold, a son of Aldo Leopold, zoologist at the University of California at Berkeley, and the chairm of the National Park Service Advisory Board on Wildlife Management.

94 George Sessions, "Ecocentrism, Wilderness, and Global Ecosystem Protection," in Max Oelschlager, ed., *The Wilderness Condition: Essays on Environment and Civilization*, San Francisco, Sierra Club Books, 1992, 93.

95 Several other small animals occur in related but distinct species that are separated by the Grand Canyon. For example, on the North Rim the long-tailed pocket mouse, bushy-tailed wood rat, and long-tailed meadow mouse correspond respectively to the rock pocket mouse, Mexican wood rat, and Mexican meadow mouse of the South Rim.

96 Zoological Society of San Diego, website, 1999.

97 Marguerite Henry, *Brighty of the Grand Canyon*, New York, Rand McNally, 1953.

98 Steven W. Carothers and Bryan T. Brown, *The Colorado River through Grand Canyon: Natural History and Human Change*, Tucson, University of Arizona Press, 1991.

99 Ibid., 67.

100 Hussein M. Fahim, *Dams, People and Development: The Aswan High Dam Case*, New York, Pergamon Press, 1981, 4.

101 A feddan, an Egyptian land area measure, is 1.038 acres.

102 After the completion of the High Dam, it generated about 50 percent of all the power used in Egypt. By the 1990s, that figure had declined to 10 percent because other power stations had come online.

103 John Waterbury, *Hydropolitics of the Nile Valley*, Syracuse, NY, Syracuse University Press, 1979, 5.

104 Edward Goldsmith and Nicholas Hildyard, *The Social and Environmental Effects of Large Dams*, San Francisco, Sierra Club Books, 1984, 1.

105 Waterbury, *Hydropolitics*, 101.

106 Gilbert F. White, "The Environmental Effects of the High Dam at Aswan," *Environment* 30, 7, September 1988, 4–11, 34–40, at 8.

107 Ali Fathy, *The High Dam and Its Impact* , Cairo, General Book Organization, 1976, 50–1; quoted in Waterbury, *Hydropolitics*, 116.

108 Fahim, *Dams, People and Development*, 165.

109 Waterbury, *Hydropolitics*, 102.

110 White, "Environmental Effects," 38.

111 Fahim, *Dams, People and Development*, 28.

112 Amin, "Safety Considerations," 40; M. El-Moattassem, "Field Studies and Analysis of the High Aswan Dam Reservoir," *International Water Power and Dam Construction* 46, 1, January 1994, 30–5; Daniel Jean Stanley and Jonathan G. Wingerath, "Nile Sediment Dispersal Altered by the Aswan High Dam: The Kaolinite Trace," *Marine Geology* 133, 1/2, July 1996, 1–9.

113 Waterbury, *Hydropolitics*, 135, mentions a rate of 29 m (95 ft) at the mouth of the Rosetta Branch and 31 m (102 ft) at Damietta.

114 White, "Environmental Effects," 9, 34.

115 H.E. Hurst, *The Nile: A General Account of the River and the Utilization of Its Waters*, London, Constable, 1957, 38–41.

116 Fahim, *Dams, People and Development*, 31.

117 Fred Pearce, "High and Dry in Aswan," *New Scientist* 142, 1924, May 7, 1994, 28–32, at 30.

118 Mahmoud Abu-Zaid and M.B.A. Saad, "The Aswan High Dam, 25 Years On," *Unesco Courier*, May 1, 1993, 37.

119 Waterbury, *Hydropolitics* , 231.

120 White, "Environmental Effects," 37

121 Fahim, *Dams, People and Development*, 34. One study said that between Cairo and Aswan, the urinary form increased from 5 percent in 1930 to 35 percent in 1972. A University of Michigan study found declines in that form in all major districts due mainly to the use of protected water supplies.

Delta	1966, 60	1978, 42	
Middle Egypt	1940, 60	"	27
Upper Egypt	1972, 38	"	25
Nubia	1958, 40	"	7

But the more severe variety, intestinal schistosomiasis, has spread and increased from 3.2 in 1935 to 73 percent in 1979.

122 Fahim, *Dams, People and Development*, 138.

123 Byron D. Cannon, "Cultural–Historical Discontinuity and Imposed Population Resettlement: Egyptian Nubia and the Aswan Dams of 1902 and 1965," paper presented at "Environmental Cultures: Historical Perspectives" Conference, University of Victoria, BC, April 1996.

124 Robert O. Collins, *The Waters of the Nile: Hydropolitics and the Jonglei Canal, 1900–1988*, Oxford, Clarendon Press, 1990, 272.

125 Fahim, *Dams, People and Development*, 93–5.

126 Ibid.

127 Waterbury, *Hydropolitics*, 139.

128 Mohammed Ali's dates are 1769–1849; he was in effective control of Egypt from 1811 to 1847.

129 Peter Mansfield, *The British in Egypt*, London, Weidenfeld & Nicolson, 1971, 115.

130 H.E. Hurst, *The Nile: A General Account of the River and the Utilization of Its Waters*, London, Constable, 1957, 50.

131 Collins, *Waters of the Nile:*, 103.

132 Willcocks, 1908, quoted in Waterbury, *Hydropolitics*, 39.

133 They were moved at the time the Aswan High Dam was constructed, as part of the UNESCO project to "save" the monuments of Nubia.

134 Herbert Addison, *Sun and Shadow at Aswan: A Commentary on Dams and Reservoirs on the Nile at Aswan, Yesterday, Today, and Perhaps Tomorrow*, London, Chapman & Hall, 1959, 43.

135 Collins, *Waters of the Nile*, 109.

136 Mansfield, *British in Egypt*, 118.

137 Norman Smith, *A History of Dams*, London, Peter Davies, 1971, 221.

138 Cannon, "Cultural–Historical Discontinuity and Imposed Population Resettlement."

139 J.N. Lanoix, "Relation Between Irrigation Engineering and Bilharziasis," *Bulletin of the World Health Organization* 18, 1958, 1011–35.

140 Waterbury, *Hydropolitics*, 135.

141 Ibid., 5.

142 Ibid., 6.

143 Fahim, *Dams, People and Development*, 160.

144 John R. McNeill, *Something New Under the Sun: An Environmental History of the Twentieth-Century World*, New York, W.W. Norton, 2000, xxi.

8   Modern environmental problems

The Grand Canyon is in a region once noted for its clear air, but in my many visits to it over the years – the first one in 1948 – I have noticed a grayish haze that increases in frequency and turbidity. Photographs from space reveal one of its sources: smog drifting eastward across the desert from the Los Angeles basin, 640 km (400 mi) away. But there are other sources even further away. Air over the Arctic Ocean has a layer of pollution that can be traced to Europe, Russia, China, Canada, and the United States. In the late twentieth century, it became clear that environmental problems affect the whole Earth. In former decades, it seemed to most people that problems affecting the natural environment were locally caused, with local impacts. A city's industries and transport polluted its own air, logging threatened a particular park or wilderness area, and sewage seemed a worry for those downstream in a single watershed. But in this period environmental impacts crossed boundaries and became international or worldwide in scope. As the magnitude of the effects of human actions increased, the size and number of the ecosystems affected by them increased. Radioactive particles, chlorine compounds that react with the ozone shield in the stratosphere, greenhouse gases such as carbon dioxide and methane, and pollutants in the sea spread worldwide and affected the largest of ecosystems, the biosphere itself.

The images of rapid environmental destruction in the late twentieth century were numerous, and information technology made possible a degree of accuracy in gathering them and an extent of dissemination that made an unprecedented impression on human consciousness. The last half of the twentieth century saw a remarkable expansion of knowledge about the workings of the biosphere, but at the same time activities that damaged the biosphere accelerated faster than ever before. Although the period covered by this chapter is shorter than any of the previous ones, it is the one in which the most rapid impacts of humans were made on the Earth, including depletion of resources and impairment of natural systems of life in the land, sea, and atmosphere. Investigation of the structure and dynamics of these communities and the damage being done to them also reached a scale unmatched before.

In 1950, many of the Earth's ecosystems had been altered by human intervention, but by the end of the century, almost every ecosystem was either degraded or seriously threatened. There were few corners of the globe without evidence of human presence and change caused by humans. Antarctica was dotted with research stations that generated waste and had to arrange for its disposal. Globules of oil and pieces of plastic foam floated throughout the oceans. Passing jet planes and their vapor trails were often visible in the sky from every part of Earth. The pressure of human numbers was pushing settlements into forests and grasslands where natural functions once were dominant. In cities, suburbs, and in industrial and agricultural zones, the works of humankind dominated the landscape. But humans are still part of, and totally dependent upon, the natural systems of the Earth. This truth was often forgotten in legislatures and company board rooms, but it was none the less crucial. Every molecule of oxygen in the atmosphere breathed by humans was produced by the photosynthesis of plants on land and in the sea. Even in the late twentieth century, most food eaten by people on Earth was the product of agriculture, and the rest came from fishing, hunting, and gathering; no appreciable amount was synthetic, and even if it were, it would have to have been processed from some natural raw material, such as petroleum, that was once living. Human activities, even in this most technological of ages thus far, depended upon and related to ecosystems, even those in distant parts of the globe.

Processes occurring in ecosystems continue to affect humans. As the proportion of the biosphere's energy taken up by human activities becomes larger, even things once considered completely natural may at least be triggered by what humans do. Climate and the intensity of storms may be subject to human influence. The spread of diseases is affected by worldwide jet aircraft transportation, and by the ubiquity of human bodies and the amount of protoplasm they present as environment for microorganisms. Even earthquakes can be human-caused, resulting from the weight of huge reservoirs or, as happened in the 1960s in Colorado, by the injection of liquid nerve-gas wastes into deep rock strata.

The visible effects of humans on ecosystems have increased greatly in number and kind in the late twentieth century. A single human action may have many results due to the complexity of ecological relationships. Some changes are within the capacity of ecosystems to absorb or compensate and still remain functioning and healthy. Others may go beyond that capacity, and erode or transform an ecosystem.

The kinds of changes inflicted by industry on ecosystems since the Second World War include some that had not been known during previous centuries. Plutonium and other radioactive wastes, non-biodegradable insecticides, chlorofluorocarbons, plastics, artificial pheromones and hormones, and many of the rest of the tens of thousands of industrial chemicals in use either did not exist or were not disseminated in major quantities until recently. If there is any judgment historians can make about technological change, it is that its pace is accelerating at a rate never previously matched, and that its environmental impacts are similarly escalating. That pace has outstripped a traditional human method of coping with environmental change through gradually altering taboos and customs. It has also outstripped the progress of carefully verified scientific research, so that damage is done before measures can be taken to ameliorate it, or even before its existence, extent, and causes are known. Late twentieth-century humans played dice as never before with the systems that support life.

The effects of acid precipitation were catastrophic in such regions as eastern Canada, New England, Scandinavia, central Europe, and parts of Russia and China.[1] Fish life perished in thousands of lakes. Millions of hectares of forests experienced dieback, known as *Waldsterben*, German for "forest death," and evidence accumulated that this was due to precipitation that in many cases demonstrated an acidity exceeding that of lemon juice. Acid rain was first noticed and named in 1872, and forest dieback was attributed to air pollution in the early twentieth century.[2] But scientists who predicted in the 1950s what actually happened in the 1980s were denounced as doomsayers. Lake Baikal in the heart of Siberia was not saved by its remoteness, nor were the wilderness areas of Labrador.

Human activities have been adding gases to the atmosphere that are known to have a warming effect; that is, they allow energy in sunlight to reach the surface of the Earth, but trap some of the heat radiation that would otherwise escape, in the so-called "greenhouse effect." Between 1800 and 2000, the concentration of one of these gases, carbon dioxide, increased by almost 30 percent, and the increase is expected to accelerate in the following decades. Concentrations of other gases with similar effects, such as nitrogen oxides and methane, are increasing even faster. The rapid rise of average temperatures observed since the late 1980s could be a result of this process. The Intergovernmental Panel on Climate Change stated in 1996, "The balance of evidence suggests that there is a discernable human influence on global climate."[3]

Another worldwide environmental impact of technology is the depletion of stratospheric ozone. It was suggested in 1974 that a group of chemicals known as chlorofluorocarbons (CFCs), used as propellants, refrigerants, solvents, and in production of foam plastics, were adding chlorine to the atmosphere and disturbing the ozone balance. The ozone layer in the stratosphere shields the Earth from ultraviolet radiation. The effects of increased ultraviolet radiation include an elevated risk of skin and eye cancer in humans and animals, and death or reduced rates of growth in plants and in organisms that live near the surface in water, such as the phytoplankton that consume carbon dioxide and provide about 70 percent of the Earth's annual production of oxygen. Losing these could accelerate the "greenhouse effect;" and to exacerbate the problem, CFCs are greenhouse gases with the same kind of effect as carbon dioxide. Recognizing these dangers, in 1978 the US, Canada, Sweden, and Norway banned CFCs in aerosol cans. Other uses continued, however, and worldwide production began to increase again. Then in 1985 scientists from the British Antarctic Survey announced the shocking news that a hole in the ozone layer had appeared over Antarctica, and this was confirmed by data from satellites and airplanes.[4] Since then, the hole has grown larger and deeper. Thirty-one nations approved a treaty in Montreal in 1987 which would reduce world production of CFCs by 50 percent in ten years. Later, scientists found a rapid depletion of the ozone layer over the Northern Hemisphere at least three times as serious as they had expected. The treaty was amended twice to make it stricter, and DuPont, the world's largest producer of CFCs, which had once dismissed these concerns, announced that it would cease production.[5] But even if all production of CFCs were to cease immediately, the concentration of ozone-destroying chlorine in the stratosphere would continue to increase as CFCs already in the atmosphere continue to make their way upward. In October 2000, the "ozone hole" over Antarctica extended as far north as the city of Punta Arenas, Chile. The ozone layer will continue to weaken in the twenty-first century.[6] One of the most important effects may be damage caused by ultraviolet radiation to agricultural crops.

Agriculture became more intensive and more productive in the late twentieth century due to trends in agricultural technology: continuing mechanization in the richer countries, the dissemination of high-yielding genetic strains of basic food crops, and the application of industrial fertilizers and pesticides. The use of farm machinery instead of human labor had already begun to decrease the agricultural work force in the United States by 1920; by 1970 it had shrunk to what it had been in 1835 with only half the land area.[7] In Europe, the same process occurred rapidly after the Second World War; the number of combine-harvesters in Denmark, for example, increased from zero in 1944 to 40,000 in 1968.[8] The Soviet Union made immense efforts to mechanize agriculture at the same time. In the developing countries, however, labor-intensive methods remained the rule. China had 1.5 tractors per 1,000 ha (2,400 acres) in 1970 as against the European average of 41, although since then China has made huge steps in mechanization. The picture was less industrial in Africa, and in relative terms has changed little since.[9] On journeys through the countryside in India and Indonesia in the mid-1990s, I saw farmers plowing with oxen or water buffaloes everywhere and tractors only rarely. In 1989–91, the United States had 225 times more harvester machines than India, which had four times the population.[10] Even so, agriculture has changed radically in those countries and throughout the world since the 1960s. The so-called Green Revolution involved the selection and rapid diffusion of high-yielding crop varieties, particularly of rice and wheat in tropical areas, and will be discussed later in this chapter.

Deforestation became more severe during the late twentieth century, especially in the tropics, but also in other parts of the world. Many nations continued to lose their natural forests, along with the biodiversity, soil and water conservation, and climate regulation they would have provided. A huge increase in world exports of tropical hardwoods began about 1955,[11] driven by demand in Japan, western Europe, and the US for products such as veneer and plywood, and by new technologies including mechanized logging, timber transport, and pulping. Timber was often underpriced, if total replacement costs are considered. Some tropical countries were small and poor, and their forest sectors had to depend on inadequately educated management. They faced powerful international corporations that could summon up huge amounts of money and numbers of employees greater than those of many governments, and sometimes even weaponry. Violence, however, was seldom necessary; large national and multinational corporations could promise jobs and other rewards. Agencies set up by governments to protect local people and resources sometimes proved amenable to bribes and pressure. Tropical deforestation proceeded at a rapid pace in the 1990s in spite of efforts to slow down and reverse it.[12] Between 1960 and 2000, the world lost at least 20 percent of all tropical forest cover. Less than 10 percent of old-growth temperate forests still stand, and their commercial exploitation continues. Everywhere, government programs to encourage exports, rising prices of timber and other wood products, and the depletion of accessible forests drove logging concerns to seek out surviving forest resources. Many of these companies have poor environmental and social records.

The dipterocarp forests of southeast Asia fell victim to the newfound usefulness of their comparatively less expensive wood. In west Africa, logging for export and local demands for wood diminished reserves. Niger, where logging and agriculture eliminated much of the standing forest, suffered a crisis in wood supply.[13] More than half of Burkina Faso's woodlands have been lost, and erosion exposing the underlying laterite prevents regeneration. In Central America, large areas of forest were removed to provide grazing for beef cattle for the fast-food market in the US and elsewhere.[14] Richard Tucker calls the Caribbean Basin "the Yankees' Tropical Woodlot."[15] Tasmania began major shipments of woodchips for pulp and paper to Japan in the 1970s.[16] Japan, whose forests had suffered from overcutting and mismanagement before and during the Second World War, but whose government subsequently enacted strict controls to preserve Japan's remaining forests, imported logs from western North America and the tropics of Asia and the Pacific in prodigious amounts. Exports from Oregon and Washington to Japan increased by a factor of six between 1961 and 1974. US law forbade export of timber from public lands, but exports from private land increased demand for federal timber. As a result, ancient forests were still declining as formerly undisturbed sections of the Pacific Northwest such as the Willamette National Forest described in this chapter were razed by clearcuts. Total forest area in the US may have increased, however, due to regeneration on cleared farmland in the eastern states. The USSR, with one-fifth of the world's forested land, depleted its forest resources by poor managerial practices, although dependable figures were hard to obtain due to policies of secrecy and over-optimistic reports from officials anxious to show good records in spite of the facts.

The technology of water use has a major impact on the environment, since less than 1 percent of the world's water is fresh and available for humans and other land organisms.

Water withdrawn from surface and groundwater sources for human use rose from approximately 1,000 cu km (240 cu mi) in 1950 to 3,500 cu km (840 cu mi) in 1980, the greater part used for agriculture.[17] Most of this, which in the latter year was 39 percent of all available fresh water, was polluted by organic wastes, fertilizers, pesticides, and industrial effluents before it flowed back into surface or ground waters. Unfortunately, much water became polluted even during precipitation, picking up acids and particulates from the atmosphere.

The problem of disposal of human excreta, the most important source of water pollution in earlier periods, had been addressed by chlorination of drinking water and sewage treatment in most industrialized countries by 1950. But in poor countries, it remained a serious source of illness and death. In 1992, almost 1 billion people had no access to safe water supplies, and perhaps twice that number lacked adequate sanitation. Waterborne infections such as cholera, dysentery, poliomyelitis, schistosomiasis, and typhoid were primary causes of infant mortality and a significant contributor to death among adults.[18] Paradoxically, construction of sewers without complete treatment facilities in places such as the Ganges River Basin might add to biological demands on the river, because most human waste, although a terrible threat to health, does not now reach the river.[19]

Other water pollution problems include inflows of toxic chemicals, fertilizers, heavy metals, and heated water from power stations, and the results of loads of these substances, such as acidification, eutrophication, and oxygen depletion. Improvements have been made in some river basins, particularly where a single political entity controls the watershed. Control of discharges to the Thames River, for example, brought water quality from an appalling state in the 1950s, with near 0 percent dissolved oxygen, to relative clarity in the 1980s. In 1974 the first salmon in 140 years were observed, and nearly 100 fish species have returned to the river.[20] Other rivers in the industrial north have not fared so well; on the Rhine, divided among a number of jurisdictions, the work of cleanup has lagged.

As an example of the danger of pollution to freshwater resources, Lake Baikal in Siberia deserves mention. The oldest, clearest, and deepest fresh water lake in the world, it contains about one-fifth of all fresh water on the Earth.[21] During 25 million years, the ecosystem of the lake evolved in relative isolation, so that of the species that occur there, 84 percent are found nowhere else. Amphipods and other tiny invertebrates consume suspended organic matter and keep the lake so clear that objects up to 40 m (130 ft) deep may be seen from the surface. Historically, it was so clean that its waters could be drunk safely without treatment. A Siberian folk song calls it "Sacred Baikal, the glorious Sea."[22] In 1958, cellulose plants were built in the Baikal basin in spite of warnings by scientists that effluents would damage the lake's ecology. The pollution of Baikal, the destruction of its unique aquatic life, and the felling of forests with attendant erosion, became issues in the Soviet Union at least as great as the proposed damming of the Grand Canyon in the United States at the same time.[23] Writers, artists, and film-makers spoke out. The government responded with protective laws, but nothing effective was done and the pollution continued. Legislation and decrees are never enough to save the environment, as people in many nations have discovered. As one Russian put it, "Paper can tolerate anything."[24]

The level of another great body of water in the former Soviet Union, the Aral Sea, which has no outlet, began to drop when its feeder rivers, the Amu Darya and Syr Darya, were dammed and the water diverted primarily to irrigate cotton in Uzbekistan. Its salinity increased, its fish died in great numbers, and fishing boats were stranded in a wind-whipped salt desert. A vast plan to divert water into its basin from Siberian rivers was shelved during the collapse of the Soviet Union. More recently, a plan has been implemented to isolate and restore a portion of the lake.

The building of large dams, like the High Dam at Aswan described Chapter 7, continued worldwide, as nations considered them to be matters of national pride. By 1988, more than 36,000 dams more than 15 m (45 ft) in height had been built.[25] In 1950, North America had almost two-thirds of the world's reservoir capacity; by 1985, the proportion had dropped to one-third.[26] The effects of these structures include the flooding of ecosystems in the reservoir areas, loss of habitat and therefore of biodiversity, altering flow, increasing evaporation, leakage to groundwater, conversion of land for irrigated crops, and provision of electric power to cities and industrial centers. Before dams were authorized, potential negative effects sometimes escaped serious consideration, since those who could commission a study were already committed to the project.[27] Brazil's Amazon Basin has several large dams that have killed vast areas of species-rich rainforest, and many more have been planned. Virtually all India's major rivers are dammed or have projects under way, although many of the latter are opposed by vocal grassroots campaigns. Reservoirs displace large numbers of people, agricultural land is lost, and by 1983, more than 16,000 sq km (6,200 sq mi) of forest in India had been submerged. In 1992, China's Congress approved the construction of the Three Gorges hydroelectric project on the Yangtze River, which would be the world's largest, generating 40 percent more electricity than the world's largest dam at present and storing 39.3 billion cu m (51 cu yd) of water.[28] It would displace more than 1 million people, endanger several species of mammals and fishes, and destroy some of China's finest scenery. In addition to flooding by reservoirs, the effects of large dams on ecosystems include fragmentation of the remaining habitat, interruption of migratory pathways, and replacement of riverine forests by reservoirs with shifting shorelines that do not encourage regrowth of vegetation.

Technology for the exploitation of fisheries was transformed in the years after the Second World War from a labor-intensive form of hunting into a mechanized and electronically sophisticated operation using sonar and satellite-assisted systems. Huge factory ships capable of processing the catch at sea, operated by crews of up to 650, were accompanied by trawlers outfitted with the most advanced technologies for finding and capturing fish. These included drift nets kilometers in length that swept great volumes of water clean of organisms beyond a certain size, including mammals such as dolphins. Several maritime nations, notably China, the USSR, Japan, Peru, the United States, Chile, and Norway, made large investments in advanced fleets. The world's total fish catch rose from 19 million metric tons in 1948 to over 60 million in 1970 and 100 million in 1989. Since then it has declined due to depletion of fish populations. More effort had to be expended for a disappointing return, and the world's fishing fleets were losing money, although government subsidies made up part of the loss. Fishing fleets plied distant seas including the Antarctic. By 1994 thirteen of the world's seventeen major oceanic fisheries were overfished.[29] The more destructive of the drift nets were banned by international agreements, although they were difficult to enforce. The potential sustainable yield of marine fish was estimated at from 62 to 87 million metric tons, a level exceeded from the 1980s onward. The sustainable level drops when it is exceeded, and more intensive fishing will not continue to increase yields.

An example of how insensitivity to natural systems can damage an ecosystem and destroy a major economic activity can be seen in the collapse of the Peruvian anchovy fishery. Beginning in the 1950s, Peru began large-scale exploitation of anchovies exported as fishmeal. The catch, less than 87,000 metric tons in 1950, reached 12.4 million tons in 1970 in spite of biologists' warnings that the maximum sustainable yield was 9.5 million tons. In 1972, an El Niño incident began, and a sharp reduction occurred in phytoplankton growth, the base of the food chain that supports the anchovy. The catch dropped disastrously, averaging

1.2 million tons per year between 1977 and 1987. "Today, the misconception persists that El Niño was responsible for the demise of Peru's anchovy fishery. Most research, however, supports the idea that although El Niño contributed to the collapse, it was unrestricted fishing that placed the resource in jeopardy."[30] Similar rapid declines occurred in related fisheries such as California sardines and Atlantic herring. The warning of the Peruvian experience was not adequately heeded; the king crab industry of the Bering Sea collapsed in the 1980s, and the cod fishery of the North Atlantic in the 1990s, both due to overfishing.

Most fishing is done in the nutrient-rich waters of continental shelves or areas of oceanic upwelling, which also tend to be near landmasses and thus vulnerable to pollution. Tidelands, including mangrove forests, provide food and spawning grounds for many marine species, but are being destroyed by coastal modification. More than half of all tropical mangrove forests disappeared between 1950 and 1990. Ecosystems are not neatly bounded by coastlines; interactions constantly occur between organisms and cycles in the sea and land, and humans who treat them as separate entities do so at their peril.

Nuclear technology may serve as a major example of a human activity affecting the biosphere. From its invention in the Second World War through the rest of the century, it had two major aspects: weaponry and power generation. These are discussed below in this chapter. The spread of radioactive isotopes through the atmosphere and the biosphere had an analogue in the dissemination of pesticides.[31] Rachel Carson's *Silent Spring*, published in 1962, marked the emergence of environmentalism into public consciousness. The book warned of the dangers to human and other life from the massive spreading of long-lasting pesticides into the environment. Birds, especially insectivorous species, were particularly vulnerable to these chemicals, which killed them directly or, acting as endocrine disrupters, interfered with their reproduction. The title referred to the fact that if birdsong were to disappear, spring would be silent indeed. The author, a biologist, had taught at the University of Maryland, worked for the US Fish and Wildlife Service, and written popular books on oceanic life.[32] The scientific argument of *Silent Spring* did not consist only in pointing out the abandon with which huge amounts of biocides were being spread across the land, sea, and air. Carson used ecological data and theory to show the particular danger posed by substances that, as they pass through food chains, accumulate in the tissue of plants and then of animals, especially those further along on the chains, such as raptorial birds. Humans, especially if they drink milk and eat meat, are high on the food chain and therefore concentrate the chemicals in their bodies. Unfortunately, as she observed, the insects that were the targets of the poisons had rapid rates of reproduction and those with genetic resistance survived and repopulated their niches, assuring that subsequent applications of biocides would be less effective against them. One of the most insidious chemicals that Carson warned against was DDT (dichloro-diphenyl-trichloroethane), which was being used not only on agricultural crops, but also in forests against leaf-eating insects[33] and in water and in cities to combat mosquitoes and flies. By the 1960s, it had been detected in mothers' milk and in the fat of penguins in the Antarctic, demonstrating its spread through the world environment. Its interference with the assimilation of calcium in egg production made birds such as the peregrine falcon and osprey endangered species, and pelican rookeries in California were not producing offspring. Carson compared the indiscriminate application of pesticides to the spread of radioactive contamination.[34]

Chemical and agricultural firms and governmental agencies in the United States and Europe attacked the book. One company even tried to stop its publication.[35] The popular press carried both appreciative reviews and intemperate attacks; *Time* magazine called Carson's argument "unfair, one-sided, and hysterically overemphatic."[36] Later its editors would honor her as one of the most important scientists of the twentieth century. The opponents raised the specter of uncontrolled insect outbreaks destroying the world's agricultural capacity if pesticides were abandoned, ignoring her clear statement that she favored use of biological and ecological controls when possible, and the careful use of biodegradable chemicals when necessary.[37]

The response to *Silent Spring* was mostly positive, illustrating how knowledge of ecological science can influence public policy. President Lyndon B. Johnson appointed a Science Advisory Committee on Pesticides, and subsequently DDT and other persistent pesticides were banned in the US. Other industrialized nations followed. Ironically, chemical firms continued to manufacture the poison and export it to countries where it was still legal.

"Ecology," although coined in the nineteenth century, became a word widely recognized by the public in the US and Europe only in the 1960s.[38] The level of concern about problems of the environment rose sharply after the middle decades of the twentieth century. Ecology emerged into wide public consciousness, and "conservationism," which regarded the sustainable use of natural resources as the basic issue, gave way to "environmentalism," which recognized a growing number of worldwide issues.[39] Knowledge and concern increased about issues such as waste disposal and pollution across national boundaries including radioactive fallout, the effects of persistent pesticides, acid precipitation, accumulation of greenhouse gases and their possible effects on global temperatures, the weakening of the ozone layer and increasing ultraviolet exposure. The decline in biodiversity, with the accelerating extinction of species, especially those in the world's rainforests that are rapidly being destroyed, received unprecedented attention from greater numbers of people. Many people saw these environmental changes as threats to the beauty and usefulness of the natural world around them, to their own health, and to their ability to continue the ways of life that supported them. Just as conservation had a public dimension in earlier times, environmentalism became a popular movement in the US, Europe, and to some extent in other parts of the world. It is not that people in the developing world loved nature less, but that economic deprivation was for them a deciding issue, and given the economic, educational, and political facts in many countries, they saw less opportunity as a public to affect decisions of governments and corporations.

An increasing number of local and national environmental efforts, including the creation of governmental environmental agencies in most nations, met with attendant successes and failures, and nongovernmental organizations with environmental concerns proliferated. Numerous protest movements as disparate as Greenpeace, India's Chipko, and America's Earth First! resisted instances of commercial exploitation that they saw as detrimental to the environment. Unfortunately, international environmental crime involving trade in illegal animals and plants and their products, poaching, and timber theft, also became a force to be reckoned with.

The emergence of environmentalism into the political sphere brought increased importance to organized groups of environmentalists and their opponents. "Green" parties emerged in Germany and several other European nations with a program emphasizing environmental values, anti-nuclear activism, economic rights for workers, and participatory democracy. Their success in the polls was moderate, generally under 10 percent, but the German party, large enough to wield a critical margin between left and right, managed to participate in a coalition government. Environmentalism also stimulated a thoughtful and increasingly sophisticated ethical analysis of the issues underlying decisions and actions, from the personal to the international scale. Religious people reexamined their traditions to find bases for environmental concerns. These recent tendencies in environmental thinking merit careful study, but also the sobering recognition that scientific, political, philosophical, and religious thought concerning treatment of the natural world does not always determine human actions. And it is human actions that increasingly determine the course of change in the community of life.

Bali: a green revolution?

The rapid ringing of the gongs of gamelan music thickens the air and gives another meaning to "heavy metal." White-masked angels advance in ordered rows along the pavement to confront the witch Rangda, with her pendulous breasts and prominent tusks. Rangda's chaotic minions, bulbous of eye and grinning gloatingly, weave the dance around her. Then Barong, the huge, fiercely friendly apotheosis of animals, charges from behind to defend the people and their crops. The battle, led by the quickening beat of drums, is loud, energetic, and colorful, but there is no victory. The two sides exhaust their magical powers upon each other, but neither wins.[40] Those who attack Rangda find their own knives turning against themselves.

Over the centuries the Balinese people created a rice agriculture that appeared sustainable, along with a worldview and associated practices that provided balanced relationship to the ecosystem. The only Hindu island in predominantly Muslim Indonesia, Bali is located off the eastern tip of Java, 8° south of the equator. It has an area of 5,600 sq km (2,175 sq mi), 1.5 times the size of Long Island, and in 2000 the population was 3 million. It is a volcanic island; the highest peak, Gunung Agung, rises to 3,142 m (10,308 ft). The soil has good texture, and is renewed by ash from eruptions. Climate is dominated by a southerly monsoon, with rains from October to April, temperatures are pleasantly warm all year, and humidity is high. The original vegetation was rainforest with some areas of dry tropical forest. Fragments of the ancient forests survive in the west, where a national park has been created. Near the artisan town of Ubud, there are two sacred groves, Sangeh and Padangtegal, containing huge ancient trees. They are commonly called "monkey forests" because they are inhabited by protected troops of long-tailed rhesus macaques that are visited and fed by local people and tourists.[41] Three species of deer live on Bali. But biodiversity on the island generally has declined. The elephant is extinct, the last Balinese tiger was shot in 1937, and the exquisite Bali mynah is endangered.[42] Banteng, the wild relatives of the domestic Bali cattle, are a vulnerable species.

Much of Bali is occupied by wet rice fields, and in the hills, terracing gives the land a sculptured look. Rice culture reached Bali in the Bronze Age, after 300 BC. The influence of Indianized rulers on Java came to Bali by the eighth century AD. At this time the traditional agricultural system was formed, combining sophisticated irrigation technology with a religion embracing elements of animism, Hinduism, and Buddhism.

Rice is the major export and staple food, the basic ingredient of every meal,[43] so important that people use the word *nasi* [rice] to mean "food."[44] Traditional strains come in three colors: white, red, and black. The food grass is grown in fields that occupy the lowlands and rise up hillsides on terraces that require the continuing labor of farmers.[45] Wet rice fields must be irrigated. High on the island rivers flow in deep canyons, so at many places water is diverted into an aqueduct, often through a tunnel to an outflow at the head of a series of terraces. After use, water goes back to the river, or by canals to fields downslope. The result is an engineering marvel: a sculptured landscape and an "artificial ecosystem that has operated for centuries."[46]

Who assured this smooth functioning? Since the tenth century, decisions on irrigation, planting, harvest, and labor have been the responsibility of *subaks*. A *subak* consists of all farmers who receive water from the same outlet.[47] Cooperation at *subak* level is essential, but does not end there. Representatives from *subaks* in the same river system meet regularly at district water temples to discuss needs and determine schedules. These temples are part of a hierarchy of water temples corresponding to the complex irrigation system, so the structure of the watershed is paralleled in a functioning religious structure.[48] Each farmer has a shrine where water enters his fields. Offerings are made there to the rice goddess and other deities. Above that, the *subak* has a temple where members make offerings. A larger Ulun Swi temple stands near a canal that feeds several *subaks*. An irrigation district has its Masceti temple. There are temples at the headwaters, lakes in the volcanic calderas which, lacking surface outlets, seep underground and feed the rivers. The Temple of the Crater Lake is most sacred, since the Balinese hold that its waters feed all other lakes and rivers on the island.[49] A selection of gods is worshipped in each temple. Larger temples offer to nature gods, such as the Earth Mother. Among the most important ceremonies is the sharing of holy water, which is collected from springs, lakes, and other sources, mixed, and used for ritual purification. The Balinese refer to their variety of Hinduism as the "Religion of Holy Water." Its sharing reflects the structure of the irrigation system. The social relationships embodied in the constellation of water temples sustain the terrace ecosystem and productive rice agriculture.[50]

The various stages of the agricultural year have rituals that go with them.[51] These are orchestrated by the *tika*, one of the world's most complex religious calendars, with weeks and months of various lengths running simultaneously in a pattern that has been compared to the interlocking rhythms of gamelan music.[52] Decisions on timing take account of the availability of water and the need to allow a fallow period so that fields can dry out and the numbers of pests will decline. In case of disagreements on water allocation between *subaks*, if not solved by discussion at local temples, the high priest of Crater Lake Temple may be invited to mediate.

The agricultural landscape operated as an artificial ecosystem with many characteristics of a natural ecosystem. It had biodiversity, although much less of it than in the tropical forest ecosystem it replaced; rice paddies were home to frogs, fish, and eels that could be caught, and with dragonflies helped keep insect numbers down. Weeds were picked and cooked as table greens. Nitrogen-fixing cyanobacteria in the water aided fertility. Rice straw was left in the fields to decompose and provide nutrients, or burned to discourage pests. Ducks consumed weeds, snails, and insects and provided fertilizer, but were controlled by duckherds to prevent them from eating rice plants. Farmers simulated natural cycles, flooding and draining rice paddies at the same time over a large district. The system was sustainable; there was no significant load of erosional materials in the runoff, little development of gullies, and no decline of fertility. Historically, food supplies were adequate; there were few crop failures over a thousand years. Long experience of trial and error was preserved in rituals and sacred calendar. As Clifford Geertz put it, "A complex ecological order was both reflected in and shaped by an equally complex ritual order, which at once grew out of it and was imposed upon it."[53]

Balinese philosophy is based on maintaining balance, and seems suited to a prudent ecological lifestyle. Many rituals express rapport with fellow beings.[54] There are celebrations for animals, plants, and rocks on auspicious days. Sacred rice plants are dressed and given offerings. In the Balinese view, the forces of the universe are counterpoised.[55] The desire is not to achieve the triumph of one over another, but to placate both and restore balance.[56] This is the theme of the Barong-Rangda dance. A similar concept is embodied in the black-and-white checkered cloth seen everywhere on images and in costumes. The colors stand for opposed forces, woven together as they are in nature, neither dominant. Offerings are made daily to keep spirits of every kind well fed.

Early rice culture was not controlled by government. Islam had entered Indonesia at least as early as the 1300s, and by the 1500s, the religion swept over and transformed much of Indonesia, but left Bali, with its isolation and cultural resistance, relatively untouched.[57] The island was divided into nine kingdoms (later eight), whose boundaries did not correspond to irrigation districts, and governments did nothing more than collect a tax.[58]

A Dutch fleet called at Bali in 1597. For decades the Dutch and English competed; Sir Stamford Raffles visited Bali in 1814, and the Dutch occupied the north coast in 1846. The invasion of south Bali did not begin until the twentieth century. It was marked by ritual suicide of the Balinese royal houses, who were slaughtered as, wielding ineffective ceremonial weapons, they deliberately attacked the well-armed Dutch. Massacres in 1906 and 1908 shocked many Dutch people and governments around the world. Perhaps repenting, colonial administrators afterwards patronized Balinese culture, excluding missionaries and tourists while making a handsome profit on opium and rice. They failed to understand the rice culture system because they confined their attention to irrigation works and regarded the water temples as a primitive "rice cult." The Dutch assumed that Balinese kingdoms historically had controlled agriculture,[59] and tried to establish colonial government as successor to the kingdoms, collecting taxes and supervising production. They never got a handle on rice culture, however, and directed their efforts to building permanent weirs, lining canals, and other improvements that did not deeply affect the traditional system. For political purposes, they reconstituted the traditional kingdoms in 1938, but less than four years later the Japanese seized the archipelago.

After the Second World War, Indonesia won independence. The first president, Sukarno, founded Udayana University in Bali's capital, Denpasar, but many of his other actions were

Figure 8.1 A Balinese farmer plowing in a rice paddy with banteng, the native cattle. Irrigated rice paddies such as this form a productive, sustainable system. The volcano Gunung Agung, considered sacred by the people of Bali, rises in the background. Photograph taken in 1994.

resented. A most disastrous event was the 1963 eruption of Gunung Agung just as Sukarno was using a major Balinese festival to launch international tourism and improve his image. Many interpreted the catastrophe as divine disapproval of presidential hubris.[60] In the country as a whole, runaway inflation and the failure of programs to raise rice production were major issues. An attempted coup in 1965 was followed by the fall of Sukarno and the slaughter of thousands of suspected Communists. Suharto, the new president, claimed the title "Father of Development" by stimulating economic growth through oil exports, encouraging tourism (not least on Bali), and signing up Indonesia for the Green Revolution.

In the 1950s and 1960s, Indonesia had an exploding population and a crisis in food production, with the lowest rate of growth in rice yields of any major producer.[61] Rice imports reached 1.7 million tons in 1964.[62] Bali was no exception; in 1967, the total fertility rate was 5.9, above the national average, and in spite of being one of the most intensive rice production areas, about 10,000 tons had to be imported to the island annually.[63] Nationally there was an effort to expand the rice cultivation area, but on Bali virtually all arable land was already producing. A two-pronged attack was launched: a family planning program to slow population growth, and the so-called Green Revolution to increase yields. The first effort was effective; the decline in total fertility in Bali, a 46 percent drop in fifteen years to 3.5, is one of the success stories of the world movement to limit population, and was combined with a decline in infant mortality and a lengthening of life expectancy, indicating improved health.[64] The strength of village organization in Bali contributed to the success.[65] Population remained high, however, and the government encouraged the Balinese to join a resettlement program to less populated islands.

The Green Revolution began with breeding studies on wheat and maize in Mexico in the 1940s and was continued by the International Maize and Wheat Improvement Center, established in 1966 by the Rockefeller Foundation in cooperation with the Mexican government. Its director, Dr. Norman E. Borlaug, received the Nobel Peace Prize in 1970 for his role in the program. A similar project for rice, the International Rice Research Institute (IRRI) was begun in the Philippines with Ford and Rockefeller foundation support. It introduced new high-yielding strains of rice with short periods of maturation, allowing more crops each year, to the rice-growing nations of tropical Asia.[66] These advantages were achieved only in conjunction with the application of chemical fertilizers and insecticides in significant amounts. The Indonesian government subsidized the chemicals, and in 1967 hired a Swiss corporation to develop a distribution system. A government agency, BIMAS (meaning "Mass Guidance"), was given the task of assuring that farmers adopted the new varieties and used them as directed by agricultural technologists. Banks offered credit for purchase of seeds, agrochemicals, and farm machinery. Initial success was spectacular, with annual production increases between 5 and 10 percent. Bali's acceptance of the Green Revolution was high.[67] In south Bali, a new rice strain was planted on 48 percent of terraces in 1974; three years later it was 70 percent. In 1979, the Bali irrigation plan, devised with help from the Asian Development Bank, envisioned complete restructuring of the island's irrigation systems and abandonment of the traditional calendar in favor of short rotation periods and virtually continuous cropping. Experts dismissed the existing system as a "rice cult." The temples lost control, and *subaks* were subjected to political and economic pressure. By 1985 irrigation scheduling was in chaos and water shortages became common in the dry season.

Other problems appeared. Genetic uniformity made crops vulnerable to disease and insects.[68] The rice strain IR-8 proved susceptible to the brown planthopper.[69] Once a minor insect, it proliferated in the monoculture in spite of massive applications of insecticide. A

Figure 8.2 The Barong, the huge, fiercely friendly apotheosis of animals, defender against evil, is represented by dancers in the town of Jimbaran on the Indonesian island of Bali. Photograph taken in 1994.

new super strain IR-26, resistant to the insect, was substituted.[70] The emergence of a plant-hopper biotype to which the new strain was not resistant forced another switch to IR-36. Unfortunately, the latter was sensitive to tungro virus, so it was replaced with PB-50, a strain that in due course succumbed to a soil fungus which in 1982 destroyed 6,000 ha (15,000 acres) of planted rice.[71] By doing away with fallow periods and other ecological measures that in traditional agriculture had controlled natural enemies, the Green Revolution enabled them to reproduce to an extent that overwhelmed even modern insecticides and fungicides. These, along with chemical fertilizers, were applied at a level that polluted the water system. Although the increase in production resumed, its rate slowed. Bali again exported rice beginning in the mid-1970s, but the problems mentioned above and the social changes accompanying the Green Revolution's conquest of Bali caused some to question the wisdom of abandoning traditional agriculture so completely.

Among these were anthropologist Stephen Lansing and systems ecologist James Kremer, who in 1987 developed a computer model of two river systems in Bali with the aid of Apple programmers Tyde Richards and Alan Peterson.[72] They compared the effectiveness of a number of systems of coordination of irrigation patterns, from one where decisions were made independently by each *subak* to one more like the typical Green Revolution arrangement, where the whole watershed had an identical cropping pattern. They found that traditional timing by water temples that coordinated *subaks* would produce the best yields, with the fewest pest outbreaks and the most efficient provision of water.

Lansing and Kremer presented the results to the Asian Development Bank, advising that restoration of the traditional water temple timing of irrigation and planting be considered.

They did not advocate a return to older native strains of rice. After first rebuffing it, the ADB gave sympathetic audience to the study and agreed to try a modified traditional system. "The water temple system in Bali, whose colorful ceremonies were never abandoned by farmers although the planting cycles were, is being reestablished."[73] In certain environmental niches farmers had continued to grow traditional varieties. Whether modifications made necessary by the shorter growing periods of the new strains can be harmonized with the old calendars and festivals remains to be seen. The response of pests to new varieties in a pattern resembling but not identical to the traditional one must be observed. However, the advantage of studying traditional agriculture in the search for sustainability has been noted beyond Bali.

In parts of Asia the Green Revolution had enormous effect; India, for example, has not had a famine since 1965–6. In Africa, however, where a Green Revolution seemed most desirable, financial resources were lacking, social and political problems interfered, and the varied growing conditions of the continent with poor soils and recurrent drought were not conducive to the high-yield genetic strains.[74] Crises of famine continued, especially in Ethiopia, Somalia, and Sudan.

In 1994 an official of the UN Food and Agriculture Organization favored using the revived Balinese water temple system as a model for programs "in other cultures to promote sustainable agricultural systems."[75] If this means study of systems of traditional agriculture that can aid sustainability, it might succeed. But if it means an attempt to copy the Balinese system, it is plainly impossible. No one else is likely to worship the rice goddess, exchange holy water, or hold Barong-Rangda ceremonies. Even resettlement of Balinese communities in Indonesia has not resulted in new productive centers of rice cultivation, due to differences in local ecosystems.[76]

The Balinese ritual water temple system is an indispensable component of an intricately engineered system of sustainable agriculture. But the religious beliefs by themselves did not produce the irrigation system, nor did the irrigation system produce the religion. They are parts of a whole. The Balinese set of cultural attitudes and religious rituals evolved together with sustainable rice agriculture as part of the same ecological process. The economic and ceremonial aspects are inseparable. To learn from the Balinese case of sustainability in an applicable sense, therefore, is simply to realize that other cultures might simultaneously create ecologically sustainable economic systems with attitudes and public rituals and decision-making processes that express and support them.

Willamette National Forest: now that the big trees are down[77]

The most trenchant comments on land management are made by the land itself. Standing on Hills Peak, looking over the valley of the Middle Fork of the Willamette[78] River in the Oregon Cascades, I saw a changed landscape. In 1995, the scene contrasted to the one I knew when I lived on this mountain for a summer as a lookout for the US Forest Service (USFS), 45 years earlier.[79] Then the slopes were covered by a green robe of ancient trees. The nearest town was 67 km (42 mi) away over a long trail and a rutted dirt road, virtually all through old growth. Today the main road is paved and mountains on every side bear the scars of a labyrinth of logging roads. The forest is a patchwork of clearcut timber sale units. Some patches are bare red-brown soil, recently scraped clear. Others are covered with low brush. Still others have bright green young trees, crowding each other for light. Sections remain of darker old growth conifers, dotted with dead snags whose hollows shelter birds and mammals. But few stands are between forty-five and 200 years old. That is because clearcutting began in the mid-1950s. Since then, this forest has seen the accommodation of the US Forest Service's principles of management to the demands of the market economy.

The Willamette National Forest embraces 7,280 sq km (2,800 sq mi) along the central Cascades Range; it is 37 percent larger than the state of Delaware. Long the most productive forest in the Pacific Northwest, it has been called the "flagship" of the national forest system. During a number of years, the timber sold to companies and removed by them from the Willamette constituted 10 percent of the annual cut from all US national forests. Its history is a prime example of the problems of federal forest management.[80] But the ecological impact is even greater. In order fully to understand the significance of events in this forest, it would be necessary to see them in the context of the biotic community of the Douglas fir and western hemlock forest ecosystem and related ecosystems, the range of the northern spotted owl,[81] and the biosphere.

The USFS was created in 1905 in the Department of Agriculture.[82] Its first chief, Gifford Pinchot, believed use and conservation were best managed on public lands by a national agency applying scientific forestry. He said, "In the administration of Forest Reserves ... all land is to be devoted to the most productive use for the permanent good of the whole people."[83] If conflict arose between various claims for use, he thought it should be resolved by the principle, mentioned above, of "the greatest good of the greatest number in the long run."[84] This policy of "multiple use" implied that watershed, logging, stock grazing, mining, hunting, fishing, and outdoor recreation would be managed to ensure their continuation.[85]

Resources such as timber are renewable. The ecosystem, as long as enough of it remains, restores what has been taken from it. This principle resulted in the theory of "sustained yield:" that it is possible to cut from a forest annually an amount of timber equal to the wood added by tree growth, minus that destroyed by fire, insects, etc. If that amount is exceeded for long, the forest will be degraded.

Pinchot and his successors were optimistic about applying these principles to American forests. But timber interests would not accept equal treatment with other forest users. From the Second World War onward, industrial forestry dominated the landscape of the national forests, as trees on large sections were sold and cut. The USFS, following congressional mandates, considered the timber industry its most important user. As Supervisor David Gibney of the Willamette remarked in 1965, "Timber is our meat and potatoes – recreation our dessert!"[86]

Most forest managers and timber companies prefer clearcutting as the method of "harvesting" primarily because it takes less labor and is cheaper than selective logging, a method that was used briefly in the Willamette National Forest during the 1930s under the administration of Regional Forester C.J. Buck.[87] After 1940, clearcutting became virtually the only method on the Willamette. For sustained yield, so-called "harvest units" would be designated in a mosaic pattern, only enough of them cut so that the same number could be cut every year. The length of the cutting cycle in Douglas fir is 100 years. In a forest so managed, there would be no old growth because attaining its characteristic form takes at least 200 years. There would be no giant trees 800 or 900 years old. Species dependent on old growth would become extinct, reducing biodiversity. Other values of old growth, such as watershed protection, supply of organic material, and the awe-inspiring size and beauty of the ancient forest, would be gone.[88]

Before 1920, timber cut in this national forest averaged 10 million board feet (abbreviated mmbf) or 23,300 cu m per year.[89] Timber companies opposed USFS sales when competition might lower wood prices. But supply on private forestland became depleted, and

Figure 8.3 View from Hills Peak Lookout eastward toward Diamond Peak in the Willamette National Forest, Oregon, in 1950, when the forest was almost continuous in this part of the Cascade Mountains.

industry increasingly looked to public lands for high-quality timber. In the late 1920s the cut rose to 50 mmbf (117,000 cu m). Although the 1930s depression reduced it to 30 mmbf (70,000 cu m), the Second World War increased demand and Congress gave priority to meeting it. The cut reached 144 mmbf (336,000 cu m) in 1944; in 1948 it was 207 mmbf (483,000 cu m), as postwar construction demanded lumber. Technology burgeoned, with chain saws, heavy machinery, and huge logging trucks on a growing system of forest roads.

Supervisor John Bruckart, in a 1949 article on the Willamette entitled "Taming a Wild Forest,"[90] stated that under sustained yield, the allowable annual cut would be 323 mmbf (754,000 cu m). This estimate was not conservative; Bruckart established clearcutting as the dominant method of "harvesting," and had reason to choose the highest figure he could defend. The actual cut surpassed it in 1952. It was twice as high in 1962, and in 1973 reached a peak of 945 mmbf (2,205,000 cu m). Timber companies active on the Willamette during 1950–5 included large concerns like Weyerhaeuser in the northern half of the forest and Westfir and Pope & Talbot in the south. Medium-sized companies and many smaller outfits were also involved. Small companies often felt shut out of deals between large firms and the USFS, whereas large companies belonged to associations that wielded political power in Oregon and nationally. Pro-industry representatives added mandatory high timber harvest targets to appropriations bills in the 1980s, forcing the USFS to maintain the high level of sales even though it was losing money on them. The average cut during 1962–89 was 708 mmbf (1,652,000 cu m). The history of the Willamette does not show a pattern of sustained yield. It shows a process of exploitation, following the demands of the market economy, far exceeding the regeneration rate of the forest. It seems headed for a crisis when

Figure 8.4 A view from Hills Peak Lookout toward Diamond Peak in 1995, in exactly the same direction as Figure 8.3, 45 years later. Clearcut sectors and logging roads are evident.

all old growth will be gone, and there will not be enough 100-year-old timber sale units to supply the supposed allowable annual cut.

After the war, the number of Americans visiting national forests for recreation increased tremendously. Many objected to the impact of logging – ugly clearcut patches and roads bulldozed across hillsides – so the USFS tried to put timber sale units where the public would not see them. Maps of "viewsheds," showing which slopes were visible from main roads, were used. Strips of uncut timber were left along highways. Today only observation from the air reveals the extent of forest removal, as I found when in preparation for writing this section. I flew in a small aircraft over the Rigdon District of the Willamette National Forest, where I had worked as a forest guard, ably piloted by Jane Rosevelt, who cooperated with LightHawk, an educational environmental flying service. I looked down over a forest checkerboarded by clearcuts. The trees around the site of my old lookout station, Hills Peak, had been cut, even though they were alpine fir and mountain hemlock, slow-growing species of little commercial value, whereas if they had been spared they would have continued to protect the watershed.

As early as the 1920s, pro-conservation groups urged preservation of wilderness areas.[91] USFS managers had little trouble designating wilderness amid tundra and timberline forests, but seldom included old growth. "Wilderness on the rocks," conservationists called it. Among those on the Willamette was the Three Sisters Primitive Area, established in 1937. In the following year, Bob Marshall, USFS recreational chief, visited the area and recommended the addition of forested land around French Pete Creek, 21,600 ha (53,380 acres) of old growth and good-quality second growth.[92] This was done, creating a policy crisis that revealed how paramount was the USFS commitment to timber production. Forest managers, regretting the decision to include French Pete in wilderness, announced their intention to open it to timber sales. The secretary of agriculture excluded French Pete from wilderness in 1957. Reaction was heated.[93] Wilderness advocates saw that administrative designation could be removed easily. French Pete helped motivate environmentalists to urge Congress to pass the Wilderness Act of 1964. The law prohibits roads, timber cutting, and motorized equipment, but permits hunting, fishing, camping, and, unfortunately, livestock grazing. Wilderness visitors have increased sharply.

French Pete remained outside the 1964 boundaries, and the supervisor proceeded with plans for sales. After industry representatives and environmentalists presented their views, he announced his intention to begin road construction and allow clearcutting. In November 1969, environmentalists in Eugene held a large peaceful rally outside the supervisor's office – the first public demonstration against the USFS in the Northwest, but certainly not the last. In years following, "tree-sitters" occupied platforms erected in trees designated for "salvage," and loggers angry at reductions in timber sales blocked public roads with their huge trucks. Oregon representatives backed wilderness designation for French Pete, which passed in 1978. Wilderness areas are not closed to public use: visitors to the Three Sisters Wilderness Area, for example, rose from 64,000 in 1965 to 193,000 in 1971.[94] In the 1990s, Opal Creek, the last intact old growth streamshed in the Willamette, received congressional designation as wilderness, largely due to the work of George Atiyeh and the Friends of Opal Creek, who received widespread public support in Oregon.

Most old growth has been invaded by logging roads and clearcuts.[95] As old growth disappeared, it became evident that species dependent on it could become extinct. The Endangered Species Act of 1973 required identification and protection of declining populations of wildlife and their habitats. It defined an endangered species as "any species that is in danger of extinction throughout all or a significant portion of its range," and a threatened species as "any species which is likely to become an endangered species within the foreseeable future."[96] The law considered species populations, not ecosystems, although the habitat provision meant that an ecosystem would be protected if one of its species were endangered. For old growth ecosystems of the Pacific Northwest, one such indicator species is the northern spotted owl. A petition to list the owl as endangered was denied by the Fish and Wildlife Service in 1986.[97] Federal courts reversed that decision, and later enjoined the USFS from most timber sales within the owl's range.[98] Since the Willamette is entirely within the range, sales and cuts plummeted. In 1994 the cut was 123 mmbf (287,000 cu m), 14 percent of the 1988 figure, and sales were even lower.

The arson-caused Warner Creek fire burned 3,600 ha (9,000 acres) on the Willamette in October 1991, much of it spotted owl reserve.[99] The USFS proposed salvage in the burned area, meaning building roads and clearcutting 40 mmbf (93,000 cu m) of timber. Environmentalists objected that salvage could encourage arson as a means of circumventing forest management. The USFS made a sale, but activists set up a roadblock and stayed for almost a year. The USFS cancelled the sale, negotiated compensation for the company, and arrested the demonstrators.

The press oversimplified the complex issue as "owls v. jobs," but jobs also declined due to technology and because the industry exports whole logs instead of processing them in the US. Volume would have declined anyway, since remaining old growth was being logged and earlier clearcuts had not regrown to marketable size. The principle of sustained yield had been honored in the USFS's official dogma but violated in practice. The USFS had announced the principle of multiple use, but treated timber sales as a higher and better use than wildlife, recreation, or watershed protection. In a fundamental error, it had treated timber as a marketable resource without considering trees as parts of a community of life. A balance between human needs and a sustainable ecosystem might be worked out, but not if the demands of the timber industry were given priority. The agency was subject to extreme political pressure. Employees who urged a change in priorities were transferred or otherwise harassed.[100]

The forest ecosystem itself must be understood and respected before policies of multiple use and sustained yield can be applied. In recent years this realization gave rise to "ecosystem management," a principle officially recognized by the land use agencies of the US government during the Clinton administration.[101] USFS Chief Dale Robertson in 1992 announced the new policy, which he called an "ecological approach in future management. It means that we must blend the needs of people and environmental values in such a way that the National Forests and Grasslands represent diverse, healthy, productive, and sustainable ecosystems."[102] The goals include maintaining viable populations of native species in their natural habitats, protecting biodiversity, maintaining ecological cycles and processes, planning over long periods of time, and accommodating human use within these constraints. In regard to the last-named goal, "Humans [are] embedded in nature. People cannot be separated from nature. Humans are fundamental influences on ecological patterns and processes and are in turn affected by them."[103] But if people are part of ecosystems, depending on them for survival and in making humans the species they are, then the maintenance of living ecosystems must be the overarching goal of management. The urgent significance of this fact has scarcely been appreciated, much less carried out in the field.[104]

In 1993, President Clinton convened a conference in Portland, Oregon to address environmental and economic needs served by federal forests of the Pacific Northwest. He asked,

> How can we achieve a balanced ... policy that recognizes the importance of the forest and timber to the economy and jobs in this region, and how can we preserve our precious old growth forests, which are part of our natural heritage and that, once destroyed, can never be replaced?[105]

He appointed a team including technical experts led by Dr. Jack Ward Thomas,[106] a wildlife biologist and subsequently chief of the USFS, asking them "to assess not only effects on individual species ... but also the likelihood that the alternatives would provide for a functional and interconnected old growth forest ecosystem."[107] The team drafted a "Forest Plan for a Sustainable Economy and a Sustainable Environment," which created Late Successional[108] Reserves to safeguard habitat for old growth related species, embracing 3 million ha (7.4 million acres), or 30 percent of federal forest lands in the owl range. Much of the rest remained open to timber sales. The plan called for worker retraining and dropped tax subsidies for exporting logs. Environmental groups objected to cutting any old growth, while the timber industry complained that allowable cut would be too low. On the Willamette, the annual cut dropped to 136 mmbf (317,000 cu m), 80 percent below the 1980s. No one believed that the former high levels could be sustained for long; the only question was whether the last bit of profit would be extracted from the remaining old growth.

The timber interests were not defeated. Their allies in Congress attached the Salvage Logging Rider to the Rescissions Bill of 1995, exempting logging in the national forests from all conservation laws through 1996. When the bill was signed, timber companies pushed immediately for sales of old growth. Conservationists who tried to block them in court found judges unsympathetic. Much timber was sold under the law before it expired, and there have been repeated attempts in Congress to renew the provision or similar ones, using as justification the recurrence of heavy wildfire years in the US.

Old growth in the Willamette is fragmentary and impaired. Unfortunately, the flagship of the US National Forest System is typical. It would be encouraging to report that the national forests, managed for ninety years under principles of scientific forestry for sustained yield, are a model for the world; but to the contrary, forests have been cut at an unsustainable rate. Profits have accrued to corporations, not the federal treasury; more tax money has been spent on managing sales and building logging roads than the USFS has received from sales. It is possible for a carefully limited number of trees to be taken every year from a forest without impairing its ability for renewal. But enough old growth must be left untouched to serve as a reservoir for the interacting species and other components of the ecosystem. The forests of the world are in need of preservation, provident use, and restoration.[109] Instead, with the worldwide triumph of the market economy, they are being liquidated for short-term economic profit.

Bryansk: the aftermath of Chernobyl

As I sat in a colleague's kitchen in Bryansk, I looked through a stack of schoolchildren's paintings. The Bryansk Region is the section of the Russian Republic that received the highest level of radioactive fallout from the 1986 Chernobyl nuclear power plant accident. One of the paintings depicted two little hedgehogs in a forest, with suspiciously dark clouds overhead. The first hedgehog had picked mushrooms, a favorite activity of Russian children. The second hedgehog asked, "Zachem ty nesyosh' gribok? On zhe radioaktivnyi." ("Why are you picking mushrooms? They're radioactive.") The first replied, "Kushat' khochetsa." ("I want to eat.") Another drawing showed a girl with a basket crying beside a sign prohibiting entrance to a forest due to radiation. There was an imaginative painting representing mutated creatures including a dragonfly with two heads. Finally, a drawing by a seven-year-old girl showed an empty school playground, with a pony looking at it and saying, "Gdye dyeti?" ("Where are the children?") The implied answer was that they had been evacuated because their homes were too radioactive to live in.

The explosion of the reactor core at the Chernobyl nuclear power plant on April 26, 1986 occurred because operators making tests while shutting down the reactor for maintenance erred in shutting off safety mechanisms, one after another. Later blunders were made in attempts to cover up earlier ones. Thus human error was at fault. The explosion injected 50 tons of nuclear fuel into the atmosphere as dispersed particles, in addition to 70 tons of other fuel and 700 tons of radioactive graphite that settled nearer the site of the accident.

Officials did not immediately warn the local population or the world. The first announcement on Soviet television came two days later, twelve hours after elevated levels of radioactivity were detected in Sweden and Finland. Fallout polluted ecosystems and human food sources in large portions of Europe and the USSR, with measurable amounts throughout the Northern Hemisphere.[110] The authorities knew what had happened, however, and took various actions. Although winds first blew the plume of pollution westward over Europe, they shifted and carried dangerous clouds toward Moscow. Reports say military aircraft were ordered to seed them to precipitate radionuclides before they reached the capital. Whatever the cause, large amounts of material fell in the western part of the Bryansk Region around Novozybkov, 177 km (110 mi) northeast of Chernobyl, where soil contamination well above 40 curies per square km resulted.[111] Bryansk city, the regional capital located about halfway between Moscow and Kiev, recorded a relatively low level of fallout.

Evacuations began near Chernobyl within twelve hours of the accident. The number of people evacuated is unclear; the Soviet government in 1987 reported 90,000, but the number for Ukraine alone in 1994 was 130,000.[112] Authorities closed an area within a radius of 30 km (19 mi) of the plant.[113] The reactor was enclosed in a concrete "sarcophagus" which was never completely sealed: a smaller release of radioactivity continued, and new cracks later appeared in the concrete.

In the Bryansk Region, thousands were evacuated from villages with high contamination. Scenes of desolation were poignant; empty houses stood with open doors. A child's doll lay on the sill of a broken window. Many evacuees were resettled in other regions and provided with jobs and housing, but a considerable number of them considered the arrangements unsatisfactory. Children from radiation districts went to new schools only to find their classmates shunning them because they were afraid they might radiate on them, calling them "glowworms." Some families returned to reoccupy their homes. A typical comment was, "It's better for us to live in the radiation zone with reasonable living conditions." But they did not appreciate the extent of danger to themselves and their children. Village folk living near a Pioneer camp that was closed due to dangerous contamination took bricks and timber from the buildings and used them to add rooms to their houses. As months passed, illnesses and deaths resulting from exposure increased among those who had stayed in the radiation zone as well as those who returned.

The official number of deaths due to the accident is 31, all workers at the nuclear plant. But the real figure of those whose lives were shortened will never be known; it is in the thousands and increasing.[114] Incidences of thyroid cancer, leukemia, and other radiation-related illnesses among the exposed population are high. Children, since their bones and other organs are growing, are more liable to accumulate radionuclides and suffer their effects. A coterie of dedicated teachers working in the radiation district reports that children are more likely to appreciate the dangers of radiation than their parents, who want to continue living as they always have and are unwilling to make behavioral changes that might lessen exposure. Of course, children do not like to be told not to eat vegetables from their gardens, play in the forest, or fish or gather berries and mushrooms there. Avoiding some exposure is next to impossible; levels of radioactivity in milk fluctuate, rising sharply in summer when cattle graze in the fields. People must either stop their normal interactions with the ecosystems within which they live, or ingest radionuclides and accumulate exposure.

A problem of living in a fallout zone is extreme variability of levels of radioactivity over short distances. My friend, Dr. Ludmila S. Zhirina, a teacher of ecology and education at Bryansk University, started an environmental NGO named "Viola." She and her associates provided schoolchildren and teachers with radiation meters and encouraged them to make maps of villages, fields, and forests, showing localized readings. Once they found that a playground had a high reading: the school paved it with a shielding layer. However, they discovered that the readings changed over time, and not just because of nuclear decay. Many radioactive particles can and do move, blowing as dust and flowing in rainwater. To burn autumn leaves makes radioactive smoke that contaminates other places. Peat, common in the region, concentrates radioactivity and spreads it when used as fuel. Dr. Zhirina wrote and distributed a pamphlet telling schoolchildren and their families how to protect themselves from radioactivity.[115] But people who live in communities that will be heavily contaminated for the rest of their lives can only be "protected" in a relative sense; most simple measures that are possible will probably prove ineffective over a long period of time.

Virtually nothing can be done to protect the local biota. The effects of radioactive contamination on an entire ecosystem, with the interaction of various forms of damage, are not well understood. A survey of soils in the Bryansk Region showed high contamination of agricultural lands over an area of 720,200 ha (1,780,000 acres), or about 40 percent of the total.[116] In addition, about 415,400 ha (1,026,000 acres) or 35 percent of the forests were contaminated. The most important radioisotopes studied were cesium-137, which behaves chemically like potassium; and strontium-90, which resembles calcium. Living tissue readily absorbs both. Cesium-137, the most prevalent long-lived pollutant, has a half-life of thirty

Figure 8.5 A young student in Novozybkov, Russia, using a radiation meter, in one of the areas with high radioactivity from the Chernobyl nuclear power accident in 1986. Photograph taken by Dr Ludmila Zhirina in 1989.

years, which means that half the amount deposited in the Chernobyl fallout will remain in 2016, and one-quarter in 2046. It is not easily leached out by water, and persists in the upper layer of soil.[117] Strontium-90, with a half-life of 28.8 years, is more mobile, and dangerous to vertebrates because it collects in bones and may cause leukemia.

Other studies showed that radioisotopes were readily absorbed by plants including forest trees. Dr. Zhirina had begun dendroclimatological studies of forests in the region before the accident, and was able to make comparisons of the situation before and after. The effects varied with species and intensity of radiation, sometimes in surprising ways.[118] Conifers such as pines suffered more noticeably than deciduous trees, which shed some radioactivity with the annual leaf loss. The common Scots pine often died; in the most contaminated zones,[119] about 40 percent died within eight years. Many surviving pines showed yellowing, loss of needles and branches, and drying of the upper crown, making them susceptible to diseases and fire. Wood vessels showed abnormal growth patterns. Trees under forty years of age fared worse than older trees, and plantations were more vulnerable than natural stands. At the same time, in about 20 percent of the pines a marked increase occurred in the width of annual rings, unexpectedly. A possible reason is less competition from other trees that had died, or it might be an effect like the accelerated growth of cancer cells. A slowing of growth occurred in common oaks, especially in spring, when the buds were observed to open later than usual. Cambium cells (the cells that provide a tree's growth) decreased markedly in the year after the accident, recovering slowly, but in plots with high radiation showed progressive deterioration. Other scientists noted malformation in oak and maple leaves.[120] Zhirina noted that some annual plants grew abnormally, possibly from radiation-induced mutations. A study by geneticists in 2000 verified this observation, reporting an elevated incidence of mutations in the DNA of wheat grown in radioactively polluted areas near Chernobyl.[121]

Long before Chernobyl, ecologists knew that radiation accumulates in food chains. Plants are at the lower end of the chain; animals accumulate higher levels of radioactivity in their tissues, with highest doses occurring in top predators. This might result in local decline or extinctions of species such as foxes, ermines, hawks, and eagles. In the Bryansk Region, biologists measured cesium-137 concentrations in fish.[122] There is an increase of two to three times for every step up the food chain, so predatory fish such as pike and perch show more radioactivity per unit weight than bottom-feeders like roach and bream.

Radiation alters genes, producing random mutations. Most of these are disadvantageous, resulting in infertility, premature death of offspring, or gross abnormalities. Bryansk newspapers published photographs of malformed births among domestic animals. Common deformities among calves and foals included absence of the anal opening, of eyes, ears, ribs, hair, or up to three legs; misshapen skull, spine, legs, or internal organs; and presence of two heads. Such births were many times more common than before the accident, with a rise from 0.07 percent of total births in 1987 to 9.9 percent in 1989.[123] Similar effects occur in wild animals of virtually every species. Unfortunately they have also increased among humans; children born after the event suffer its consequences and will suffer them for generations to come.

The Bryansk Region offers one example of a problem of worldwide dimensions that will continue to affect the history of the community of life in future centuries. Chernobyl was by no means the only major injection of radioactive material into the environment. Since the first atomic test in 1945, the biosphere has been subjected to pollution by radioisotopes that have raised background radiation above naturally occurring levels. Historically, increasing radioactivity has had an as yet unmeasured effect on the functioning of ecosystems, which mostly evolved in the presence of low background levels. A significant change in the rate of genetic mutation will have unpredictable effects on the functioning of the ecosystem and on the humans who are part of it. Eventually there may be adaptations by plants and animals, and ecosystems also, to conditions of higher radioactivity, but just what those adaptations might be will not be known for decades. Available evidence suggests serious disruptions of the community of life.

Environmental problems are worldwide, and the Soviet Union is far from a unique example of a system that produced environmental destruction. But it is interesting to consider how it happened there. In the ideology that prevailed in the Soviet Union, human beings were considered to be wholly social creatures, human essence being determined by the system of social relations. This led to the conclusion that nature, external to culture, has no effect on human development. Economic and political considerations, therefore, always prevailed over environmental ones. After the mid-1960s, however, ecological problems were embraced as weapons of propaganda, providing evidence for the superiority of the "socialist" economic order, which was assumed to provide for the well-being of the people. The ecological crisis that alarmed people in the West was proclaimed an inevitable part of the general crisis of the capitalist system. The possibility of environmental crisis in the Soviet Union was rejected because of the rational character of the ostensibly socialist economy, allowing planners to foresee the results of industrial development and to prevent such crises. But ecologically, good intentions remained good intentions while economic and political priorities took the upper hand.

The issue of radioactive pollution had emerged in the 1940s and 1950s, as atmospheric testing of weapons by the US, the USSR, the UK, and later France and China produced fallout of radioisotopes around the world, causing concern about the effects of radiation on humans and lesser concern about effects on other organisms. An American bomb test in 1954 exposed 236 Marshall islanders and 23 Japanese fishermen on the boat *Lucky Dragon* to high levels of radiation, causing at least two deaths.[124] Other boats were contaminated. Radioactive fish were found in the Pacific, but the effects on marine ecosystems are little understood. Governments were secretive about nuclear information, especially when public knowledge might have produced political repercussions. Radiation damage to people and livestock in Nevada and Utah was hushed up for years.[125] However, the agencies with oversight were interested in discovering effects on local ecosystems near test sites and production facilities. In the United States, for example, the Atomic Energy Commission hired ecologists to study and report on these effects.[126] Various experiments were conducted, including the placement of radiation sources in forests, and putting domestic animals and plant materials within test sites. The toll among wildlife in test sites around the world, and the contamination of island, desert, and arctic ecosystems with nucleotides, was severe, although natural recovery was also noted. Mutations of genetic material certainly occurred. As an undergraduate student in genetics, I worked with maize seeds derived from some that had been exposed to one of the tests at Bikini Atoll, and observed seedlings that grew, pale white in color, several centimeters in height before any chlorophyll appeared. This behavior had never been noted before; maize seedlings are ordinarily green from the time they emerge.

Not a test, but an accident involving release of radioactivity was a fire at the Windscale military reactor in Britain in 1957. This was followed by a terrible accidental explosion of buried radioactive materials early the next year in the Soviet Union at Kyshtym in the Urals. This contaminated forests, farms, and cities, but was kept secret and the extent of damage to ecosystems and the human population is still unknown.

A ban on testing in the atmosphere, oceans, and space was proposed by leading scientists and public figures, but a conference of experts failed to reach agreement. Then in 1962, with the American blockade of a Soviet attempt to place nuclear missiles in Cuba, war between nuclear powers came close. People around the world were aware that such a war would have effects on them as radioactive particles would be carried by currents in the atmosphere and deposited by precipitation. This had happened with more than 500 tests already held. There was concern among scientists that radioactivity was damaging the genetic material of humans and other life forms, and producing cancers and other illnesses. The Atmospheric Test Ban Treaty of 1963 was signed by the US, the USSR, the UK, and more than a hundred other nations. France and China, both of whom wished to continue tests, refused to sign. Underground tests, permitted under the treaty, continued for a number of years. A non-proliferation treaty of 1968, intended to prevent the spread of nuclear weapons to other nations, was ratified by most nations, but avoided by those most likely to join the nuclear club. India, a non-signer, conducted an atmospheric test in 1974, and both India and Pakistan tested in 1998. Several nations kept their capabilities secret; even some signatories were suspected of trying to get their own bombs. Other treaties between the United States and the Soviet Union, and the breakup of the USSR in the 1990s, promised to reduce the danger of nuclear war, but weapons remained, along with the possibility of use by extremist national leaders or terrorists, or of a renewed confrontation between the great powers. In 1999, the US Senate rejected a nuclear arms limitation treaty.

In the "Cold War" period in the 1980s, certain scientists emphasized the potential world-wide environmental effects of nuclear war, predicting that they might constitute some of the most catastrophic results of human activity on Earth that can be conceived.[127] Not only blasts and radiation needed to be considered, they warned, but also the tremendous quantities of dust and smoke particulates which would enter the atmosphere from the explosions and the firestorms they would produce in cities and forests. Some scientists predicted that these particulates might block out the Sun's heat and light, killing plants and animals in a "nuclear winter."

Nuclear technology was also used for power generation. Electricity was experimentally generated in 1951, and commercial power later became available. Nuclear energy seemed safe and inexpensive, without some of the pollution problems of fossil fuel. By 1987, there were 417 plants in operation in twenty-seven nations, generating 17 percent of the world's electricity, with 120 additional units planned.[128] The nations with the highest capacity were the United States, France, the Soviet Union, Japan, Germany, Canada, and the UK. However, orders for new plants had decreased. There had been no new licenses in the United States since 1978. Costs had been higher than expected, the problem of storage of long-lived radioactive wastes – there is no way to dispose of them – was troublesome, and the number of accidents involving core damage was disturbingly high. Equipment failure and human error produced an accident in 1979 at the Three Mile Island plant, Pennsylvania, which destroyed 35 percent of the reactor core and caused release of radioactive material to the environment. Although damage to humans and the ecosystem was small, a potential danger was recognized by the public.

Some of the radioisotopes have exceptionally long half-lives and will remain dangerous after thousands of years. "There is no precedent in technology for the long periods of time for which risk assessments are required in radioactive waste management ... or the amounts of radioactive materials that should be permitted to enter the biosphere in future millennia."[129] Radioactive wastes have been treated in a number of unsatisfactory ways: storage on-site, injection through wells into deep rock formations, and dumping in containers onto the seabed with the possibility of rupture of the containers and contamination of the hydrosphere. International restrictions now forbid dumping at sea, but are difficult to enforce. At present, the recommended method is storage in underground chambers excavated in stable formations of rock or salt. There are difficulties with estimating future problems of earthquake faults and groundwater pollution, and one limiting factor in democratic societies has been the unwillingness of people to allow such facilities, not to mention the plants themselves, to be located near their homes; the acronym used for the phenomenon is "NIMBY" ("Not In My Back Yard").

The experience of Chernobyl, discussed in this section, and a dozen other accidents prompted a pause in the growth of the nuclear power industry around the world, except for France and a few other nations that remain firmly committed to it. More recently, some have advocated a reevaluation of nuclear power as an alternative to sources using fossil fuels that generate carbon dioxide effluents.

How did it happen that modern humans decided to introduce active substances into the ecosystems of which they are an inextricable part, substances which are degradable only over long periods of time and for which organisms and natural systems lack resistance? The usual answer is that fear and competition on both sides of the international political divide drove nations to do so. Another answer is that humans thought of themselves as separate from the rest of the biosphere, so that they would be protected by distance or by dilution of dangerous substances. But radioactive products were carried in the atmosphere to every part of the Earth. Yet another answer is that they intended to isolate radioactivity within safe containers such as reactor core protection systems, concrete sarcophagi, or safe buildings at plutonium production plants, all of which, in some times and places, have ruptured or leaked. Every form of technology experiences accidents from time to time. The nature of human beings is to learn by trial and error, but eventually if unpredictably to make errors. Inescapably, Pandora will open the box.

Denver: a sense of place

I lean on the railing of my tenth-story balcony and try to sense my place, the city where I live. The Rocky Mountains form an irregular horizon to the west. Many people think of Denver as being in the Rockies, but the city is on the High Plains and looks almost flat from here. Although much land is paved, most of what I see is green; in this older residential section the urban forest flourishes in spite of the inroads of Dutch elm disease. Strictly speaking, it is not a forest. When the leaves fall, they are swept up to be carried away by a Public Works Department contractor, not left to decay and form soil. Most trees are exotics: elms and maples planted by homesick easterners, or Colorado blue spruce and aspens giving the illusion of a Rocky Mountain environment, though their true habitat begins at an elevation 300 m (1,000 ft) above the city. The only large native tree is the cottonwood, which will not grow far from water. Neither will a city, in the Mountain West.

I can glimpse a watercourse from where I stand: Harvard Gulch, a minor feeder of the South Platte River, itself a tributary of the Platte, Missouri, and Mississippi.[130] I can't hear the little stream over the irregular noise of traffic on University Boulevard, the occasional airplane headed for Denver International Airport, and the distant hum of Interstate 25. Are there any sounds not of human origin? Yes, the west wind in the trees, the buzz of cicadas, and the trill of a finch on a neighbor's balcony. Is my city, Denver, an ecosystem, or part of an ecosystem?[131] In what ways?

The answers would have been easier 140 years ago, when Utes, Arapahoe, and Cheyennes lived east of the Front Range. Then this was High Plains habitat, the short grass prairie ecosystem, the western shore of a sea of grass.[132] It was a complex community of plants dominated by perennial grasses such as buffalo grass, western wheatgrass, bluestem, blue grama, wiregrass, switchgrass, sand dropseed, needle-and-thread; and other tough species, including yucca, mallows, yellow-rayed composites, and cactuses. Most was plowed up or overgrazed decades ago, invaded by introduced weeds like cheatgrass, bindweed, thistle, and prickly lettuce.

Ecologically, the city is the result of an historical process of change from the ecosystem that flourished in this place before Euro-American settlement. In the early days, the short-grass prairie was a veritable Serengeti; the dominant herbivore was the American bison, always called buffalo here. There was a buffalo wallow on the present site of North Lake in Washington Park. The Denver Zoo acquired buffalo in 1898, a few months after the last wild herd in the state was killed; the captive herd thrived by 1908.[133] In pre-settlement times, there were antelope in tens of thousands. A remnant survives at the Plains Conservation Center, a stretch of the High Plains that long provided environmental education, but now is surrounded by subdivisions and too valuable (land is a commodity) to keep in its natural state; land managers are searching for a substitute further out. In the past, there were elk and bighorn sheep, and still are mule deer and white-tailed deer. Beaver live along streams and build dams; they are now considered pests because they cut down trees with their ample teeth. Predators then were wolf and grizzly bear, now missing. Rarely, black bear or mountain lion get in as far as our part of town, but Animal Control finds them, tranquilizes them, and takes them to be released in some unspecified spot in the mountains or, failing that, shoots them dead. Coyotes are prevalent. Smaller wildlife still exist in surprising numbers inside the city, along with opportunistic introduced species that live in urban environments elsewhere: mice, rats, pigeons, and starlings.[134] During walks along Harvard Gulch, especially before the vest-pocket cattail marsh was removed to "improve" drainage, I have seen muskrats, foxes, and a beaver (and know there are skunks and raccoons), bats swooping over the stream in the evening, and birds – mallards, Canada geese, western tanagers, magpies, saw-whet owls, kingfishers (there are tiny fish, frogs, and leeches in the water), and others too numerous to list. Compared to the biological richness of the nineteenth-century Great Plains, what remains is fragmentary, but the fragments reassert themselves whenever permitted. I have seen kestrels nesting on a tower at the University of Denver. Restoration of species that have been lost is a possibility. The Denver Museum of Natural History co-sponsored a successful project with the Colorado Division of Wildlife to release peregrine falcons on high-rise buildings in downtown Denver.

A spectacular illustration of the ability of the ecosystem to repair itself is Rocky Mountain Arsenal, an area of 8,059 ha (19,915 acres) between Denver's old and new airports.[135] Before the 1940s, it was farmland dotted with lakes, but was taken by the military during the Second World War for production of munitions and chemical agents, which continued until 1969. Afterwards, through the 1970s until 1985, the arsenal was used as a site to destroy munitions and chemically related items. Coincidentally, from 1946 to 1982 the Army leased facilities to private industries such as Shell Oil Company for production of pesticides and herbicides. Pollution took place on the surface and also in the geological strata underground, since a deep injection well for toxic fluids, drilled to a depth of 3,671 m (12,045 ft), was used from 1961 to 1966, when it became apparent that it was triggering numerous earthquakes, and its operation ceased. Due to the release of a wide variety of contaminants on the surface, soil and water became so toxic that animals and birds died from contact with them, and human access had to be restricted. Nonetheless, wildlife infiltrated and prospered in less polluted zones. Deer, raptors, white pelicans, and songbirds proliferated; up to 100 bald eagles established nests. Perhaps 50,000 prairie dogs (the object of eradication in much of the rest of the city)[136] lived in the arsenal along with burrowing owls, badgers, coyotes, and ferruginous hawks. More than forty-six species of mammals and 176 of birds have been identified. In 1992 much of the land was designated a potential National Wildlife Refuge and, although cleanup has been slow and is now scheduled for completion in 2011, the Fish and Wildlife Service operates guided tours and a visitor center. Photographs of deer herds with high-rise downtown Denver in the background are reminiscent of Nairobi National Park in Kenya. Adding to the wildlife scene, a herd of bison was transferred from the National Bison Range in Montana to the Rocky Mountain Arsenal in 2007.

Despite proximity, Denver residents are no longer closely dependent on the local ecosystem for food. Restaurants such as the Buckhorn Exchange downtown, and the Fort in nearby Morrison, serve old-time fare like buffalo and elk steaks, pheasant, and Rocky Mountain oysters (bulls' testicles), but some of the meat may be imported from Canada, and the last two dishes named are from introduced species. Anyway, few can afford game very often, even if they shoot the animals themselves. The affluent majority breakfast on cereals grown in Iowa and packaged in Michigan, oranges from California, and, in winter, peaches from Chile. They live in houses built of Oregon Douglas fir timber, wear shirts sewn in Bangladesh, and use electricity generated from Colorado coal, but supplemented by a grid spanning the

Figure 8.6 Denver's downtown skyline from Broadway, next to the state capitol. Although a photograph such as this may seem to record only the works of humans, even the center of a modern city is inhabited by many other species and continues to be an ecosystem, even if much changed, impacted, and altered. Photograph taken by Dr M.D. Subash Chandran in 1996.

US.[137] Like all modern cities, Denver is largely inhabited not by "ecosystem people" who interact mainly with the local environment, but by "biosphere people" who import and export resources as components of the world market economy.[138]

In the early twentieth century, people with lung diseases came to Denver to recuperate in the clean air. But coal smoke from industries and home heating prompted a smoke abatement ordinance in 1911.[139] Improvements due to replacement of coal by natural gas were wiped out by adoption of the automobile. Today the air quality is among the worst in the US. Like Los Angeles, Denver suffers from temperature inversions that trap pollutants in a basin near the mountains, and, like Mexico City, it is at a high elevation where internal combustion engines operate less efficiently. Major causes of air pollution are motor vehicle operation, power generation using coal, industrial processes, and wood burning in fireplaces. Mandatory wood-burning restrictions go into effect on winter days when there are inversions. Controls on emissions from stationary sources, anti-pollution devices and inspections of motor vehicles, and use of oxygenated fuels have reduced levels of carbon monoxide, ozone, and particulates, but the increase in numbers of vehicles, even with control devices and less polluting fuels, may cause air quality to deteriorate again in future decades.[140] Public transportation in the form of buses and an expanding light rail system with park-and-ride centers has helped to moderate the increase in automobile traffic. Pollution is not only damaging to human health, but also affects other parts of the ecosystem. Trees help to ameliorate air pollution, but many species such as pines are weakened or killed by it.

Denver's need for water has visibly rearranged the region's hydrology.[141] The first project, City Ditch, was begun in 1859, the year of settlement. Ground water was plentiful but subject to pollution, and its level fell, so Denver exploited its river. The first masonry dam rose in 1900; today there are more than 780 dams and reservoirs in the South Platte drainage. The Denver Water Department (DWD) soon realized that the South Platte would be inadequate to supply agriculture and urban growth, so it began to acquire rights over the Continental Divide in the Colorado River watershed. Water flowed from the Western Slope through the Moffat Tunnel in 1936, and Dillon Reservoir, able to store 310 million cu m (254,000 acre-feet), doubled Denver's supply in 1963. The other side of the coin was reduced stream flow and wetland depletion in the mountain tributaries that were siphoned into aqueducts. In the 1980s the DWD proposed a $500 million project to store water for accelerating growth: Two Forks Dam, planned to rise 187 m (615 ft) and flood 48 km (30 mi) along the river valley. The Environmental Protection Agency (EPA)[142] withheld approval due to potential violation of the Clean Water Act and probable effect on wildlife habitat. In June 1996 a federal judge upheld that ruling.

In the 1980s the DWD supported mandatory and voluntary measures to reduce lawn watering, which uses more than half the supply brought into the city. The latter fact underlines an important ecological effect of water transfer to the urban area. It creates an artificial ecosystem, an oasis in the arid high plains, with planted trees, shrubs, and grasses. Aggressive introduced trees such as green ash, Russian olive, and Chinese elm crowd out the native cottonwoods. Eastern birds move in and hybridize with, or replace, native species, and there are English sparrows and starlings to contend with. The urban forest is undergoing ecological succession, and it is not always the succession people want. A program aimed at reversing the trend is xeriscaping: landscaping with plants that require less water, especially High Plains natives that can survive on local rainfall. The DWD conducts seminars and maintains a xeriscape demonstration garden. Although xeriscaping is still rare on residential streets, new corporate buildings have used it.

What happens to the water that goes down the drain, and what it contains? The Metro district returns 630 million litres (140 million gallons) of water a day to the river. Before the 1980s, the river below the foaming sewage outlet at Northside was a biological desert, its fish killed by ammonia, nitrogen, and a deficiency of oxygen. But the Environmental Protection Agency assessed fines and ordered major changes. Now the effluent has improved so much that the river below the plant can be used for recreation, and 85 percent of the dry sludge removed from it is used as fertilizer; the rest goes to landfills along with the city's solid waste. Methane gas from the treatment process is used as an energy source at the plant. Downstream, 90 percent of the river's volume is treated urban effluent, and boaters between newly replanted riverbanks float on reclaimed sewage. Metro Wastewater and EPA operate a laboratory to improve water recycling technology, and have demonstrated that treated water could be cleaner than water now coming out of taps in Denver. The DWD hopes that 20 percent of the water shortfall expected by 2045 can be supplied by recycling.

Residents are never far from open space; Denver has one of the most extensive public park systems in the US – 206 city parks, plus parkways and bicycle paths, as well as a constellation of mountain parks west of the city. Varying from traditional parks with manicured lawns and flowerbeds to natural forest, they provide refuges within the ecosystem. Historically, the evolution of the park system seesawed between ambitious projects and cautious penny-pinching. Curtis Park, the first one, was donated by a real estate developer in 1868. Fourteen years later, Mayor Richard Sopris purchased 128 ha (320 acres) from the state for City Park, which soon had an artificial body of water, Duck Lake. The Denver Zoo was established in City Park in 1896 with the purpose of displaying only native Colorado wildlife. Within two years the zoo acquired bison; as noted above, the last wild bison herd in the state was slaughtered during those two years, so the zoo participated in saving a captive population of an endangered species before that became a major announced purpose of zoos.[143] Display of non-native species began when monkeys proved to be a drawing card for children; today the zoo exhibits everything from okapis to snow leopards. The Museum of Natural History, built not far from the zoo in 1904, has become one of the leading institutions of its kind in the US.

Robert W. Speer, mayor from 1904 to 1912, led Denver in doubling the area of its parks, building an eighteen-mile network of parkways with tree-clad medians, and gaining voter approval for creation of a system of mountain parks and drives.[144] Frederick Law Olmsted, Jr., creator of the first curriculum in landscape architecture in the US at Harvard, was commissioned to design it.[145] By 1935, Denver owned 8,500 ha (21,900 acres) of mountain parkland close enough to be enjoyed in a day's outing, including Winter Park Ski Area. But the war years cut travel, and the costs of providing infrastructure for a burgeoning postwar population meant neglect of the parks.[146] The Mountain Parks property tax was dropped, and the city divested itself of more than a third of the mountain parks area, slashing staff and budget. Parks inside the city also suffered. A number of volunteer non-governmental groups stepped into the breach, including the Parks People and the Denver Urban Forest Group, organizing public support that succeeded in voting $59 million for parks in a 1989 bond issue, including renewal of the South Platte riparian corridor and a greenway plan. The 108 km (68 mi) High Line Canal has become a tree-lined recreational amenity with a trail for hiking, jogging, biking, and horseback riding. Concern for biodiversity was shown by two programs to reestablish species: the Denver Zoo's help in the Species Survival Plan by breeding the rare Bali mynah and thirty other species,[147] and, closer to home, the release of peregrine falcons not only in the mountains, but also on high-rise buildings in downtown

Figure 8.7 McWilliams Park, on Harvard Gulch in Denver, in winter. One of more than 200 parks in the city, it offers open space and amenities such as a bicycle path and playground. Wildlife seen here includes many birds such as kingfishers and tanagers, and mammals such as foxes, muskrats, and even occasionally a beaver.

Denver.[148] The latter project was co-sponsored by the Colorado Wildlife Federation, the Denver Museum of Natural History, and the Colorado Division of Wildlife.

My city is only one example of urban ecosystems around the world. There are other cities whose settings resemble Denver's: Calgary stretches out on the high plains of Alberta with a view of the Canadian Rockies, and Alma Ata in Kazakhstan is correspondingly placed below the snow-capped Tien Shan. But the similarity of the ecological processes at work in all cities calls for emphasis.

The character of a locality may be a reason why a city appeared there. But the original ecosystem altered as the city expanded. Native species disappeared as their habitats, the forests or grasslands, shrank and were replaced by power poles and paving. Usually the new human residents did not tolerate the larger animals, especially predators. Pollution killed organisms or weakened their resistance to diseases. The chemical balance of the air changed, and fish died in contaminated waters.[149]

The original ecosystems did not simply disappear. They were transformed step by step into urban-specific ecosystems. The climate altered: as a rule, urban environments have higher temperatures and lower humidity than the surrounding countryside, along with weaker winds, less sunshine, more clouds, and higher precipitation. Some of these phenomena result from the "heat-island" effect of large cities.[150] An urban forest may replace the former plant cover, but in some parts of the city there may be almost no vegetation. Some native species can adapt to these conditions: in India, predatory pariah kites soar in city skies, ready to swoop down and grab whatever morsel may present itself. Then there are interstices, protected parks that provide refuges within the ecosystem, or neglected fragments, often called "wasteland," containing some of the earlier assemblage of plants and animals. Vulnerable to invasions and extinctions, they also demonstrate that not every part of the city is subject to human planning.

Another universal fact about urban ecology is that the ecosystem is not contained within the city limits.[151] City, countryside, and wilderness are parts of a mutually dependent system.[152] Like other cities, Denver stimulated suburbs, first by cable cars and trolleys, then automobiles. Suburban malls threaten to eclipse downtown. Rural landscapes alter as highways generate "strip cities," in Brazil and India as well as the US. Cities import water and energy over hundreds, and food over thousands, of kilometers. Forests are felled because cities need fuel, paper, and timber. No wilderness is so isolated as not to feel the influences of cities, from acids in the air and pollutants in the water to the noise of jet planes. City folk no longer depend only on local or regional resources; they are involved with the ecosystems of the Earth.

City planning must increasingly take the biotic community into account, and work toward sustainable urban ecosystems.[153] The urban forest requires holistic management no less than the national forests. In the final paragraph of *The City in History*, Lewis Mumford wrote, "The final mission of the city is to further man's conscious participation in the cosmic and the historic process."[154] To that, I would add the ecological process.

Conclusion

Humans in the first half of the twentieth century did things to the natural environment that were quite new, compared to what went before. In the second half, they produced changes that were truly revolutionary. Processes that were previously regarded as of "natural" origin and beyond human influence except possibly to ameliorate their effects are now seen to have human activities involved in their causes. Some of these processes are climatic change, the chemical composition of rainfall and the atmosphere, the abundance and availability of fresh water, variations in the ozone layer and ultraviolet radiation received from the Sun, the stimulation of earthquakes, the emergence and spread of diseases, the genetic evolution of species, and the radioactive decay of elements. This does not mean that humans have achieved their control; far from it. What it does mean is that human activities, now of unprecedented dimensions and power, have had unintended effects for good and ill upon the systems of the Earth, effects that we are beginning to understand. We have found it extremely difficult to moderate the undesirable effects. We cannot yet clean up the radioactivity after an accident like Chernobyl, and the prospects of slowing global warming are truly daunting. Once we might have thought that the Earth was too vast to be changed significantly by humankind; now we see that we have changed the Earth, but in ways that may threaten us. Still, humans can be intelligent and creative. What achievements with potential for guiding change have we made in science, new technology, and in worldwide institutions? This question will be addressed further in Chapter 9.

Notes

1   T. Schneider, ed., *Acidification and Its Policy Implications: Proceedings of an International Conference Held in Amsterdam, May 5–9, 1986*, Amsterdam, Elsevier, 1986; National Research Council, *Acid Deposition: Long-Term Trends*, Washington, DC, National Academy Press, 1986.

2 Robert Angus Smith, *Air and Rain: The Beginnings of a Chemical Climatology* (1872); Ellis B. Cowling, "Acid Precipitation in Historical Perspective," *Environmental Science and Technology* 16, 2, 1982; "An Acid Rain Chronology," *EPA Journal*, June/July 1986, 18–19; Elisabeth Johann, "The Impact of Industry on Landscape and Environment in Austria from the Second Part of the 19th Century to 1914," unpublished, presented at "Forests, Habitats and Resources: A Conference in World Environmental History," Duke University, Durham, NC, 30 April–2 May, 1987.

3 Intergovernmental Panel on Climate Change, *Climate Change 1995*, 3 vols, Cambridge, Cambridge University Press, 1996, Vol. 1, 5.

4 F. Sherwood Rowland, "A Threat to Earth's Protective Shield," *EPA Journal*, December 1986, 4–6; Richard S. Stolarski, "The Antarctic Ozone Hole," *Scientific American* 258, January 1988, 30–6; Tim Smart and Joseph Weber, "An Ozone Hole Over Capitol Hill," *Business Week*, April 4, 1988, 35.

5 Philip Shabecoff, "DuPont to Stop Making Chemicals that Peril Ozone," *New York Times*, March 25, 1988, 1, 9; William Glaberson, "Behind DuPont's Shift on Loss of Ozone Layer," New York Times, March 26, 1988, 17, 19.

6 Timothy H. Quinn *et. al.*, *Projected Use, Emissions, and Banks of Potential Ozone-Depleting Substances*, Santa Monica, CA, Rand, 1986; Kathleen A. Wolf, *Regulating Chlorofluorocarbon Emissions: Effects on Chemical Production*, Santa Monica, CA, Rand, 1980.

7 D.B. Grigg, *The Agricultural Systems of the World: An Evolutionary Approach*, London, Cambridge University Press, 1974, 180.

8 Ibid., 173.

9 Ibid., 55.

10 World Resources Institute, *World Resources 1994–95*, New York, Oxford University Press, 1994, 294–5.

11 Jan G. Laarman, "Export of Tropical Hardwoods in the Twentieth Century," in *World Deforestation in the Twentieth Century*, , Durham, Duke University Press, 1988, 147–63, 149.

12 Fred Gale, *The Tropical Timber Trade Regime*, London, MacMillan, 1998, 2.

13 James T. Thomson, "Deforestation and Desertification in Twentieth-Century Arid Sahelian Africa," in John F. Richards and Richard Tucker, eds, *World Deforestation in the Twentieth Century*, Durham, NC, Duke University Press, 1988, 70–90, at 84.

14 Michael Williams, "Forests," in B.L. Turner, ed., *The Earth as Transformed by Human Action*, Cambridge, Cambridge University Press, 1990, 179–201, at 191.

15 Richard P. Tucker, *Insatiable Appetite: The United States and the Ecological Degradation of the Tropical World*, Berkeley and Los Angeles, University of California Press, 2000, 347.

16 John Dargavel, "Changing Capital Structure, the State, and Tasmanian Forestry," in Richards and Tucker, eds, *World Deforestation in the Twentieth Century*, 189–210.

17 J.W. Maurits la Rivière, "Threats to the World's Water," in *Managing Planet Earth*, New York, W.H. Freeman & Co., 1990, 37–48, at 41.

18 Adam Markham, *A Brief History of Pollution*, New York, St. Martin's Press, 1994, 53; figures from World Health Organization, *Our Planet, Our Health*, Geneva, 1992.

19 Harry E. Schwarz *et al.*, "Water Quality and Flows," in Turner, ed., *Earth as Transformed by Human Action*, 253–70, at 264.

20 Alwyne Wheeler, *The Tidal Thames; The History of a River and Its Fishes*, London, Routledge & Kegan Paul, 1979, 144–55.

21 George St. George, *Siberia*, New York, David McKay, 1969, 162–81.

22 Ibid., 162.

23 Charles E. Ziegler, *Environmental Policy in the USSR*, Amherst, University of Massachusetts Press, 1987, 53–7; Boris Komarov, *The Destruction of Nature in the Soviet Union* White Plains, NY, M.E. Sharpe, 1980, 3–19; Craig ZumBrunnen, "The Lake Baikal Controversy: A Serious Water Pollution Threat or a Turning Point in Soviet Environmental Consciousness?" in Ivan Volgyes, ed., *Environmental Deterioration in the Soviet Union and Eastern Europe*, New York, Praeger, 1974, 80–122; Marshall I. Goldman, *The Spoils of Progress: Environmental Pollution in the Soviet Union*, Cambridge, MA, MIT Press, 1972, 178–209; Philip R. Pryde, *Conservation in the Soviet Union*, Cambridge, Cambridge University Press, 1972, 147–51.

24 Heard by the author at a conference in the Pribaikalskaya Lodge in August 1987.

25 International Commission on Large Dams (ICOLD), *World Register of Dams, 1988 Updating*, Paris, ICOLD, 1989, 25–7.

26 Mark I. L'Vovich *et al.*, "Use and Transformation of Terrestrial Water Systems," in Turner, ed., *Earth as Transformed by Human Action*, 235–52, at 239.

27 Gilbert F. White, "The Environmental Effects of the High Dam at Aswan," *Environment* 30, 7, September 1988, 4.

28 Shen Gengcal, "Three Gorges Needs to Power Ahead," *China Environment News* 32, March 1992, 4–5.

29 United Nations Food and Agriculture Organization (FAO), *Marine Fisheries and the Law of the Sea: A Decade of Change*, FAO Fisheries Circular 853, Rome, 1992, 4–7.

30 World Resources Institute, *World Resources 1994–95*, New York, Oxford University Press, 1994, 187.

31 Ralph H. Lutts, "Chemical Fallout: Rachel Carson's *Silent Spring*, Radioactive Fallout, and the Environmental Movement," *Environmental Review* 9, 3, Fall 1985, 210–25.

32 Linda Lear, *Rachel Carson: Witness for Nature*, New York, Henry Holt, 1997.

33 I witnessed an incident in 1952 when, as part of a U.S. Forest Service training program, an airplane demonstrated the spraying of DDT on a Douglas fir forest to kill spruce budworms. There may or may not have been spruce budworms in the trees that were sprayed, but the trainees, including myself, were certainly exposed to the DDT, which was at that time considered harmless to humans.

34 Rachel Carson, *Silent Spring*, Boston, MA, Houghton Mifflin Co., 1962, 6.

35 Frank Graham, Jr., *Since Silent Spring*, Greenwich, CT, Fawcett Publications, 1970, 59–63.

36 Paul Brooks, *The House of Life: Rachel Carson at Work*, Boston, MA, Houghton Mifflin Co., 1972, 297, citing *Time*, September 28, 1962.

37 H. Patricia Hynes, *The Recurring Silent Spring*, New York, Pergamon Press, 1989, 18.

38 Arthur George Tansley, "The Use and Abuse of Vegetational Concepts and Terms," *Ecology* 16, 3, 1935, 284–307, at 299. Noted in Frank Benjamin Golley, *A History of the Ecosystem Concept in Ecology: More Than the Sum of the Parts*, New Haven, CT, Yale University Press, 1993, 8.

39 Samuel Hays, "From Conservation to Environment: Environmental Politics in the United States since World War Two," *Environmental Review* 6, 2, 1982, 14–41; and Samuel Hays, *Beauty, Health, and Permanence: Environmental Politics in the United States, 1955–1985*, Cambridge, Cambridge University Press, 1987.

40 Jane Belo, *Bali: Rangda and Barong*, Monographs of the American Ethnological Society 16, New York, J.J. Augustin, 1949.

41 Bruce P. Wheatley, *The Sacred Monkeys of Bali*, Prospect Heights, IL, Waveland Press, 1999.

42 Tony Whitten, Roehayat Emon Soeriaatmadja, and Surayat A. Afiff, *The Ecology of Java and Bali*, Singapore, Periplus Editions, 1996, 227–31.

43 Albert Ravenholt, "Man-Land–Productivity Microdynamics in Rural Bali," *American Universities Field Staff Reports, Southeast Asia Series* 21, 4, 1973, 1.

44 Fred B. Eiseman, Jr., *Bali: Sekala and Niskala*, 2 vols, Singapore, Periplus, 1990, Vol. 2, 285.

45 J. Stephen Lansing, *Priests and Programmers: Technologies of Power in the Engineered Landscape of Bali*, Princeton, NJ, Princeton University Press, 1991, 12.

46 Jane E. Stevens, "Engineered Paradise," *Earth: The Science of Our Planet* 3, 6, November 1994, 47.

47 I Made Ponti of Tambanan, Bali, "Rice Culture," talk and interview, February 27, 1994.

48 Lansing, *Priests and Programmers*, 54.

49 Ponti, "Rice Culture."

50 Lansing, *Priests and Programmers*, 130.

51 Ibid., 65.

52 Ibid., 67.

53 Clifford Geertz, *Negara: The Theatre State in Nineteenth Century Bali*, Princeton, NJ, Princeton University Press, 1980, 82.

54 Ponti, "Rice Culture."

55 Carol Warren, *Adat and Dinas: Balinese Communities in the Indonesian State*, Kuala Lumpur, Oxford University Press, 1993, 37.

56 Ponti, "Rice Culture."

57 A.J. Bernet Kempers, *Monumental Bali: Introduction to Balinese Archaeology*, Berkeley, Periplus Editions, 1991, 33–49.

58 Fredrik Barth, *Balinese Worlds*, Chicago, University of Chicago Press, 1993, 70.

59 Colonial administrators may have been misled by a scholarly theory of ancient irrigation systems which held that they arose with the state and, in particular, with the institution of kingship. According to this

theory, only "oriental despots" could command the employment of enough workers to construct and maintain canals and aqueducts.

60 J. Stephen Lansing, *The Three Worlds of Bali*, New York, Praeger, 1983, 135–7.

61 Scott Pearson *et al.*, *Rice Policy in Indonesia*, Ithaca, NY, Cornell University Press, 1991, 11.

62 Randolph Barker *et al.*, *The Rice Economy of Asia*, Washington, DC, Resources for the Future, 1985, 249.

63 Sisira Jayasuriya and I. Ketut Nehen, "Bali: Economic Growth and Tourism," in Hal Hill, ed., *Unity and Diversity: Regional Economic Development in Indonesia since 1970*, Singapore: Oxford University Press, 1989, 331; Eiseman, *Bali: Sekala and Niskala*, Vol. 2, 285.

64 Kim Streatfield, *Fertility Decline in a Traditional Society: The Case of Bali*, Indonesian Population Monograph Series 4, Department of Demography, Canberra, Australian National University, 1986, 9.

65 Carol Warren, "Adat and Dinas: Village and State in Contemporary Bali," in Hildred Geertz, ed., *State and Society in Bali: Historical, Textual and Anthropological Approaches*, Leiden, KITLV Press, 1991, 233.

66 Michael Lipton, *New Seeds and Poor People*, Baltimore, MD, Johns Hopkins University Press, 1989.

67 Gary E. Hansen, *Agricultural and Rural Development in Indonesia*, Boulder, CO, Westview Press, 1981, 65–7.

68 James J. Fox, "Managing the Ecology of Rice Production in Indonesia," in Joan Hardjono, ed., *Indonesia: Resources, Ecology, and Environment*, Singapore, Oxford University Press, 1991, 67.

69 Lansing, *Priests and Programmers*, 112.

70 Barker *et al.*, *The Rice Economy of Asia*, 64–5.

71 Lansing, *Priests and Programmers*, 114.

72 Ibid., 117–26; see also the film by Dr. André Singer, "The Goddess and the Computer," 1988.

73 Stevens, "Engineered Paradise," 53.

74 World Resources Institute, *World Resources 1986*, New York, Basic Books, 1986, 43, 57.

75 Stevens, "Engineered Paradise," 54.

76 Oekan Soekotjo Abdoellah, *Indonesian Transmigrants and Adaptation: An Ecological–Anthropological Perspective*, Center for Southeast Asia Studies Monograph 33, Berkeley, Centers for South and Southeast Asia Studies, University of California at Berkeley, 1993.

77 The subtitle was suggested by a line, "Who will hold the sky up, now the big trees are down?" in the song by Douglas Wood and Edith Rylander, lyrics, "The Big Trees Are Down," in Douglas Wood, *Earth Songs*, NorthSound NSAC 27124, Minocqua, WI, NorthWord Press, 1992.

78 For readers unfamiliar with Oregon pronunciation, it should be noted that Willamette is pronounced to rhyme with "God *damn* it!", not "Want to *bet*?"

79 I had a very similar experience on Logger Butte, the lookout station where I worked during the following summer in 1951. A road was bulldozed to Logger Butte in 1966, and one also reached Hills Peak. Both lookout buildings have since been removed (Hills Peak in 1966; Logger Butte in 1969) except for wooden platforms. See Donna Marie Hartmans, "Historic Lookout Stations on the WNF: Management Plans for Preservation," Master's Thesis, University of Oregon, 1991.

80 David J. Brooks, "Federal Timber Supply Reductions in the Pacific Northwest: International Environmental Effects," *Journal of Forestry* 93, 7, 1995, 29–33.

81 In the US, the range of the northern spotted owl extends from the eastern slopes of the Cascade Mountains to the Pacific coast in Washington, Oregon, and parts of northern California. It also occurs in British Columbia.

82 Harold K. Steen, *The U.S. Forest Service: A History*, Seattle, University of Washington Press, 1976, 74.

83 USDA, FS, *The Principal Laws Relating to Forest Service Activities*, Agricultural Handbook 453, Washington, DC, GPO, 1978, 138–9, quoted in David A. Clary, *Timber and the Forest Service*, Lawrence, University Press of Kansas, 1986, 22.

84 Ibid.

85 In the Multiple Use–Sustained Yield Act of June 12, 1960. See J. Michael McCloskey, "Natural Resources – National Forests -- The Multiple Use–Sustained Yield Act of 1960," *Oregon Law Review* 41, December 1961, 49–78.

86 Gibney File, WNF Historical File, April 1, 1965, quoted in Richardson, Elmo, "Willamette National Forest: The History of a Public Enterprise," Report to the USDA Forest Service, MS, July 1982, 167. A copy of the latter is available at the Library of the Forest History Society, 701 Vickers Ave., Durham, NC 27701.

87 Richardson, "Willamette National Forest," 85.

88 Larry D. Harris, *The Fragmented Forest: Island Biogeography Theory and the Preservation of Biotic Diversity*, Chicago, University of Chicago Press, 1984, 108–13. The distance of the "edge effect" averages two to three times the average tree height; in old-growth Douglas fir forest, that could easily be 200–250 m (600–750 ft).

89 Available figures from the period 1915–24.

90 John R. Bruckart, "Taming a Wild Forest," in US Department of Agriculture, *Trees: The Yearbook of Agriculture*, Washington, DC, US Government Printing Office, 1949, 326–34.

91 Arthur H. Carhart, *Timber in Your Life*, Philadelphia, PA, J.B. Lippincott, 1955, 141.

92 The best study of the French Pete issue, on which this account is based, is Gerald W. Williams, "The French Pete Wilderness Controversy, 1937–1978: A Leadership Case Study," unfortunately unpublished at present but possibly available from the USFS Pacific Northwest Region, P.O. Box 3623, Portland, OR. A good, even if biased, description can be found in Rakestraw, *History of the Willamette National Forest*, 111–15. My uncle, Emmett Underwood Blanchfield, a landscape architect and recreational planner for the USFS, assisted Marshall in this survey.

93 Roy A. Elliott, "A Year to Remember," *Timberlines* 11, May 1957, 3–4.

94 Rakestraw, *History of the Willamette National Forest*, 145.

95 Peter H. Morrison, *Old Growth in the Pacific Northwest: A Status Report*, Washington, DC, The Wilderness Society, 1988, quoted in Elliot A. Norse, *Ancient Forests of the Pacific Northwest*, Washington, DC, Island Press, 1990, 243–51.

96 16 USC 1532. See Richard B. Watson and Dennis D. Muraoka, "The Northern Spotted Owl Controversy," *Society and Natural Resources* 5, 1, 1992, 85–90.

97 Jeb Boyt, "Struggling to Protect Ecosystems and Biodiversity Under NEPA and NFMA: The Ancient Forests of the Pacific Northwest and the Northern Spotted Owl," *Pace Environmental Law Review* 10, 2, 1993, 1009–50.

98 United States Department of Agriculture, Forest Service, and United States Department of the Interior, Bureau of Land Management, *Record of Decision for Amendments to Forest Service and Bureau of Land Management Planning Documents Within the Range of the Spotted Owl*, US Government Printing Office, 1994-589-111/00003 Region No. 10, April 13, 1994, 38–9.

99 Interviews, Neal Forester, US Forest Service; and Doug Heiken, Oregon Natural Resources Council, August 1995.

100 Paul W. Hirt, *A Conspiracy of Optimism: Management of the National Forests since World War Two*, Lincoln, University of Nebraska Press, 1994, 285–8.

101 One of the best discussions of the term is R. Edward Grumbine, "What is Ecosystem Management?", *Conservation Biology* 8, 1, March 1994, 27–38. See also Robert B. Keiter, "NEPA and the Emerging Concept of Ecosystem Management on the Public Lands," *Land and Water Law Review* 25, 1990, 43–60; and Jeff R. Jones, Roxanne Martin, and E.T. Bartlett, "Ecosystem Management: The U.S. Forest Service's Response to Social Conflict," *Society and Natural Resources* 8, 2, 1995, 161–8.

102 F. Dale Robertson to Regional Foresters and Station Directors, "Ecosystem Management of the National Forests and Grasslands," USDA Forest Service Memorandum, June 4, 1992, 1, quoted in Hirt, *A Conspiracy of Optimism*, 287–8.

103 Grumbine, "What is Ecosystem Management?", 31.

104 Ibid.

105 Margaret A. Shannon and K. Norman Johnson, "Lessons from the FEMA," *Journal of Forestry* 92, 4, April 1994, 6.

106 Jack Ward Thomas, "Forest Ecosystem Management Assessment Team: Objectives, Process and Options," *Journal of Forestry* 92, 4, April 1994, 12–19.

107 *Record of Decision for Amendments to Forest Service and Bureau of Land Management Planning Documents Within the Range of the Spotted Owl*, 5.

108 "Late successional" is an ecological term more or less equivalent to "old growth." It means that the natural processes of change and growth have proceeded for a long period of time without destructive incidents such as fire, avalanche, landslide, or clearcutting.

109 Douglas E. Booth, *Valuing Nature: The Decline and Preservation of Old-Growth Forests*, Lanham, MD, Rowman & Littlefield, 1994.

110 Nigel Hawkes *et al.*, *Chernobyl: The End of the Nuclear Dream*, New York, Random House Vintage Books, 1987; David R. Marples, *Chernobyl and Nuclear Power in the USSR*, New York, Free Press, 1986; Valeri A. Legasov, *The Lessons of Chernobyl are Important for All*, Moscow, Novosti Press Agency, 1987.

111 G.T. Vorob'yev, N.I. Volkova, and V.K. Zhuchova, "A Landscape-Based Radiation Map for Bryansk Oblast," *Geochemistry International* 31, 6, 1994, 116–24, map on 117.

112 Lida Poletz, "Paradox of Power: Chernobyl Eight Years Later," *Surviving Together* 12, 2, 1994, 23.

113 Vladimir Blinov, "The Zone," *Surviving Together* 12, 2, 1994, 26–7.

114 Murray Feshbach and Alfred Friendly, Jr., *Ecocide in the USSR: Health and Nature under Siege*, New York, Basic Books, 1992, 146–7, includes estimates by competent experts familiar with the aftermath of the accident.

115 L. Zhirina, *Chelovyek, spasi sebya sam ot radiatsii (Person, Save Yourself from Radiation)*, Bryansk, Pressa i Reklama, 1991.

116 Vorob'yev *et al.*, " Landscape-Based Radiation Map for Bryansk Oblast," 116–24, at 116.

117 V.A. Poiarkov, A.N. Nazarov, and N.N. Kaletnik, "Post-Chernobyl Radiomonitoring of Ukrainian Forest Ecosystems," *Journal of Environmental Radioactivity* 26, 1995, 259–71.

118 Ludmilla S. Zhirina and A.V. Kryshtop, "Use of the Dendroclimatological Method in Bioindication: Effects of Chernobyl," May 20, 1994, paper read at the International Conference on Tree Rings, Environment, and Humanity: Relationships and Processes, Tucson, Arizona.

119 That is, zones with a gamma background level of more than 40 curies per sq km. One curie per sq km corresponds to an irradiation dose of 0.01 milliroentgen per hour.

120 Vladimir Shevchenko, illustration published in Zhores A. Medvedev, *The Legacy of Chernobyl*, New York, W.W. Norton & Co., 1990, opposite 179.

121 Olga Kovalchuk, Yuri E. Dobrova, Andrey Arkhipov, Barbara Hohn, and Igor Kovalchuk, "Germline DNA: Wheat Mutation Rate After Chernobyl," *Nature* 407, 6804, October 5, 2000, 583–4.

122 David G. Fleishman, Vladimir A. Nikiforov, and Agnes A. Saulus, "137Cs in Fish of Some Lakes and Rivers of the Bryansk Region and North-West Russia in 1990–1992," *Journal of Environmental Radioactivity* 24, 1994, 145–58.

123 V.M. Chernousenko, *Chernobyl: Insight from the Inside*, Berlin, Springer-Verlag, 1991, 233–5.

124 Paul Craig and John A. Jungerman, *Nuclear Arms Race: Technology and Society*, New York, McGraw-Hill, 1986, 25.

125 Stewart L. Udall, *The Myths of August: A Personal Exploration of Our Tragic Cold War Affair with the Atom*, New York, Pantheon Books, 1994, 205.

126 Frank Benjamin Golley, *A History of the Ecosystem Concept in Ecology: More Than the Sum of the Parts*, New Haven, CT, Yale University Press, 1993, 72–4.

127 Scientific Committee on Problems of the Environment (SCOPE), *Environmental Consequences of Nuclear War*, Vol. 1, A. Barrie Pittock *et al.*, eds, *Physical and Atmospheric Effects*; Vol. 2, Mark A. Harwell and Thomas C. Hutchinson, eds, *Ecological and Agricultural Effects*, Chichester, John Wiley & Sons, 1985–6.

128 E.G. Nisbet, *Leaving Eden: To Protect and Manage the Earth*, Cambridge, Cambridge University Press, 1991, 92; figures from International Atomic Energy Agency annual report, 1987, GC (XXXII), 835.

129 Merrill Eisenbud, "The Ionizing Radiations," in B.L. Turner, ed.,*The Earth as Transformed by Human Action*, Cambridge, Cambridge University Press, 1990, 455–66, at 462.

130 See Robert Michael Pyle, *The Thunder Tree: Lessons from an Urban Wildland*, Boston, MA, Houghton Mifflin, 1993, for an informed evocation of another Denver watercourse, the High Line Canal.

131 Ian Douglas, *The Urban Environment*, London, Edward Arnold, 1983; Michael Hough, *Cities and Natural Process*, London, Routledge, 1995.

132 Gleaves Whitney, *Colorado Front Range: A Landscape Divided*, Boulder, CO, Johnson Books, 1983, 67.

133 Andrew C. Isenberg, "The Returns of the Bison: Nostalgia, Profit, and Preservation," *Environmental History* 2, 2, 1997, 179–96, gives an account of various motives for saving, and using, bison in the United States.

134 Frederick R. Rinehart and Elizabeth A. Webb, eds, *Close to Home: Colorado's Urban Wildlife*, Boulder, CO, Roberts Rinehart, 1990.

135 Chris Madson, *When Nature Heals: The Greening of the Rocky Mountain Arsenal*, Boulder, CO, Roberts Rinehart, 1990.

136 Susan Jones, "Becoming a Pest: Prairie Dog Ecology and the Human Economy of the Euroamerican West," *Environmental History* 4, 4, October 1999, 531–52.

137 Mark H. Rose, *Cities of Light and Heat: Domesticating Gas and Electricity in Urban America*, University Park, Pennsylvania State University Press, 1995.

138 Raymond Dasmann, "Toward a Biosphere Consciousness," in Don Worster, ed., *The Ends of the Earth*, Cambridge, Cambridge University Press, 1988), 277–8.
139 Lyle W. Dorsett, *The Queen City: A History of Denver*, Boulder, CO, Pruett Publishing, 1977, 159.
140 Joseph Conrad, ed., *Colorado Environmental Handbook: The State of the State*, Denver, Colorado Environmental Coalition, 1996, 24–51.
141 James L. Cox, *Metropolitan Water Supply: The Denver Experience*, Boulder, CO, University of Colorado, Bureau of Governmental Research and Service, 1967.
142 Edmund P. Russell III, "Lost Among the Parts Per Billion: Ecological Protection at the United States Environmental Agency, 1970–93," *Environmental History* 2, 1, 29–51, reviews the history of this bureau.
143 Carolyn Etter and Don Etter, *The Denver Zoo: A Centennial History*, Boulder, CO, Roberts Rinehart, 1995, 34.
144 Lyle W. Dorsett, *The Queen City: A History of Denver*, Boulder, CO, Pruett Publishing Co., 1977, p. 148.
145 Stephen J. Leonard and Thomas J. Noel, *Denver: Mining Camp to Metropolis*, Boulder, University Press of Colorado, 1991, 262.
146 W.E. Riebsame, H. Gosnell, and D.M. Theobald, "Land Use and Landscape Change in the Colorado Mountains I: Theory, Scale, and Pattern," *Mountain Research and Development* 16, 4, 1996, 395–405.
147 Etter and Etter, *Denver Zoo*, 169.
148 Jerry Craig, "Peregrine Recovery in Downtown Denver," in Frederick R. Rinehart and Elizabeth A. Webb, eds, *Close to Home: Colorado's Urban Wildlife*, Boulder, CO, Roberts Rinehart, 1990, 115–32.
149 Joel A. Tarr, *The Search for the Ultimate Sink: Urban Pollution in Historical Perspective*, Akron, OH, University of Akron Press, 1996.
150 Laurance C. Herold, "The Urban Climate," in Elizabeth A. Webb and Susan Q. Foster, eds, *Perspectives in Urban Ecology*, Denver, CO, Denver Museum of Natural History, 1991, 35–44.
151 Michael Hough, *Out of Place: Restoring Identity to the Regional Landscape*, New Haven, CT, Yale University Press, 1990.
152 William Cronon, *Nature's Metropolis: Chicago and the Great West*, New York, W.W. Norton, 1991, see especially 5–25.
153 Orie L. Loucks, "Sustainability in Urban Ecosystems: Beyond an Object of Study," in Rutherford H. Platt, Rowan A. Rowntree, and Pamela C. Muick, eds, *The Ecological City: Preserving and Restoring Urban Biodiversity*, Amherst, University of Massachusetts Press, 1994, 49–65.
154 Lewis Mumford, *The City in History*, New York, Harcourt, Brace & World, 1961, 576.

9   Present and future

While histories do not often concern themselves with the future, it is appropriate for a world environmental history to look at the trends active in the present that are likely to persist into the future and will continue to affect the worldwide picture. Some of these were discussed in Chapter 8. In this introduction, I will comment briefly on three kinds of change that are particularly salient and which promise to shape the future in positive and negative ways. These are high technology, including space technology; the world market economy in relation to natural capital; and the reduction of biological diversity.

A pervading transformation that seems certain to dominate human interaction with the environment is the continuing spread of high technology and its rapid series of innovations, so radical as to merit the historian's designation as a new technological revolution. Machines with greater power and sophistication in making environmental changes will be created. The speed and spread of the reach of communication will continue to accelerate. Information of many kinds, including the facts of environmental change, will be more easily available. At the same time, governments will be able to watch social developments, gather information on their citizens, and possibly control their actions as never in the past. Satellites and other instruments in space will provide ever more detailed knowledge about the Earth's environment, and information on processes of change that will aid in making judgments about sustainability and the advisability of various kinds of projects. The purposes to which such knowledge will be put remains in question.

In the latter half of the twentieth century, the exploration of the universe beyond the Earth provided a series of startling insights. People everywhere saw images of the Earth photographed from the Apollo 8 space capsule on the way to the Moon, as a single planet, undivided by borders, a small island of life in a sea of space. Some have dated the beginning of prevalent modern environmental concern from that glimpse of our planet; as the poet Archibald MacLeish put it,

> For the first time in all of time men have seen the Earth with their own eyes – seen the whole Earth in that vast void as even Dante never dreamed of seeing it ... It may remake our lost conception of ourselves ... To see the Earth as we now see it, small and blue and beautiful in that eternal silence where it floats, is to see ourselves as riders on the Earth together ...[1]

The view back toward the home planet, with the incredible detail of its environment that could be discerned in every part of it, is an aspect of the various national space programs with long-term value. In fact, that was the justification given by the United States and the USSR in 1955 when both nations announced that they would launch earth-orbiting satellites as part of their participation in the International Geophysical Year sponsored by the United Nations, which covered the eighteen months from July 1957 to December 1958.[2] The Soviets launched two satellites, Sputnik I and II, in late 1957, and the Americans followed with Explorer I in early 1958. Within the following three decades, eight nations had placed satellites in orbit, and others had participated in some of these flights. Many of the experiments were surveys intended to observe the Earth's atmosphere, geology, ecology, oceans, and land use. Also graphically revealing were time-lapse images of clouds and precipitation, now routinely used in forecasting and television weather programs. The patterns of globe-circling atmospheric systems resembled organic circulation and may have suggested the renewal of the idea that the Earth (Gaia) is alive.[3] But much more than weather can be observed from space. Applications of satellites like the American Geodetic Earth Orbiting Satellite (GEOS) and Landsat include forecasting crop production, assisting in soil and forestry management, locating energy and mineral resources, and measuring urban population densities. Landsat management was privatized in 1984, and much of the information gathered is available for commercial as well as scientific use.

While national investments in space programs slowed late in the century, it was nonetheless possible to gather a tremendous body of data not only about the solar system and the galaxies, but also about the process of environmental change on the Earth. Ecosystems could be inventoried and the process of change within them measured. The pace of deforestation and of atmospheric change, to give two examples, could be monitored. The UN has no satellites of its own, but coordinates information from member states. The United Nations Environment Programme (UNEP), described in more detail later in this chapter, collects remote sensing data as part of a cooperative effort called Global Environmental Monitoring System (GEMS), and has assembled a Global Resources Information Database (GRID) as a major data management program. UNEP programs have been concerned with constructing natural resource databases, monitoring changes in tropical forest cover and desertification, assessing soil erosion and lake sedimentation, making analyses of watersheds, and testing the applicability of high-resolution satellite data to urban management and planning.[4] The European Space Agency has an Earth observation program, with environmental studies and monitoring of resources among its objectives, and there are several other intergovernmental organizations with similar tasks.

The International Geosphere–Biosphere Program was proposed in the late 1980s and carried out in the 1990s. This was a global research effort to study the interrelated processes of the atmosphere, hydrosphere, geosphere, and biosphere, necessary for a comprehensive understanding and evaluation of the effects of human activities on the environment. It had the support of the International Council of Scientific Unions and, in the US, the National Science Foundation (NSF) and the National Aeronautics and Space Administration (NASA).

Two satellite photographs of the Brazilian state of Rondônia placed side by side, one taken in the early 1980s and the other in the early 1990s, show a startling pattern of deforestation. A branching system of long, straight roads with bare fields along them has spread across and fragmented an expanse that was once unbroken rainforest. In ten years, more than half a million settlers were brought into this frontier region and had managed to clear about 25 percent of the land, or 60,000 sq km (23,000 sq mi) out of 243,000 sq km (94,000 sq mi).[5] The pace of blight continued even after the rate of immigration fell, because when the farms first opened lost their fertility, many of their tenants moved further into the forest to eke out a living in new clearings. At the same time, though almost invisible in the photographs, native South American Indians who lived as hunter–gatherers within the rainforest found the means of their subsistence destroyed, and in many cases were killed or driven into isolated tracts where they were given little protection even when these areas were designated as reserves for them. Rubber tappers, who had earned a living over the decades by gathering latex sap from the widely separated rubber trees in the natural forest, found their jobs in peril, and when they organized to resist, their leader, Chico Mendes, was murdered.[6] The environmental history of the Amazon basin is the subject of a later section of this chapter. The images of rapid environmental destruction in the late twentieth century are numerous and striking, and information technology has made possible a degree of accuracy in their gathering and an extent of dissemination that makes an unprecedented impression on human consciousness.

One of the most far-reaching ways in which human impacts on the natural environment are augmented in present times is the growing world market economy. This is true because industrialization and intensive agricultural production increase demands for land and resources and generate pollution, and trade accelerates economic growth. Demand in one region can be met by impacting the environment in a distant part of the world. For example, urban North Americans who want fruits in the winter can import them from Chile in the Southern Hemisphere, where seasons are reversed. Japan prefers to import timber from the tropics rather than increase the pressure on domestic forests. This distancing of the consumer from the sources of resources makes ecological awareness difficult. Where people depended on what local ecosystems could supply, they were aware of environmental worsening and anxious to reverse it. But the world market economy transfers resources from the region where they were produced to a second region where they are consumed, and may dispose of the wastes in a third region. As Gilbert Rist analyzes it,

> Everything undertaken in the name of expanding international trade allows production to be dissociated from consumption and consumption from disposal (that is, from conversion into visible or invisible waste). This spares the consumer-polluter from realizing that he is involved in using up resources and accumulating waste, as the trade circuit obscures what is actually taking place. Transnational companies favor this dilution of responsibility, operating as they do in many different places at once and constantly splitting creation from destruction of resources. The "polluter pays" principle does not do away with pollution, but implies that those with the means can reserve the right to pollute.[7]

As mentioned in Chapter 7, the world system of "free trade" which gives a degree of unrestricted operation to multinational corporations is facilitated by a number of supranational agreements and organizations including the International Monetary Fund (IMF) and the World Bank (officially, the International Bank for Reconstruction and Development, or IBRD), which emerged from the Bretton Woods Conference of 1944. The General Agreement on Tariffs and Trade (GATT) was negotiated subsequently. At first focused on the need to help Europe recover from the Second World War, these agencies later concentrated their efforts on encouraging economic growth in the less industrialized countries and world trade generally. The organization intended to oversee GATT was the International Trade Organization (ITO), which was fairly weak, but was succeeded in 1995 by the more effective World Trade Organization (WTO) as a result of the "Uruguay Round" of trade negotiations.[8] The WTO, with a membership of over 150 nations, can make a claim to universal oversight. It is committed to ceaseless growth in trade and the world economy. These organs of the international financial system have eroded the traditional sovereignty of the nation-state, and their effects on the biosphere have yet to be measured.

Dominant economic thought today presents a neoclassical model that treats the environment as a factor of production, a subset of the human economy, instead of what it is: a biophysical system which embraces the human economy and makes it possible.[9] Market economists discount the importance of natural resources, maintaining that the market and human technology will find substitutes for whatever we run out of. Living organisms and their diversity are attributed no intrinsic value in their calculations, which become so mathematically abstract that they usually ignore human values as well. An attitude that treats the natural world not as a series of ecosystems that include human beings, but as a set of resources and commodities separate from humankind, is dangerous. Unfortunately, this way of resigning from the community of life has been embodied in institutions of the world economy which have nullified some national laws including several intended to protect endangered species in international trade. Some economic theorists regard environmental regulations as an unnecessary restraint of trade. Fortunately, there is a growing number of environmental economists who argue for sustainability, conservation of resources, and the protection of biodiversity, such as Herman Daly, Robert Costanza, and Robert Goodland.[10]

The WTO provides very limited support to measures for environmental improvement. It permits its member nations to enforce laws necessary to protect the life and health of humans, animals, and plants, and to conserve natural resources, but does not address the broader area of environmental protection. A landmark case was brought by Mexico in 1991 before a GATT panel. The US, under its Marine Mammal Protection Act, had decided that Mexican-caught tuna would be excluded unless Mexican fishermen used methods that would spare the thousands of dolphins that were being destroyed in their nets. GATT decided that this was an improper attempt by the US to impose its own environmental regulations on Mexico, and ordered the US to accept tuna that was not "dolphin-safe."[11] Critics of the decision pointed out that an appointed panel had negated a law passed by the democratically elected government of a member state, a state which is not among the weaker ones economically or politically. GATT also has determined that a Canadian law to conserve fisheries, a Thai limitation on cigarette imports, and US measures that use taxes on oil and chemical feedstocks to pay for cleaning up hazardous wastes are unfair obstacles to trade.[12]

Neoclassical economists oppose on principle such measures as the ban on trade in ivory, while the global economy seems designed to assure by inflating prices on rare commodities that the trade will continue until the last tusker is harvested. Living forests are conceived as economic abstractions, which means clearcutting to save on labor costs, not careful selective silviculture. The subsidy the economy has been taking from wild nature may be near an end,[13] as the last wild places yield to the inexorable advance of tree farms, industrial agriculture, strip mines, power plants, and urban encroachment. Pollution carried by air and water to formerly distant regions affects even protected wilderness.

The emerging world trading system ignores an ecological principle, namely the limiting factor. Ecologists point out that any organism can increase in number and total biomass, and spread geographically, to the point where it encounters an environmental factor that prevents further increase. Liebig's law of the minimum means that growth is limited by the least available factor.[14] That factor may be another species in the ecosystem, or water, a chemical substance, or physical space. Obviously none of the limiting determinants is infinite in availability, so that every organism, every species, faces a limit to its growth.

The principle of the limiting factor obviously runs counter to the present doctrine of economists, who regard unbounded growth not only as a possibility, but as the preferred solution to poverty. Since they contend that environmental quality is a "luxury good" desired only by people whose basic needs for food, shelter, and economic security are already met, they believe that economic growth is the best way to achieve environmental improvement.[15] The world market economy seeks to escape from local limitations by tapping resources around the globe, but fails to recognize the limits of the Earth itself. Looking at the beginning of the new millennium, neoclassical economist C. Fred Bengtsen predicted,

> standards of living will rise sharply almost everywhere, even as the global population rises to between 12 billion and 15 billion, as technology continues to expand exponentially and virtually all regions adopt the policy reforms that began to proliferate in the late 20th century.[16]

This "rosy view," which posits China and India with more than 2 billion apiece with a living standard equal to or greater than that of the present-day US, pays no attention to the finite dimensions of fossil fuel supplies and the ultimate constraints of the laws of thermodynamics.

Ecologists point out that the environmental degradation contingent upon the resource use required for such growth would interfere with meeting the basic needs of the vast new human population. One recalls the images of settlers in desolate, logged, and burned-over stretches of the Amazon basin, or the workers in polluted districts of Romania covered with soot. By 1985, the proportion of the world's population living in cities was more than 40 percent, and in the twenty-first century, more than half of all humans will live in large urban concentrations. Cities in the less industrialized countries are growing most rapidly, and their slums make up most of this growth.

There are limits to human population. Malthus advanced one: the availability of agricultural land. But we should look for other factors that may come into play even before we run out of arable soil. An independent group of economists, scholars, and industrialists called the Club of Rome met for the first time in 1968, appropriately in the Accademia dei Lincei, home of a society to which Galileo Galilei belonged. They launched a program to determine with the aid of computer analysis when the world economy might run out of essential non-renewable materials and reach what their first report called *The Limits to Growth*.[17] Their computer models indicated that a complex disaster of resource shortages, overpopulation, and massive pollution would happen during the twenty-first century unless drastic and unlikely counter-measures were taken. Economists attacked the methodology and conclusions of the report, with some accuracy, since a number of the deadlines set by the report have already passed without the debacles it predicted.[18] The Club of Rome undertook a revised study in light of the criticisms and in 1992 concluded,

> The human world is beyond its limits. The present way of doing things is unsustainable. The future, to be viable at all, must be one of drawing back, easing down, healing. Poverty cannot be ended by indefinite material growth; it will have to be addressed while the material human economy contracts.[19]

Meanwhile, in 1987, the World Commission on Environment and Development, created by the United Nations and chaired by Prime Minister Gro Harlem Brundtland of Norway, issued a report, *Our Common Future*, which brought to the fore the concept of "sustainable development," two words which were to resound in the halls of international organizations and receive the endorsement of numerous conferences. This idea represents the hope that economic improvement and environmental protection can go hand in hand. According to the report, "sustainable development is development that meets the needs of the present without compromising the ability of future generations to meet their own needs."[20] Such a definition implies living within limits, but the report mentions only "limitations imposed by the state of technology and social organization on the environment's ability to meet present and future needs." But there are limits other than those set by our abilities to use resources; there are limits set by natural systems. Humans cannot use more biomass than photosynthesis produces. We cannot generate more heat than the atmosphere can dissipate. We are limited by deterioration of soils, loss of biodiversity, and degradation of the ozone layer.[21] Unfortunately, in many discussions of the world market economy, sustainable development is taken to mean indefinitely continued growth, which evades the question of limits and is intrinsically impossible.

Herman Daly and other economists envision a "steady state" economy that would operate within the constraints set by the biophysical environment.[22] This would mean that the size of the human population would stabilize, perhaps at a level lower than at present. Use of non-renewable resources would slow and eventually depend entirely on recycling, while use of renewable resources would remain below the replacement level. At present, these ecological economists, however discerning they may be, have little influence on the course of the world market economy. Still, even many of those friendly to expanding markets have been forced to admit, "There's not much point in growth that completely lays waste to the environment."[23]

Salient among the processes of change that will extend into the new century is the impact of changes caused by human intervention in natural ecosystems, including habitat destruction, extinction of species, and loss of biological diversity, often called biodiversity. Although it is perhaps most often used to indicate the number of species in an ecosystem, biodiversity is "the variety of living organisms at all levels, from genes to species, populations and communities, including ... habitats and ecosystems."[24] Evolution seems to foster biological diversity by its innate tendency to variation, producing forms of life to occupy every available niche in the environment. These forms in turn offer niches for additional forms, i.e. "other bugs to bite 'em ... ad infinitum." Many ecologists contend that a high degree of biodiversity helps to maintain the balance and productivity of an ecosystem in reaction to moderate stress, according to the analogy that when one of its strands is broken, a net or web can continue to hold better if it has many strands instead of a few.

We can appreciate the value of diversity to ecosystems as a manifestation of life itself. There is also value to humans, since we are inevitably part of and depend upon functioning ecosystems. Historians of science and medicine have pointed out the importance of biodiversity in the discovery of drugs and other useful substances. The synergy of the diversity of human cultures with biodiversity has provided much of the knowledge necessary to this discovery. Societies living in close contact with abundant ecosystems have complex ethnobotanies and ethnozoologies, so that humankind as a whole can potentially gain from them and has an interest in preserving both biodiversity and indigenous peoples.

The destruction of both kinds of variety is a notable fact of world history, especially in the last century or two. Habitat destruction, with attendant extinction of species, and pollution with toxic substances are among the ways in which biodiversity is being diminished, along with human cultural diversity. E.O. Wilson conservatively estimates the rate of decline of biodiversity by comparing the present rate of extinction caused by worldwide human interference, 27,000 per year, or seventy-four per day, with the "background" extinction rate over millions of years in the past, and finds that the rate today is between 1,000 and 10,000 times higher than the natural rate.[25]

In the twentieth century, more powerful technologies, increasing exploitation of natural resources, and an expanding human population led to an accelerating destruction of other forms of life by humans. A hundred years ago, large sections of the continents were still teeming with wildlife. There seemed to be no end to the bounty of the sea. By the 1990s, extinctions had occurred on a scale matched only by catastrophic events of the geological record.[26] Wildernesses shrank to isolated retreats, and few were safe from destructive invasions. Varieties of frogs and other amphibians inexplicably disappeared in ecosystems around the world.[27] India had 4 million blackbuck antelope in 1800; only 25,000 remained in 1990. One of the blackbuck's major predators, the cheetah, vanished from India. Similar declines have been recorded for other animals around the world. Wild ecosystems shrank and their component species declined in number or disappeared.

Technology provided humans with immense power to fracture ecosystems and to alter the environment. Assault weapons designed for use in war came into the hands of poachers. In open-pit mines, bulldozers and excavating machines large enough to dwarf the dinosaurs stripped away vegetation, soil, and underlying rock. Giant dams impounded reservoirs that flooded extensive lowlands, former homes of many forms of life. Ancient forests fell to clear-cutting so rapidly as to threaten their disappearance before the twenty-first century is half over. The unparalleled ecological richness of the rainforests, with the genetic record of millions of years of evolution, gave way to agricultural and mining projects of questionable long-term value. Their removal means a crisis of extinction. Between the mid-1970s and mid-1980s, the timber extracted legally from the primeval forests of the Brazilian Amazon rose more than 270 percent, from 10.36 to 28.10 million cu m (13.5 to 36.5 cu yd).[28] In addition, entrepreneurs and settlers cut and burned much larger amounts, and the amount taken illegally can only be guessed. By 1980, according to the UN Food and Agriculture Organization, 78 percent of Ghana's forests had been logged, and Costa Rica was cutting 4 percent of its forests annually.[29] When a single ridge top in Peru was cleared, more than 90 plant species known only from that locality were lost.[30] At the same time, the original forests of giant trees in the northwestern United States and western Canada, and the vast taiga of the Soviet Union, were being logged faster than the Amazon. In the United States in the 1960s, Congress had enacted laws to protect endangered species, but there is as yet no law to protect endangered ecosystems. International concern appeared over the imminent extinction of single species: the panda in China, the tiger in India and Siberia, and the elephant in Africa. These are highly visible indicator species, but the real problem in each case is the diminishment of the ecosystem to which each of them belongs. It is a process often called "habitat destruction," but in fact it is the fragmentation of communities of life.

Biodiversity was on the agenda of the United Nations Conference on Environment and Development, the Earth Summit at Rio in 1992. Most discussion was not on the need to preserve species and ecosystems, but their usefulness for sustainable economic development, and the demands of industrializing nations to distribute the gains realized from the development of biological resources more equitably. A primary document produced by the conference was the Convention on Biological Diversity.[31] Its goals are the conservation and sustainable use of biodiversity and fair trade and compensation involving products made from the genetic resources of nations. It charges each signatory to make plans to protect habitats and species, and provides for aid to developing countries to help them do this. There are regulations concerning biotechnology. The treaty was signed at the conference by 153 nations of 178; only the US voiced refusal to sign, on grounds that the financial obligations were open-ended and insufficiently supervised. President Clinton later signed it, but when it was submitted to the Senate for ratification, although the Foreign Relations Committee approved it, a sufficient number of conservative senators to defeat it expressed opposition (a treaty must be approved by a two-thirds vote), and consequently a vote was not scheduled. Theoretically, it could still be revived. As John Rodman remarked, "The ecology movement, to the extent that its central worry is the rapid extinction of ecological diversity, is essentially a resistance movement against the imperialism of human monoculture."[32] Some environmental non-governmental organizations, therefore, objected to the Convention on Biological Diversity because it assumes that non-human forms of life on Earth are the property of nation-states. It forbids interference in the way any nation chooses to protect or exploit species within its borders. But national frontiers rarely coincide with ecosystems, and the welfare of life on the whole planet is of concern to all. Of course, no other species had representation at Rio, nor did any ecosystem.

If the cultural attitudes of the modern industrial age remain the determiners of human actions in regard to the ecosystems of which humans are part, while the human population continues to increase, or remains at its present excessive level, an unprecedented crisis of survival is likely in this century. Humankind is subject to change as a result of the impact of a rapidly diminishing biosphere. As E.O. Wilson put it, "We are in the fullest sense a biological species and will find little ultimate meaning apart from the remainder of life."[33] It is the community of life in its many forms, not humankind alone, that made us what we are.

Amazon: threats to biodiversity

A canopy walkway gives access to the treetops in the rainforest near the Amazon Center for Environmental Education and Research (ACEER) in Peru. Usually a visitor to the ancient tropical forest must peer upward through many layers of foliage that grow as trees strive to reach and use every bit of available light, along with the epiphytes: bromeliads, orchids, and other plants that perch on the tree trunks and branches. But the canopy walkway ascends by a series of wooden stairways and long, hanging bridges suspended between the most massive emergent trees, reaching a level almost 60 m (200 ft) above the forest floor. From that height, above the early morning mist, I looked out over the unbroken rainforest in a biosphere reserve, the largest remaining refuge in the western Amazon basin, the least disturbed part of a sea of trees.

The variety of life in that place strikes anyone who looks for it. The trees are of scores of different species; standing in a platform on one, the observer may have to look far to find a second one of the same kind. I saw one tree covered with bright yellow leguminous flowers, and never saw another. There are at least 60,000 species of plants in the Amazon basin, and there can be hundreds in a single hectare. Then there are the birds; different ones in the canopy from those near the ground. A friend also staying at ACEER was a bird-watcher who had a life-long list of those he had seen. In one week in this forest, he recorded many species, including toucans, macaws, oropendolas, woodcreepers, antbirds, and curassows, to name a few, more than the total number that have ever been seen in my home state of Colorado. We also saw the archaic-looking hoatzins, but only after a long hike in search of them. Insects were extremely numerous, of many different kinds and striking forms and colors; with the exception of ants, I never found myself surrounded by many of the same species. Scientists believe that the number of species of insect in the Amazon numbers in the millions, most of which have never been described and named.[34] I photographed a remarkable lizard with a reddish-brown head and a blackish body; later on, a researcher at ACEER said that this species has never been noted anywhere else.

High species diversity is characteristic of moist tropical forests, and nowhere is it more notable over a large area than in the Amazon. Most major groups of living things there exhibit an amazing number of varieties, including mammals, birds, reptiles, amphibians (the frogs around ACEER are numerous and of kaleidoscopic colors and patterns), fresh water fish, insects, spiders, snails, flowering plants, and ferns.[35] Each species represents a long history of evolution in this warm, moist environment, in competition and in cooperation with others. The information contained in the DNA of any one of those species represents a priceless fund of biotic information that will be lost if it becomes extinct. ACEER is located in an area drained by the Río Napo, a tributary of the Amazon, which has been identified in several scientific surveys as a Pleistocene refuge. That is, during the temperature changes and desiccation of the Ice Ages, the rainforest and its species survived there, and spread to reoccupy the entire basin. The Forest Development and Research Project of the UN's Food and Agriculture Organization designated it as the largest First Priority Conservation Area in the Amazon.[36] If the area near ACEER had been logged, my lizard and her kind might have disappeared forever along with unknown numbers of beetles and other species. Such things are unfortunately happening every day along the Amazon and its tributaries, and around the world in tropical forests.

The Amazon is the largest river in the world, in volume. It drains an area of 7 million sq km (2.7 million sq mi), and the outflow to the Atlantic Ocean is 175,000 cu m (6.25 million cu ft) per second,[37] an amount exceeding the total discharge of the next ten largest rivers. This outpouring carries fresh water into the sea for scores of kilometers. The river is often 11 km (7 mi) wide, and ocean-going ships can navigate upstream 3,700 km (2,300 mi) to Iquitos, Peru. The volume of fresh water is due to rainfall in the Amazon basin averaging 2,300 mm (90 in) annually, reaching 3,600 mm (142 in) in the northwest. More than half the water that falls as rain is returned to the atmosphere by evaporation and transpiration, and much of it falls again as rain as the air masses that carry it move westward toward the Andes. The existence of the rainforest and the physiological processes within it increase the rainfall upon which it depends. The activities of life improve the conditions for life. But "if the forest is destroyed, the system will regress from the current dynamic equilibrium to a ... state characterized by lower annual precipitation, which would represent a climatic change."[38]

The richness and variety of animal and vegetable life in the Amazon rainforest, and the enormous biological mass which it contains per unit area, led potential exploiters to assume that the basin had fertile soil, and that if the forest were cleared, rich crops would grow there. But experience has proved otherwise. Soils in much of the area are poor in minerals and organic matter, some of them extremely so. Luxuriant rainforests grow on, not in, these infertile soils. Almost all the organic material is in the forest, not the soil, and an efficient system of recycling keeps it there. There is a thin layer of decomposing material on the forest floor. Tree roots spread out in thick mats on the surface to absorb the available minerals; many trees are buttressed to give them support in the absence of deep root systems. Often roots will climb up the trunks of adjacent trees to absorb nutrients leached from those trees. Every leaf that falls represents valuable nutrients and is quickly reabsorbed by the living portion of the ecosystem.

The removal of a large section of rainforest is a catastrophe for the ecosystem. Initially, the organic material left on the surface of the ground, or the minerals in the ashes left by fire, may fertilize the soil enough for a crop or two afterwards, but that will be all. Erosion is rapid as torrential rains beat down on fragile soils. The forest will not return quickly, and may be replaced by grasses. Also, soils found commonly in the tropics, when exposed to downpours

Figure 9.1   A lizard in a rainforest tree on the canopy walk at the Amazon Center for Environmental
Education and Research near the Rio Napo, Peru. According to a researcher at
ACEER, this remarkable lizard with a reddish-brown head and blackish body
represents a species that has never been noted anywhere else. High species diversity
is characteristic of moist tropical forests, and nowhere is it more notable over a large
area than in the Amazon Basin. Photograph taken in 1995.

and heat, turn into a bricklike substance called laterite and lose their productivity.[39] The
popular image of the aggressive jungle invading open country may be true for small clearings
in the forest, such as those native people made for swidden agriculture, but it is a myth
insofar as the wide swaths made by fire, bulldozers, and logging machinery are concerned.[40]

The present state of the Amazon basin must be viewed in perspective of the history of
human occupation. Indigenous people occupied the lands around the river for perhaps
12,000 years before Europeans arrived.[41] They displayed a great variety of cultures, many of
them village societies depending on hunting, gathering, fishing, and swidden agriculture.
Recent archaeological research has shown that some of these peoples had more complex
societies and more sophisticated agriculture than had hitherto been expected, and artifacts
including pottery and large earthworks. Students of prehistory in this area have raised their
estimates of the size and density of native population, and the scale of human effects on the
rainforest and its denizens.

The routes of European exploitation followed the rivers, and their first settlements were
in the floodplains. The Spanish explorer Vicente Yáñez Pinzón sailed into the Amazon's
mouth in 1500, and Francisco de Orellana descended the river by ship in 1541–2, reaching
the Amazon from Ecuador by the Río Napo. The Portuguese gained the upper hand against
Spanish, British, French, and Dutch interests in the lower and middle sections of the river.
Pedro Texeira ascended the Amazon and Napo and eventually reached Quito, reversing
Orellana's route, in 1638. Europeans sought gold, unsuccessfully at first, although it would

eventually be found. Their relations with the native peoples were ambivalent; they founded missions to civilize them, but also made efforts to defeat, enslave, and destroy them. The Indians contracted diseases brought by Europeans and died in great numbers. The Jesuits, who were authoritarian and suppressed tribal customs, but who tried to protect their native converts, were expelled from Portuguese and Spanish dominions in 1759 and 1767, leaving the Indians in the hostile hands of secular authorities. Native communities suffered extreme reductions, and many were wiped out, although not without fierce resistance.[42] Settlers experimented in establishing plantations of sugarcane, cotton, tobacco, and rice; and in the collection of forest products including valuable woods, nuts, oils, and flavorings such as cacao (chocolate), the most important export during colonial times, vanilla, root beer, and substitutes for clove and cinnamon.[43] Clearings were limited to lowlands near riverbanks, but these were the areas that had been most densely occupied by native peoples. By the 1750s there were few Indians left there. As late as 1840, however, the vast interior of the Amazon forest was relatively intact, due more to barriers to travel than to governmental conservation efforts.[44]

Then came the Great Rubber Boom. Charles Goodyear perfected vulcanization of rubber in 1839, and latex from wild trees was in demand for hoses, belts, shoes, and raincoats.[45] Tens of thousands of men were recruited as rubber tappers. The trees that could be bled for latex were widely distributed, so tappers had to travel long distances. They came into conflict with tribes in the inner forest, and genocide and slavery spread.[46] There were serious impacts on flora and fauna. An upsurge of population followed as laborers flocked in from many parts of Brazil, especially the impoverished northeast, and from abroad. River traffic increased. The population of the Amazon basin increased by a factor of ten from 1820 to 1910. Manaus grew from 5,000 in 1870 to a city of 50,000 in 1910, boasting an opera house and public library. Iquitos was founded as a port for rubber export in 1864.[47] But the rubber balloon burst as rubber plantations came into production in Malaysia and south India.[48] The price of rubber dropped, and it was no longer profitable to send men into Brazilian forests on long collecting trips. Prosperity disappeared, livestock production dropped, and the economy and population stagnated between 1920 and 1940. Henry Ford started rubber plantations in Brazil, but leaf blight swept through the monoculture and his attempted modernization of production was a financial failure.[49]

The Second World War began the period of greatest environmental change in the Amazon forest. With the Japanese occupation of southeast Asia, the United States turned again to Brazil for rubber. But then it became possible to manufacture synthetic rubber from petroleum, and the market for Amazon wild rubber shrank once more. Other economic factors brought in more population, denuded forest land, and caused a crisis of such proportions that the survival of the rainforest became an international issue. Most of these large-scale changes have occurred since 1970.

One agricultural incursion, especially in the Peruvian and Colombian headwaters, was coca, the raw material for cocaine.[50] A multimillion dollar illegal business, it is protected by an international crime syndicate. The area carved from forest for coca production in Peru rose from 16,360 ha (40,425 acres) in 1964 to 200,000 ha (500,000 acres) in 1990.[51] The poor are often forced into actions which are ecologically destructive through no fault of their own, but necessary for survival. This is the case in the Brazilian province of Rondônia, where thousands of square kilometers have been stripped of their trees and countless animal species by destitute people from overcrowded cities, who have been promised land in the interior, but who lack the resources to exploit it to support them, if indeed it ever could. After destroying the environment they had come to farm, many are forced to hunt and fish for subsistence, or move back to the cities worse off than before. Meanwhile, cattle ranching by large landowners returned to the Amazon, driven by demand from fast-food companies in the US and elsewhere. Livestock raising involves clearing vast areas of forest, largely by burning. It exposes the soil to erosion and its productivity declines rapidly, causing further cycles of deforestation.

Timber corporations in the Amazon have selected the most valuable species to the extent that merchantable examples of mahogany, cedar, podocarpus, etc. are rare or nonexistent. Logging to meet demand in industrialized countries no longer takes high-grade trees only, since even low-grade wood can be used for wood chips, pulp, and paper. Clearcutting has become a practice in the Amazon, facilitated by a far-reaching network of highways.

In earlier times, rivers were the avenues of invasion into the rainforest. Now they have been augmented by roads cut through blocks of wilderness. Brazil signaled its intention to exploit the interior by establishing a new capital, Brasilia, in 1960, halfway between the old coastal capital, Rio de Janeiro, and the Amazon. A highway was pushed from Brasilia to Belém, followed by work on the Transamazon Highway cutting east to west through the heart of the rainforest, with plans to connect through Peruvian highways to the Pacific, bisecting the continent. Meanwhile, other roads pushed into Amazonia from the south, bringing ecological impacts in their wake.

Hydroelectric dams destroyed expanses of rainforest. The Amazon descends only 55 m (180 ft) in elevation from the Peruvian frontier to its mouth; since the gradient of the Amazon and its tributaries is so low, even a low dam will impound a reservoir covering a vast area. Many dams have been proposed, and if built would result in a greater loss of biodiversity than anywhere else on Earth.[52] They would also displace indigenous people. The Tucuruí Dam, with a length of 1.2 km (0.75 mi), impounds the Rio Tocantins. Five major dams and a number of smaller projects have been built. Balbina Dam near Manaus drowned 250,000 ha (618,000 acres) of rainforest and two native towns, generates little power at high cost, and represents a public works fiasco. Reservoirs interfere with migration of fish populations and provide a breeding ground for the disease-spreading mosquito.

El Dorado, the lure of gold, was a myth for early explorers of the Amazon, but in the gold rush after 1979 it became reality. The international price of gold had reached phenomenal heights. Prospectors and wildcat miners in the hundreds of thousands probed every part of the basin, and many of them found what they were looking for. The richest find was Serra Pelada, in the Carajás mountains of Pará state, where swarming miners scratched out 40 tons of gold by 1986. Mining damages the rainforest by tearing up the soil, exacerbating erosion, and most seriously by causing mercury pollution. The effect on the native people is devastating. The Yanomamö people, one of the few remaining tribes in the Amazon that maintain their traditional ways, live in northern Roraima state and adjoining Venezuela.[53] A gold rush beginning in 1987 brought in 40,000 miners – there were only 20,000 Yanomamö. Violent clashes occurred. The Yanomamö are fierce warriors, but the miners, better armed, slaughtered more than a thousand of the natives, raped women and forced many into prostitution. Drugs, venereal diseases, malaria, and tuberculosis took a heavy toll. The Brazilian government vacillated between colluding with the mining interests and declaring that the rights of the Yanomamö would be protected, but with patchy enforcement. In contrast, another warlike people, the Kayapó, who live on the Xingu River near Serra Pelada, have become familiar with non-Indian law and politics and have used their knowledge adroitly to gain native rights and title to their land.[54] Petroleum and natural gas have been located in the western Amazon, and oil operations in the Ecuadorian rainforest have caused excessive pollution and damage to native people.[55]

Figure 9.2 A fisherman casting a net into the River Amazon below Iquitos, Peru. The people in this area depend on fish for much of their food. Many of the species of fish subsist on fruit from the rainforest trees, which are being cut down, especially in the river floodplain. An ecological cycle is being broken. Photograph taken in 1995.

Roads, dams, mines, oil wells, and cattle ranches have contributed to a population explosion in the Amazon. Cities from Belém to Iquitos have mushroomed; in the years from 1960 to the present, Manaus has grown from a city of 200,000 to a sprawling agglomeration of more than 1 million, many of the new inhabitants living in makeshift shacks, a symptom of prevailing unemployment.[56] In 1970, the population of Brazilian Amazonia was 4.5 million, of whom nine-tenths lived in Belém and Manaus. President Emilio Medici announced a policy of providing "land without people for people without land," but poor people from the Brazilian northeast had little success in becoming landholders in the interior. Wealthy ranchers moved in, making it "land with cattle for men with capital," in the words of John McNeill.[57] In 1992, the population had increased to 20 million and its growth had not slowed.[58] Environmental changes in the Amazon rainforest in the last quarter of the twentieth century exceeded by far everything seen before.

Forest removal is presently at an annual rate of 405,000 ha (10 million acres). During fourteen years, 1975–88, 24 percent of the Brazilian state of Rondônia was deforested, and the process continues. In 1995 the Amazon Treaty Organization set guidelines for the sustainable management of tropical forests by the nations of the Amazon basin, and listed indicators for judging progress, but its effect has been minimal.[59] Logging receives state subsidies. The Peruvian Amazon is being subjected to massive mechanized deforestation. But a larger area of forest is burned off during the dry season. The total loss of forest cover from the Amazon basin by 2000 as revealed by images taken by satellites was 15 percent, and the forest shrinks noticeably every year. Only 2 percent of the Brazilian Amazon is within designated parks or reserves, and these areas are not well protected; in many cases, farmers and ranchers have located within the boundaries, and miners and loggers have invaded.[60] Experiments have demonstrated that when "islands" of forest remain in a cleared landscape, the forest ecosystems continue to lose species even in the absence of hunting and other disturbances.[61]

Animal species are being reduced to rarity as rapidly as the tree species that are in highest demand. Not only do residents shoot monkeys and other animals for food, but commercial hunters kill anything in the forest that people will buy in city markets. Rare species are captured for sale to unscrupulous collectors, and are shipped illegally to northern countries. Most die in transit, and even those that survive represent a loss to Amazon ecosystems. But more devastating than the death and removal of individual animals is the destruction of the forest habitat, which means that all forms of life adapted to it disappear.

Mammals such as jaguars, tapirs, anteaters, and armadillos were abundant, but now are seldom seen. Monkeys have been decimated by a combination of deforestation and hunting. Sloths, easily caught, are declining in numbers. Birds, especially those adapted to the floodplain, suffer from hunting and loss of nesting sites. Parrots and songbirds are frequently trapped, caged, and sold. Reptiles are exploited for food; turtles are killed and eaten and their eggs collected wherever they can be found. Between 1951 and 1976, Colombia exported 10 million caiman hides.[62]

Fish constitute the main protein in the diet of Amazonian people; at least 200,000 tons are taken from the river every year. Their numbers have been decimated by dams and pollution, and fewer fish mean malnutrition for impoverished residents. The greatest reduction in fish population results from the removal of forest from the riverbanks. Many fish are dependent on fruits, other vegetable matter, and small animals that drop into the river. In the undisturbed Amazon forest, trees hang over the water, but as I traveled along the rivers near Iquitos, I noticed that they had been everywhere cleared some distance back from the water's edge. By cutting these trees, people are reducing the number and size of the fish they want to catch. If fishing is to remain viable, substantial areas of floodplain forests must be preserved.[63] Many colorful tropical fish in demand for aquariums come from the Amazon's tributaries; collecting them is illegal, but a major trade continues.

Use of poisons in agriculture destroys many non-target organisms. It troubles me that a typical method of studying rainforest biodiversity is to fog a tree with insecticide, killing all the insects and then identifying and counting them. It is quite possible that this could make species extinct just before they were first noted by biologists.

The Amazon rainforest is the largest remaining tropical forest on Earth, and contains the planet's leading reservoir of biodiversity. Because deforestation is proceeding even more rapidly in Indonesia, southeast Asia, and Africa, the Amazon's green robe will in all probability be the last of the great tropical forests to disappear in the course of the twenty-first century. When it disappears or shrinks to a few protected forest remnant "islands," the

Earth will be impoverished. The loss of millions of tons of moisture formerly sent into the atmosphere by transpiration will reduce precipitation, and therefore the volume of the Amazon River.[64] As a result, hydroelectric dams will generate less electricity. The replacement of the bulk of the rainforest's carbon-rich biomass by the less voluminous vegetation resulting from human interference may result in an increase of carbon dioxide in the atmosphere of about 8 percent, adding to the greenhouse effect and contributing to global warming.[65] But the greatest deprivation will be the extinction of millions of species, many of which will never have been seen by human eyes, and the reduction of the world's most complex and balanced ecosystems to a simplified and impoverished condition. Some of the lost species might have had important medical or economic uses. The economies of the Amazonian nations, and the world market economy which they in large part serve have received a fleeting "subsidy from nature"[66] in degrading the ecosystems of the Amazon rainforest, but one which, at present rates of destruction, cannot continue for much longer.

New Orleans: causes of environmental disaster

In August 2005, New Orleans was a historic city, a cultural center, the birthplace of jazz music, and one of the world's busiest ports. It attracted 10 million tourists in a year.[67] It was also a city with most of its surface located below sea level, and in the path of a storm that would bring heavy rains and a surge of water from the sea on the 29th of that month. I saw New Orleans eighteen months later, after the disastrous Hurricane Katrina had breached the levees and flooded more than four-fifths of the city with toxic, sewage-laden water as deep as 6 m (20 feet) in some places. Much of the debris had been cleared away, but the city was still suffering from its wounds. On the shore of Lake Pontchartrain there was a lighthouse smashed over at an angle, and bare pilings in the water where there had been restaurants and theaters. I went to the Lower Ninth Ward with university professors who have studied the city, to see first hand something of what had happened. There were whole blocks where houses had been swept away by water exploding from a breach in the floodwall. Some houses were left smashed and crushed, and a few turned upside down. When I looked at the ground and the sidewalks, I saw shells from the canal along with knives, forks, scissors, and children's toys. Block after block were empty houses still erect but often windowless and gutted, stained with mold. All were marked next to the front door with a large X and the necessary data: the date inspected, the inspecting agency, and the number of dead found. In most cases, the latter figure was zero, but Katrina and its aftermath had killed about 1,500 people in New Orleans. In spite of all that, there were signs of hope: someone had scratched in the concrete at the base of a re-erected electric pole, "Ninth Ward Lives!" I met people wearing T-shirts with the motto, "Re-New Orleans!"

The driver in the taxi that took me from Baton Rouge to the New Orleans airport was an African-American resident of New Orleans who had been at work in a K-Mart the day Katrina struck. She had brought her nine-year-old son with her. When the floodwall broke, the water quickly rose to chest height. She put her son on an inflated air mattress and managed to get to a bridge. The other side was not flooded, but that was a predominantly White-inhabited neighborhood and police with guns stopped the people from crossing. Helicopters later evacuated old people and children, including her son; eventually she made her way out to Baton Rouge and was reunited with him. Like our driver, more than half the population of New Orleans resettled elsewhere, in Baton Rouge, Houston, Atlanta, and smaller numbers across the US. Approximately 200,000 of an original population of

440,000 now live in New Orleans, and the African-American proportion has dropped from 67 percent to 47 percent. Certainly the poor suffered more than the affluent. It took a week to evacuate all of the 122,000 people who were stranded in the Superdome and the Convention Center.

Ari Kelman, my colleague who wrote a fine book on the earlier history of New Orleans entitled *A River and Its City*,[68] said that the idea was:

> Move the homeless, the elderly, the impoverished, the unlucky, all those poor souls who couldn't get out of New Orleans in time to avoid Hurricane Katrina; move them into the city's cavernous domed football stadium. Anyone who has seen a disaster movie could have predicted what would happen next: Katrina slammed into the Superdome, ripped off the roof, and knocked out the power, cutting off the drinking water and the air conditioning. Those trapped inside had to be moved again – to Houston's Astrodome, of course. If it's not too callous to say so, the stadium mishap is an apt metaphor for New Orleans' environmental history. The sodden city has long placed itself in harm's way, relying on uncertain artifice to protect it from predictable disasters.[69]

If Egypt is the gift of the Nile, similarly New Orleans and all southern Louisiana are the gift of the Mississippi River. Each was formed of the debris deposited by a great river from a vast watershed draining part of a continent. The Mississippi, before the dams and diversions, carried water and sand, silt, and mud from 40 percent of the land area of what became the United States, and a smaller area in Canada, over many thousands of years. Without human interference, the river would continue to add to its vast, flat delta, flooding and shifting from one channel to another. Most of New Orleans today is below sea level, because the alluvial soil compacts as it accumulates, and further shrinks when it is drained. While in Louisiana, I took another field trip into the wetlands and bayous, where forests of huge bald cypresses and tupelos flourished in the fresh water brought down by the river. Those wetlands, along with grassy marshes, and the barrier islands formed further out on the edge of the Gulf of Mexico, are rich in fish, alligators, water birds, and other wildlife. They provide a stopping place for 70 percent of the migratory birds in the Great Mississippi Flyway. In the past they also formed insulation against hurricanes. New Orleans has been hit by a major hurricane every few decades, but the earlier ones tended to do less damage due to the protection offered by these millions of acres of wetlands, forests, and barrier islands. They were like a series of speed bumps against the storm surges of salt water. It was a case of healthy ecosystems serving as natural defenses. But much of it has disappeared and the rest is endangered, and the reason why can be explained by the environmental history of the region. That protection has been stripped away in large part by human projects, not by natural processes alone.

Hurricanes, like other natural disasters such as floods, tsunamis, and volcanic eruptions, present a problem for historians. Are they events that happen to people without choice, often without much warning, and make humans helpless victims? To what extent can humans control them, or at least modify them and guard against them? And in what ways are humans responsible for the destruction they and their works suffer from natural disasters because of the choices they do make? The damage caused by a hurricane depends not just on the force of the storm, but also on what people have done to the land. That includes city planning and activities that weaken and destroy natural entities that might protect them, such as the wetlands of the Mississippi delta.

Figure 9.3 New Orleans: One of the houses overturned by a surge of water from the collapsed floodwall of the Industrial Canal during Hurricane Katrina, August 2005. This is in the Lower Ninth Ward, where flooding was deep and widespread, and almost no houses escaped damage. Photograph taken in 2007.

For European colonists determined to exploit North America in the early eighteenth century, the delta of the Mississippi River was an ideal location, a portal to the immense interior along a myriad tributaries large and small. It was, however, an environment that did not offer an ideal site for a city among its intermittently flooded and unhealthy swamps. The French explorer and governor of Louisiana, Jean-Baptiste Le Moyne, Sieur de Bienville, after making settlements at several other places, discovered a crescent-shaped bend where the river had built a natural levee that seemed to him relatively safe from tidal waves and hurricanes, and in 1718 decided to establish a city there, naming it *La Nouvelle-Orléans*. It became the capital of the Louisiana colony in 1722, a year which foreshadowed future events when in September a hurricane struck, blowing most of the houses down. The site of today's French Quarter was on slightly elevated ground (3.7 m, or 12 ft above sea level) near the river, and therefore somewhat protected from flooding. Nonetheless, there were floods; the river always presented a threat and the colonists added an artificial levee (a dike or embankment) 1.2 m (4 ft) high on top of the natural one formed by the river, beginning a process that would last at least two centuries of building barriers against river floods and raising them increasingly higher. During the French period and the Spanish period that intervened from 1763 to 1800, the main flood threat was from the Mississippi River, and landowners were required to build levees in the sectors of land that they owned. North of the new settlement, in an area that was to become part of the modern city, cypress swamps and grassy marsh stretched to the shores of Lake Pontchartrain. This almost completely flat expanse was crossed from east to west by the Metairie and Gentilly ridges, ancient natural levees of no great elevation along a former path of the river. These ridges were separated by Bayou St. John, a sluggish watercourse debouching into the lake.

New Orleans became part of the United States as a result of the Louisiana Purchase from Napoleonic France in 1803. The city soon began to use boat-mooring fees to compensate landowners for levee building. The new Territory of Louisiana, and the State of Louisiana organized in 1812, lacked a central flood authority, and although levees ran along many kilometers of both banks of the river, the protection they offered was uneven and failed to prevent floods that occurred frequently, often only a few months to two or three years apart. Frequently breaches in the levees that caused the floods were some distance above New Orleans and entered the lower parts of the city from behind. In 1849 the Sauvé Crevasse flood penetrated New Orleans, including part of French Quarter, and the river undercut levees, making extensive repairs necessary.[70] The state, and increasingly the federal government, took greater roles in flood protection as the value of development in New Orleans and the Mississippi delta became apparent. Congress passed the Swamp Land Acts, which allowed the state to sell federal land for money to construct levees. Meanwhile, the US Supreme Court had ruled that the commerce clause of the constitution made the federal government responsible for maintaining navigation on rivers, and Congress delegated that job to the Army Corps of Engineers, whose role grew to a leading one in later years. The Corps built levees on the lower river leading to the Gulf, and it dominated the Mississippi River Commission created by Congress in 1879. After the Great Flood of 1927, the Corps was given oversight of flood control and navigation works on the entire Mississippi and its branches, and, after Hurricane Betsy in 1965, direction of the hurricane protection system of all southeast Louisiana. Unfortunately, the projects undertaken for flood control also had the effect of changing the hydrological regime of the floodplain to prevent the deposition of new soil which formerly maintained the bottomland of southern Louisiana.

The attempt to control flooding, it is clear, depended almost entirely on building levees, and as this effort continued, the river was gradually contained between two sets of levees from far above New Orleans to the Gulf of Mexico. By constricting the river, the levees raised its level further above the surrounding land, with the result that floods that did occur had the potential to be higher and more destructive. They also shunted the river's load of fertile soil to the Gulf, effectively stopping the renewal of land in the delta.[71] In addition, the Atchafalaya River, a lower branch of the Mississippi, was straightened and contained between high levees that convey floodwaters rapidly to the Gulf, where the erosional materials they contain are dissipated.[72] Most of the Mississippi's flow, however, was not allowed to enter the Atchafalaya and was directed past New Orleans. Several engineers pointed out the advantages of gates and spillways that could give river water and river mud alternate pathways to the Gulf, relieving pressure in the main stem and helping to build up the delta. There were experiments of this kind, but for the most part the policy was "levees only," and as far as the city's defense against the river was concerned, the policy was a success; from the late nineteenth century onward, the river levees usually held. In the Great Flood of 1927, the bowl of the city filled when 355 mm (14 in) of rain fell and the pumps failed. The Army Corps broke a levee downstream at Caernarvon to relieve pressure on the levees next to the city.[73] This action flooded St. Bernard Parish, but it may have been unnecessary. In 2005, none of the levees along the Mississippi River above or in New Orleans failed.

The main flood danger to the center of the city had shifted to another front: the south shore of Lake Pontchartrain.[74] New Orleans grew rapidly in the nineteenth century as a result of its busy port and the success of steamboats in making swift upriver travel possible. The relatively high ground near the river soon filled up with structures, and the obvious direction to expand was northward into the swamps toward the lake. For that to happen, the landscape would have to change. Largely to be cleared were the majestic bald cypress trees; their wood was durable timber, so resistant to rot that it was called "the wood eternal."[75] Cypress forests were exploited in Louisiana from early in the French period,[76] but although sizeable forest survived toward the lakefront, it was increasingly depleted and the maximum period of removal was from the 1890s through the 1930s, after which there were few trees left that were accessible and of useful size.[77] Floods only made felling the trees by boat and floating them out easier, but this was wasteful because they were sawn off 4–5 m (12–15 ft) above root level and the tall stumps remained.[78] Fortunately in the late nineteenth century the city and state created two large parks, Audubon Park and City Park, and with plans developed in part by John Charles Olmsted, scenic versions of the original landscapes of the city were preserved.[79] The next step necessary before streets could be laid out and houses built was drainage; the accepted way to do that in the early nineteenth century was to dig canals, but the extremely low gradient toward the lake rendered canals ineffective in getting the water out, and in addition provided a way into the city for water driven by storms from the lake. Steam-driven waterwheels were installed to lift water into the canals, but drainage lowered the water table and the ground level, and left the canals, provided with their own levees, higher above the city. Electric pumps and new canals provided more effective water removal in the years before the First World War, when there were seven pumping stations and 112 km (67 mi) of canals and the water table had been lowered by as much as 3 m (9 ft).[80] Development proceeded toward the lake, from south to north, and the pumping stations stayed where they were first placed, toward the south ends of the canals, away from the lake, instead of at the lakeshore levees, where they could have pumped water directly into the lake and served as barriers. As a Dutch engineer remarked after Hurricane Katrina had demonstrated the weakness of the canal levees, "Why in the world would you invite the enemy deep inside your own camp?"[81] The major construction of lakefront protection levees took place in 1922–34. In the 1970s, the lakefront levees were raised to 4.25 m (14 ft).

Then in the eastern part of the city the Industrial Canal, opened in 1923, connected the Mississippi River to Lake Pontchartrain. About halfway along that canal, it is joined by two other canals from the east, the Gulf Intracoastal Waterway, a navigable inland channel that parallels the Gulf coast from Florida to Texas, and the Mississippi River–Gulf Outlet (MR-GO), a navigation channel 122 km (76 mi) long completed in 1965 to connect the Port of New Orleans directly to the Gulf of Mexico, shortening the distance that ships would otherwise have to traverse along the curves of the river. These two waterways, with their levees, form a "funnel" leading directly toward their T-shaped junction with the industrial canal, setting a scenario for disaster when hurricanes drive surges from the Gulf.

Never a stranger to hurricanes, New Orleans had discovered that the high winds could also drive waters from Lake Pontchartrain into the city. A hurricane in 1947 tossed waves over the lakeshore levees and caused significant flooding. By the 1950s, hurricanes received female names bestowed in alphabetical order, an idea taken from George R. Stewart's novel, *Storm*, in which meteorologists called a Pacific low-pressure system "Maria." Consequently, New Orleans suffered Hurricane Flossy, which burst gaps in the canal levees in 1956, and Hurricane Hilda in 1964. In the following year, memorable Hurricane Betsy brought winds of 260 kph (160 mph) and breached the Industrial Canal, damaging 7,000 homes and 300 industries. Hurricane Camille also breached the Industrial Canal in 1969. Incidentally, the idea that hurricanes should be given feminine names was recognized as an instance of gender discrimination, and in 1978 the official lists of names began to contain alternately masculine

Figure 9.4 Bald cypress in the Edenborn Brake, Louisiana. Wetland forests like this help to counteract the flood surges of hurricanes; their removal exposes homes and businesses to greater damage. Photograph taken in 1916. Courtesy of the Forest History Society, Durham, NC.

and feminine names. It was the luck of the draw that gave New Orleans two female hurricanes in 2005, Katrina and Rita.

Meanwhile, the oil and gas industry in the Louisiana wetlands was expanding to become the largest source of crude oil and the second largest source of natural gas in the United States. Oil platforms were also constructed to access petroleum reserves under the Gulf of Mexico, and oil tankers required delivery to the port at New Orleans, so canals were built through the wetlands to provide access. Combined with those excavated for operations on land, the total distance of cuts and canals in the wetlands has been estimated at almost 13,000 km (8,000 mi). These canals allowed salt water to flow into the wetlands, killing trees and other vegetation and eroding away the land. The longest navigation canal was MR–GO, mentioned above, leading directly from the Gulf of Mexico to the heart of New Orleans, providing a channel for oil tankers to come into the port, but also providing a potential funnel for hurricane surges. This is exactly what happened during Hurricane Katrina. Some engineers predicted this danger and recommended building gates on MR–GO that could be closed in case of a storm, but this was not done because of cost and other objections, some of them environmental.

In July 2004, the Federal Emergency Management Agency (FEMA) brought together emergency officials from fifty parish, state, federal, and volunteer organizations for a five-day exercise held at the Louisiana State Emergency Operations Center in Baton Rouge to help officials develop joint response plans for a catastrophic hurricane.[82] It involved a simulated, computer-generated event called Hurricane Pam. Hurricane Pam was assumed to be category 3, bringing sustained winds of 200 kph (120 mph) and up to 510 mm (20 inches) of rain in parts of southeast Louisiana and a storm surge that would top levees in the New Orleans area. More than 1 million residents were targeted for evacuation and Hurricane Pam was expected to destroy 600,000 buildings. According to the scenario, only a third of the population was predicted to leave New Orleans before the storm hit. This was recognition of the fact that much of the city's population lived in relative poverty, with approximately 127,000 in households in which no one owned a car. Hurricane Pam turned out to be an almost exact prediction of Hurricane Katrina, the first of the two hurricanes that actually hit New Orleans just over a year later. I mention it because it indicates that no one should have said that Katrina was unpredictable. There was earlier warning, indeed, from scientists, historians, engineers, and a newspaper, the New Orleans *Times-Picayune*, which had carried a series of stories in 2002 on the very subject, also predicting a disaster very much like that which actually occurred. In spite of this, the federal administration's unpreparedness is epitomized by the slow response of FEMA when the emergency occurred.

The large and destructive hurricane designated Katrina reached the northern coast of the Gulf of Mexico on August 29, 2005, four days after it had crossed the Florida peninsula. When it reached the Mississippi delta in Louisiana, it possessed sustained winds of 204 kph (127 mph). It produced a storm surge approximately 9 m (30 ft) high, with waves up to 16 m (52 ft), which did catastrophic damage when it smashed into coastal areas of Louisiana, Mississippi, and Alabama.[83] Considered as a single event, it was one of the deadliest and most costly weather-related disasters up to that time in the history of the United States, causing more than 1,833 deaths and about $125 billion in economic impact. New Orleans suffered most of the deaths and a major share of the damages.

The storm surge of Hurricane Katrina found weak points in the system of levees set up by engineers to protect the city. A surge into Lake Pontchartrain, pushed by winds from the north as Katrina passed by, entered the drainage canals and breached the floodwalls in several places, flooding the northern and central parts of the city. The surge from the east piled through the "funnel" into the Industrial Canal and breached its floodwalls in three major places, pouring violently into East New Orleans, the Lower Ninth Ward, and parts of the city to the west.

Oil- and sewage-laden water flooded more than 80 percent of the city as power failed and the pumps ceased to function. In some places the water level rose more than 6 m (20 ft) above the ground. People who were still in the city were trapped in their attics as water came up, unless they could break through and seek rescue on the roofs, rescue that was all too often very slow to arrive by boat or helicopter, if it arrived at all. Indeed, the slowness of government agencies to respond to the disaster, on the federal, state, and local levels, constitutes a lasting disgrace associated with the Katrina disaster. As Kelman expressed it,

> Nor do I suggest that this was a natural disaster. It was not. It was an outgrowth of a host of bad decisions, large and small, of miscalculations, of ignorance, even of hubris. It was, in sum, a byproduct of the city's and the nation's environmental history.[84]

Past efforts to protect New Orleans were in the main attempts to control nature, and although nature time and time again has demonstrated its forces to be uncontrollable, those efforts are parts of plans for the future. New Orleans will not be abandoned any time soon, although if a radical change in the way its problems have been addressed does not take place, that might become an unavoidable alternative. But if not abandoned, it must be defended by working technology. Other hurricanes, some of them more powerful than Katrina, will inevitably hit the city sooner or later, and the danger is exacerbated by the expected incremental rise in sea level caused by global increases in average temperature. Projects have been advocated, some of them already under way, to repair the breaches, raise and strengthen the levees and build new ones regionally, put pumps where the canals meet the lake, and complete a lock system on the Industrial Canal. The Army Corps of Engineers has proposed to close the Mississippi River–Gulf Outlet. Perhaps using aspects of the Dutch experience with dikes and gates, at the entrance to Lake Pontchartrain and elsewhere, will help. Projects like these are necessary, but they will not be enough.

The effort to keep New Orleans alive does not need to be only a struggle against nature; it can be a venture to cooperate with nature. The city is inextricably part of the Gulf Coast wetland ecosystem, and that ecosystem has historically operated to insulate the city against a number of dangers. The area of delta marsh that becomes open water every year is variously estimated at 60–100 sq km (25–50 sq mi),[85] but Katrina destroyed 250 sq km (100 sq mi), and chopped up the Chandeleur Islands, a barrier island chain.[86] Many planners advocate a regional system of coastal and wetland restoration.[87] This would involve diversion of some of the flow of the Mississippi so that sediment can build up wetland where it has been lost.[88] It would mean establishing new protected areas and limiting development within them, especially the excavation of canals. Cypress trees would be protected, including a ban on cutting them for chipping into mulch for suburban gardens, which is today the most common form of exploitation. The barrier islands could be strengthened, extended, and vegetated. Such a program would be very expensive, although perhaps not compared with the damage that has occurred by the neglect of the ecosystem.

Nairobi and the world: the United Nations Environment Program

The world headquarters of the United Nations Environment Program (UNEP) is an attractive campus located on the outskirts of Nairobi, Kenya in a suburb called Gigiria. Most of the buildings are of one or two stories, set among artificial streams and waterfalls, with carefully tended plantings of flowering shrubs, trees, and succulents. The colorful birds of East Africa make themselves at home, and visit outdoor tables at the restaurants provided for visitors and staff. Conferences take place in ample meeting halls provided with electronic facilities. I visited with an environmental study group, and we were impressed by a painting, "A Tree for Every Child," showing a group of children from many nations dancing in a circle around a newly planted tree, symbolizing environmental concern for the next generation. We saw films on the organization of UNEP and on the dangers of ozone depletion, and received printed information on many projects, including one on the status of efforts to save the African elephant.[89]

UNEP's mandate is "to provide leadership and encourage partnership among nations in caring for the environment, by inspiring, informing, and enabling nations and peoples to improve their quality of life without compromising that of future generations."[90] It was planned "as a catalyst, rather than an implementer or enforcer."[91] At the same time, it became the first major UN agency centered in a country in the developing world. Jomo Kenyatta, the first president of independent Kenya, invited the international body to locate the headquarters for UNEP in his country, and his offer was accepted over several others. The choice of Nairobi, located so far from its sister UN agencies, created problems for UNEP, whose mode of operation was to coordinate efforts of other agencies on environmental issues. It made contacts with many governments, environmental movements, and the mass media more difficult. UNEP Nairobi keeps in contact with New York, Geneva, and other UN "capitals," as well as its own regional offices in places such as Bangkok, Bahrain, and Mexico City through the Mercure satellite communications network, acquired in 1997 from the European Space Agency.[92] Particular responsibility has rested on UNEP's Regional Office for North America in New York, which keeps contact with other major UN offices and bodies there, and the diplomats who visit them. From 1972 to 1995, the regional director at the UN in New York was Noel Brown, a citizen of Jamaica with a doctorate from Yale University, who represented UNEP at many important international conferences.

UNEP was an outgrowth of the United Nations Conference on the Human Environment (UNCHE), the landmark international environmental meeting of the century, convened in Stockholm in June 1972, which included representatives of 113 nations, nineteen intergovernmental agencies, and 134 non-governmental organizations (NGOs).[93] That meeting was the first major modern international gathering on the theme of human activities in relationship to the environment. It marked a major step in awareness among nations that many environmental problems are worldwide in scope, and must be addressed on an international level. Representatives of industrialized and developing countries attended, and the issues that divided those two groups were subjects of searching discussion.

Stockholm 1972 was organized and chaired by Maurice F. Strong, a Canadian industrialist who subsequently became the first executive director of UNEP. Unlike the 1992 meeting in Rio de Janeiro which was its successor, Stockholm 1972 was not an "Earth Summit." The only heads of state present were the host, Sweden's Prime Minister Olaf Palme, and Indira Gandhi of India, who served as an articulate spokesperson for views shared by many developing countries. Some developing-world representatives noted that environmentalist views were most vocal in the industrialized world, in the very nations which had reached their economic pinnacles by using natural resources from around the Earth and producing the major proportion of the planet's pollution. Would measures for resource conservation and reduction of pollution limit the development of poorer countries while leaving the richer countries in relative affluence? Was the environmental movement a colonialist conspiracy to thwart development by imposing upon them extra costs and prohibitions that developed states had not faced in the nineteenth and early twentieth centuries in implementing their pollution-generating, resource-consuming industrial revolutions?[94]

Indira Gandhi had a more measured view that environmental concern should accompany a desire for development: "Are not poverty and need the greatest polluters?" she asked, adding, "The inherent conflict is not between conservation and development, but between environment and the reckless exploitation of man and earth in the name of efficiency."[95] A major contribution of developing countries to the discussion was to insist that environmental problems always be considered along with issues posed by basic human needs.

Concerned environmentalists from industrialized countries had to recognize the justice of these arguments, and at the same time made the point that a livable environment is a critical dimension of successful development. The principles approved by the Stockholm conference reflected a compromise on which the two major groups of nations could agree. These principles included recognition of the fundamental right of people to live in an environment of a quality "that permits a life of dignity and well-being."[96] They urged the conservation of natural resources and the limitation of pollution. Industrialized countries, it was agreed, should aid other nations in development guided by environmental concern. Scientific and technological research and education should be undertaken to promote environmental protection. These environmental efforts should proceed through international cooperation that respects national sovereignty. After Stockholm, there was a more widespread belief that development and sound environmental management are not incompatible, but that both are necessary to create a sustainable society.

UNCHE at Stockholm has been called the single most influential event in the evolution of the international environmental movement. It heralded a period of deliberations in which several new treaties on environmental issues would be negotiated. And it laid the foundation of UNEP, which would forward environmental programs within the structure of the United Nations. UNEP was authorized by a UN General Assembly resolution in December 1972, and was charged with coordinating efforts to implement the recommendations approved at Stockholm. It was, however, to be a policy and information center that would assist and coordinate the activities of other agencies. Many other UN bodies had ongoing environmental protection activities which they would continue. For example, the UN had recently created a Fund for Population Activities (UNFPA), so population was not included in UNEP's mandates. The new agency would be hampered by a small budget and staff.[97] The administrative expenses of its Council and Secretariat come from the general UN budget, which is seriously limited, and its programs depend on an environment fund that is financed with voluntary contributions from UN member states. UNEP chronically lacks enough money to finance urgent plans. Two-year contributions in 1997 totaled $140 million, less than three-quarters of the amount spent to make the motion picture *Titanic*, an ironic fact since the situation of people "on board" the Earth has been compared with that of the passengers of the great ship headed for an environmental collision. UNEP has a "big soul in a little body."[98]

Maurice Strong was the first executive director of UNEP. Dr. Mostafa Kamal Tolba, who received his Ph.D. degree in micobiology from Imperial College, London, and served as education minister in the Egyptian government, took over in 1975, and charted the course of UNEP for seventeen years.[99] He was succeeded by Elizabeth Dowdeswell of Canada in 1993, and Klaus Topfer, former German environment minister, in 1998. Achim Steiner, former head of the World Conservation Union (IUCN), was selected executive director in 2006 for a four-year term. A German, he was born in Brazil and studied at Oxford, the University of London, and Harvard.

The achievements of UNEP were in three major areas. First, UNEP maintained an information-gathering and retrieval program called Earthwatch. Second, it gave diplomatic support to the evolution of a body of international environmental law, and served as a secretariat to several important treaties. Third, it educated and inspired nations and peoples to realize the critical importance of environmental problems, and the necessity of efforts to address them.

One of the most useful aspects of UNEP's work has been to facilitate the negotiation of international treaties and agreements. Dr. Tolba believed that UNEP could be an "honest broker" between various interest groups such as nations, businesses, non-governmental agencies, and scientists. With diplomatic skill, UNEP generally proceeded by seeking to discover a scientific consensus on the environmental problem being considered, then trying to find a strategy that would be effective in addressing it.[100] Many delegates found they could trust UNEP's legal and scientific expertise, as well as its willingness to take their interests into account. UNEP gained recognition, and eventually a mandate to work for the development of international environmental law.[101] Its negotiations led to the adoption of more than thirty treaties, conventions, and other agreements.[102] Most international agreements depend on voluntary compliance by signatories. UNEP has little ability to provide economic incentives to encourage observance.[103] Thus hampered by having neither a stick nor very much of a carrot, it must depend on persuasion.

Among major agreements in which UNEP has played a role are the Convention on Conservation of Migratory Species of Wild Animals (CMS), and the Convention on International Trade in Endangered Species of Wild Fauna and Flora (CITES). These treaties protect more than 35,000 endangered species. One of the achievements of CITES has been to restrict the trade in ivory to combat the disastrous decline of the African elephant population due to poaching. UNEP has taken the lead in getting together nations that abut on regional seas to take measures to stop pollution and protect sea life. One of the most successful of these efforts produced the 1976 Barcelona Convention for the Protection of the Mediterranean Sea against Pollution, an amazing achievement considering the often antagonistic dispositions of the nations around that inland sea toward one another. UNEP negotiated the Convention on the Control of Transboundary Movements of Hazardous Wastes and Their Disposal, signed in Basel, 1989.

Perhaps the most impressive achievement of UNEP's labors is the drafting and negotiating of the Vienna Convention for the Protection of the Ozone Layer and the 1987 Montreal Protocol, aimed at reducing the production of chlorofluorocarbons and other chemicals that weaken the Earth's atmospheric shield against harmful ultraviolet radiation.[104] This is one of the most successful international environmental agreements, with excellent compliance around the world.[105]

A raising of environmental consciousness in the United Nations has been one of UNEP's finest contributions. But UNEP has also helped to educate governments and peoples on the importance of environmental issues to all nations. World Environment Day, set on June 5 to mark the anniversary of UNCHE in Stockholm, 1972, is celebrated in many countries. With the United Nations Educational, Scientific and Cultural Organization (UNESCO), UNEP has worked to encourage environmental education, training educators and providing materials for schools. It has sponsored publications and films.[106]

Figure 9.5 Administrative building on the campus of the United Nations Environmental
            Program in Nairobi, Kenya. Located far from other UN facilities, this UNEP
            headquarters uses electronic media to keep in touch with them. UNEP's mandate is
            to provide leadership and encourage partnership among nations in caring for the
            environment, by inspiring, informing, and enabling nations and peoples to improve
            their quality of life without compromising that of future generations. It was the first
            major UN agency to be centered in a Third World country. Photograph taken in
            1989.

Another UN program that deals with environmental issues is the Man and the Biosphere
Program (MAB), an initiative of UNESCO.[107] Launched in 1971, MAB studies "the
interrelationships between natural ecosystems and socio-economic processes."[108] It evolved
the idea of biosphere reserves as representative samples of significant ecosystems and habitats of plants and animals including rare and endangered species.[109] Biosphere reserves are
proposed by the governments of the nations in which they are located. Some cross national
boundaries; a large one includes Glacier Bay in the US (Alaska) and adjacent parklands in
Canada. The biosphere reserve system is an international recognition that ecosystems have
a claim on their human stewards to be preserved intact. It is an international network of
areas intended to conserve genetic resources and ecosystems and to assist in the maintenance of biological diversity, with associated scientific and educational efforts. A key principle of biosphere reserves is a zoning pattern combining a core area that is to be strictly
protected, a delineated inner buffer zone, and an outer buffer zone or transition area.[110] The
core areas are strictly protected, and as a result can serve as benchmarks for observation of
long-term changes in the biosphere and in the ecosystems they represent. The biosphere
reserve concept foresees a mutually beneficial relationship with the local human societies
surrounding the reserves.[111] Traditional economic activities are encouraged in the buffer

zones, to provide indigenous peoples means of survival that use but do not deplete the resource. There were 531 biosphere reserves in 105 countries in 2008. Almost all were areas already set aside as national parks or nature reserves.

To give an example, Amboseli National Park is one of five biosphere reserves in Kenya. It had been part of a tribal and game reserve since 1900, and Kenya made it a national park in 1971.[112] The biosphere reserve, designated in 1990, covers 39,200 ha (151 sq mi). Situated at the foot of the magnificent volcano, Kilimanjaro (which is on the other side of the border in Tanzania), Amboseli contains an impressive remnant of the abundant wildlife that once ranged over much of East Africa. It is home to many Maasai people, whose traditional way of life is cattle herding. Among the MAB research projects approved for Amboseli are studies of changing swamps, the effects of agriculture, ecosystem restoration, and long-term observation of baboons, vervet monkeys, and elephants. Unfortunately, the status of the biosphere reserve is in question because in 2005, President Mwai Kibaki decided to turn over control of the park to a county council and the Maasai tribe, a move that has been challenged in court in Kenya.

Does the biosphere reserve program hold the promise of becoming an effective world network of diverse protected fragments of the Earth's living systems? Taken individually, few of the reserves are large enough to protect all of the species needed to comprise a full ecosystem in the long run.[113] UNESCO can only urge the governments involved to protect the reserves adequately; there is no UN budget item to assist in managing them. Even though the network continues to grow, those designated do not represent all of the Earth's major biogeographical types.[114] But even as fragments of an endangered whole, they serve as indicators of what a more adequate system of protection might be like.

In 1987, the Brundtland Commission recommended a second world conference on ecological and economic problems. The UN approved the idea, and the United Nations Conference on Environment and Development (UNCED), often called the "Earth Summit" because so many heads of state attended, met in Rio de Janeiro during June 1992, on the twentieth anniversary of Stockholm. Delegations from 178 countries, heads of state of 105 countries, and representatives of more than 1,000 non-governmental organizations (NGOs) attended. The conference was covered by 8,749 accredited journalists, about half of them from Brazil.

A separate Global Forum held at the same time in Rio's Flamengo Park attracted members of almost 8,000 NGOs. Some of these independent people perceived a narrowness at the governmental level and drafted alternative treaties that attempted to express wider concerns.[115] Representatives of indigenous peoples from eighty-five countries, denied an official place in UNCED, convened their own Earth Parliament, appropriately held in a small village outside Rio called Kari Oca.

The presence of so many presidents, prime ministers, and monarchs assured that the conference would receive attention around the world from the first gavel to the last. The issues separating the developing countries from the developed countries occupied even more attention than at Stockholm, although by the time of Rio they were better defined.[116] Fernando Collor de Mello, president of Brazil and also of the conference, opened the proceedings by declaring that the main enemies of the environment and of sustainable development are poverty and lack of opportunity. In his opening address, he called for a new global partnership to ensure the common future of all people, and stressed the need for affluent nations to help the poor achieve the goals that would be set at the conference. The US president, George H.W. Bush (the first President Bush), announced his intention to attend the summit at a late date. His speech was received without enthusiasm by delegates who knew that the

US delegation opposed some and had succeeded in weakening others of the documents being considered at Rio.

UNEP was involved in the pre-conference preparation of most of these. The five principal agreements were: the Rio Declaration, Agenda 21, the Statement of Forest Principles, the Convention on Biodiversity, and the Framework Convention on Climate Change.

The Rio Declaration was a statement of principles emphasizing sustainable development. As approved, it affirmed: "Human beings are at the center of concerns for sustainable development. They are entitled to a healthy and productive life in harmony with nature."[117] It went on to say that states have the right to exploit their own resources, a principle that might be open to question on ethical grounds.[118] It called for priority to be given to the needs of developing countries, women, youth, and indigenous peoples. Environmental protection was considered an integral part of the development process. The document asked for use of peaceful means in resolving environmental disputes and declared that war is destructive of sustainable development. Nowhere in this, or any other UNCED document, is there a recognition that development, and human population, might eventually face limits.[119] Since it is not a treaty, it has moral force only. Although the US agreed to the document, it was the only country to issue a written statement of dissenting opinions. Among these, the US objected to principles which stated that developed countries have obligations to developing countries; and that trade should be subject to environmental measures.[120]

Agenda 21 is a broad statement of goals and potential programs related to sustainable development; it has forty chapters occupying more than 800 pages. Like the Rio Declaration, it is not legally binding, but UNEP regards it as a thematic guideline for its own programs. It confirms UNEP's mandate and states the need for its "enhanced and strengthened role."[121] It affirms UNEP responsibility in the process of formulating international environmental law, regional cooperation, technical environmental advice, and environmental emergency planning.[122]

The Statement of Forest Principles is a non-binding agreement on preservation and management of the Earth's remaining forests. Since deforestation is one of the most destructive processes affecting loss of biodiversity and the deterioration of the atmosphere (forests constitute a carbon reservoir almost equal to the amount of carbon in the atmosphere, and their destruction contributes a significant proportion of the release of carbon dioxide into the atmosphere) it was hoped that a treaty on the subject could be approved at Rio. However, nations treat their forests as internal resources and often object to outside interference. Industrialized countries called for regulations to stop the destruction of tropical rainforests, and developing countries countered by pointing out the ongoing loss of temperate forests in such nations as Russia, Canada, and the US, and demanded that it be limited. The gap between the two sides could not be bridged, so the principles approved were not binding. They speak of the need for sustainable forest management, but affirm the right of each nation-state to utilize its forests in accordance with its development needs, and call for free international trade in forest products. In 1995, the UN Commission on Sustainable Development created the Intergovernmental Panel on Forests to implement the statement.

The Convention on Biological Diversity is a binding international treaty aimed at strengthening national control and preservation of biological resources.[123] The weight of discussion, however, was not on the need to preserve species and ecosystems, but the desirability of assuring sustainable economic development for nations, and to distribute equitably the gains realized from the development of biological resources. One argument for preservation of ancient forests was that they are storehouses of species producing substances that might prove of use to humankind as medicines or in other ways. This is certainly true; researchers derived many healing drugs from tropical rainforests, and taxol, a derivative of the yew tree in northern forests, proved valuable in treating ovarian cancer. Biodiversity, the world suddenly realized, had economic value, and the discussion had changed its tenor. The goals of the treaty expressed in the final draft were the conservation and sustainable use of biodiversity and fair trade and compensation involving products made from the genetic resources of nations. It charges each country to make plans to protect habitats and species, and provides for aid to developing countries to help them do this. The Convention on Biological Diversity had its headquarters in Montreal beginning in 1996.

The Framework Convention on Climate Change was an international agreement that sought to limit or reduce emissions of gases, mainly carbon dioxide and methane, with the potential to exacerbate global warming. It and the later negotiations to establish specific goals for reductions to give it effect are discussed below in the section on global warming.

On the final day of UNCED, Maurice Strong openly speculated about its accomplishments: "Our experience in Rio has been as historic and exhilarating as the road that brought us here. The road from Rio will be long, exciting, challenging. It will open a whole new era of promise and opportunity for our species if we change direction, but only if we start now."[124] Some present expressed cautious optimism; Mahathir Mohamad of Malaysia said that he would consider the conference "a success if there emerged a better understanding of the enormity of the problems we face and the need for us to cooperate on an equitable basis."[125] Others were less pleased with the course Rio had taken. Wagaki Mwangi, a Kenyan Youth delegate, exhorted:

> Those of us who have watched the process have said that the Earth Summit has failed ... Multinational corporations, the United States, Japan, the World Bank, the International Monetary Fund have got away with what they always wanted ... The Summit has ensured increased domination by those who already have power. Worse still, it has robbed the poor of the little power they had. It has made them victims of a market economy that has thus far threatened our planet ... Few negotiators realized how critical their decisions are to our generation. By failing to address such fundamental issues as militarism, the regulation of transnational corporations, the democratization of international aid agencies and the inequitable terms of trade, my generation has been damned.[126]

Many of these comments assumed that the success or failure of Rio would be measured by its effect on the distribution of the world's resources between rich and poor countries. It is more difficult to find statements from observers who would ask whether, after the conference, both rich and poor people would continue the destruction of habitats and ecosystems in a competitive search for wealth, or in a desperate struggle for survival in the short term. Genuine sustainability would require the continuation of the community of life.

Indeed, national interest is an inescapable component of international agreements, no less in the field of the environment than in any other. "Nation-states, despite Rio, still consider themselves as the only important actors on stage."[127] Nation-states have always taken a self-serving political approach to problems that affect the entire world community. The historians Thucydides in ancient Athens and Machiavelli in Renaissance Florence pointed out that states are never motivated entirely by ethics, and today it might be added that they are seldom motivated by science. By the end of the twentieth century, however, it had also become clear that nations are not the only entities that must be taken into account in reaching international agreements. Transnational corporations are often richer than many nations, and employ numbers larger than the working classes of smaller countries. Organs of the world market economy, such as the World Bank, International Monetary Fund, and World Trade Organization, can override national laws and command far more resources than the United Nations.[128] Leaders in many developing nations are convinced, rightly or not, that the burden of debt prevents them from taking certain environmental steps. This must not prevent those who work for positive outcomes in organizations such as the United Nations from using arguments based on ethics and science, but it will constrain them to a certain realism. They must be "wise as serpents and harmless as doves."[129]

Global warming: an environmental historical perspective

When I entered the Main Storage Facility of the US National Ice Core Laboratory (NICL),[130] I experienced the coldest temperature I had ever felt: -36°C (-33°F). There, on shelves filling a vast space of 1,540 cu m (55,000 cu ft), lay an archive of the history of the Earth's climate and atmosphere in the form of cores of ice taken from the ice sheets of Antarctica and Greenland, and from mountain glaciers. Many cores were drilled from the ice surface to bedrock, and some approach 4 km (2.5 mi) in length, so that one drilled hole is represented by thousands of cylindrical ice segments on the shelves. Looking at one of those segments removed from its silver sleeve, I could see layers one above the other representing the annual snowfall of a series of years many centuries in the past. These layers can be dated exactly, using electrical conductivity, visual layer counting, and stable isotope analysis – much more accurately than, say, a method such as radiocarbon allows. They represent unbroken records stretching back as far as 250,000 years in Greenland and 450,000 years in Antarctica. I was amazed to see a disk of ice that had not melted for a quarter of a million years.

The value of a physical archive of past snowfalls is immense, since it contains not only samples of the water that fell, but also atmospheric gases captured and retained as bubbles; dust including fallout from forest fires, industrial pollution, volcanic eruptions, cosmic particles, and nuclear explosions; and sea salts carried by wind onto the ice sheets.[131] This valuable record of the history of the Earth's environment is kept in a building whose concrete barrel vaults remind me of ancient Egyptian grain storage vaults near Luxor. But NICL is located in the Denver Federal Center, west of Denver near the foothills of the Rocky Mountains. There are similar facilities in other countries including Australia, Denmark, India, and Japan. NICL makes core samples available to researchers for study, which can begin in the Exam Room, next to the Storage Facility and kept at a relatively balmy -25°C (-13°F). Among the many kinds of information about the past that can be learned from the samples are temperatures and weather patterns; the composition of ancient atmospheres including, for example, levels of carbon dioxide and other greenhouse gases; sea level and biological productivity of the ocean surface; and size and frequency of natural catastrophic events such as volcanic eruptions. Especially interesting is the fact that the ice cores show that the climate of the Earth has undergone many changes, some of them long and cyclical and some, unexpectedly, extremely rapid. For example, the end of the last phase (the "Younger Dryas," 12,800 to 11,500 years ago) of the most recent ice age appears to have included a rise of about 8.3°C (15°F) in the average temperature within a period of a few years, perhaps only one year, not generations or centuries.[132]

It is fortunate for human knowledge that the ice cores are being preserved and studied, since Greenland, much of Antarctica, and mountain glaciers are threatened by a worldwide phenomenon during the last century or so: the rapid retreat of ice. Not all glaciers are retreating, but there is an unmistakable pattern of loss of volume and shortening almost everywhere that measurements are taken, and photographs taken over decades show spectacular losses.[133]

Melting glaciers are only one form of evidence that the Earth's surface temperature, taken as a whole and as a general trend, has been rising since the mid-nineteenth century. Weather stations and satellites indicate an average increase of 0.6–0.7°C (1–1.25°F) over the past 100 years. The upper-level ocean temperatures have risen about 0.3°C (0.5°F) in sixty years. In the years between 1979 and 2004 the minimum annual extent of ice in the Arctic Ocean declined by 8 percent, and its average thickness lost 1.3 m (4.25 ft), or 40 percent. The loss continues. In August 2000, open water was reported at the North Pole; this was not necessarily the first time it had occurred there in recent years, and most probably not the last time it will occur. The Greenland ice cap is losing volume rapidly. During the twentieth century, global sea levels rose at a rate of 1–2 mm (0.04–0.08 in) per year, partly as a result of the expansion of warming seawater and partly due to meltwater from ice loss on the continents. Alterations in the ranges and migrations of many species imply that they are reacting to warmer temperatures. The majority of climatologists who study all these changes judge that the Earth is in an episode of rapid global warming.

Climate changes, whether warming or cooling, have occurred throughout history and prehistory. During the last few hundred thousand years, a series of glacial stages, or ice ages, has alternated with warmer interglacial stages. During recorded human history, there have been warm periods like the Medieval Climatic Optimum in the eleventh through thirteenth centuries, and cold periods such as the Little Ice Age of the fifteenth through mid-nineteenth centuries. In trying to explain these phenomena, scientists in the last two centuries have developed a number of hypotheses of causation. An excellent history of ideas about global warming and the science of climatic change can be found in Spencer R. Weart, *The Discovery of Global Warming*.[134] While it is clear that climate changes of thousands of years in prehistory, at least up to the end of the most recent ice age, had causes that were entirely natural,[135] there has been discussion and disagreement about the extent to which human activities, in particular those that generate greenhouse gases, have contributed to climate change during historic times, particularly during the past two or three centuries. As will be further noted in this section, climate scientists have reached a broad consensus that the most important cause of the present rapid warming of the overall temperature of the Earth is the accumulation of greenhouse gases such as carbon dioxide by human activities, including especially the combustion of fossil fuels (coal, petroleum, and natural gas) which has increased exponentially since the beginning of the Industrial Revolution. Recently the paleoclimatologist William F. Ruddiman has argued that other human activities, including deforestation and agriculture including irrigated rice production that is a prime source of methane, have had a net warming effect over the past 8,000 years that has reversed what would otherwise be a natural cooling trend.[136] Ruddiman recognizes that the human-caused warming since the late nineteenth century represents a major acceleration above earlier centuries.

Probably the first person to maintain that global warming could be caused by increased levels of carbon dioxide (CO_2) produced by human industrial activities was the Swede Svante Arrhenius (1859–1927), a Nobel Prize-winning physical chemist, in an article published in 1896.[137] He based this on the theory of Joseph Fourier (1768–1830) that the atmosphere slows heat loss and that without an atmosphere the Earth would be much colder.[138] Fourier had experimented with a glass-paneled black box and sunlight to show

Figure 9.6 Margerie Glacier, a tidewater glacier in the Tarr Inlet of Glacier Bay, Alaska: the height of the glacier front here is about 75 m (250 ft) above water and 30 m (100 ft) below water. The retreat of the Grand Pacific Glacier has left this fast-moving tributary isolated. Most glaciers in this area are retreating, and over the last century the loss of ice has been spectacular. Global warming is expected to cause this process to continue. Photograph taken in 1998.

how heat could be retained in what Arrhenius called a "hot-house" and is now called the "greenhouse effect." In addition, in 1859, John Tyndall (1820–93) had discovered that methane and carbon dioxide, as well as water vapor, like other atmospheric gases are transparent to sunlight but unlike other gases are opaque to heat rays and therefore capture heat that might otherwise be reflected from the Earth's surface into space. Arrhenius thought that human-caused production of carbon dioxide by combustion of fuels could raise the proportion of carbon dioxide in the atmosphere and cause a significant rise in temperature. He attempted to calculate the amount of this increase but, although the principle was right, his calculations were flawed. He proposed that a doubling of carbon dioxide would take 3,000 years; it is now estimated at possibly 100 years. He thought the higher temperatures would be beneficial to humanity, making agriculture more productive at higher latitudes and preventing a new ice age. That there might be damaging effects of higher temperatures including rising sea level, desertification, and increased heat energy for violent weather systems did not occur to him.

Several natural causes for climate change were put forward in the following years. One of the most interesting of these was the proposal of Milutin Milankovitch (1879–1958), advanced in the 1920s, that ice ages are related to variations in the parameters of the Earth's orbit around the Sun that affect insolation, that is, the amount of solar energy that reaches various parts of the Earth's surface.[139] These variations occur in cycles of different lengths that reinforce or weaken one another. Simply, the longest cycle is a swing from a nearly circular orbit to a more elliptical one that takes about 100,000 years. Another cycle, about 41,000 years in length, changes the tilt of the Earth's axis. The precession cycle, 19,000–23,000 years long as it affects seasonal insolation, is caused by the change in the direction of the Earth's axis of rotation relative to the fixed stars, a kind of circular wobble. These three cycles can be combined into curves representing changes in potential heating due to total solar radiation for various latitudes in the Northern and Southern Hemispheres. Milankovitch's predicted variations do seem to correspond roughly with past ice ages. Interestingly, they suggest that the Northern Hemisphere should now be slowly cooling. Some scientists take this as evidence that the observed rapid warming over the past century and a half is in large part the effect of processes initiated by humans.

Another proposed natural cause of climate change is sunspot activity; when there are few sunspots, less solar energy reaches the Earth and temperatures may cool. There is an eleven-year cycle of sunspot frequency, first noticed in the nineteenth century, which seems to influence weather. But the Sun sometimes has an irregular phase when sunspots are rare. Edward W. Maunder (1851–1928) studied historical records of sunspots and discovered a period now called the Maunder Minimum, 1645–1715, during which sunspots dropped to as low as one-thousandth their present numbers. That corresponds to the coldest part of the Little Ice Age, suggesting that it is one of the causes of that phenomenon.

Explosive volcanic eruptions can introduce into the high atmosphere massive amounts of sulfur dioxide aerosols, which can circulate around the Earth. These cause cooling by reflecting sunlight before it reaches the Earth's surface. For example, Mount Tambora in Indonesia erupted in 1815, apparently causing 1816 to be the "year without a summer," with snows in June in Europe and North America, crop failures, and famine. After the cataclysmic explosion of Krakatoa in 1883, average world temperatures briefly dropped by 1.2°C (2.2°F). Notable cooling also followed the 1991 eruption of Mount Pinatubo in the Philippines. Looking further back in history, ice core evidence indicates that an equatorial eruption in AD 633–6 created an atmospheric dust veil that caused a severe cooling of worldwide temperatures.[140] This happened during the reign of the Roman Emperor Justinian, and may have weakened the Mediterranean population just before the advent of the Great Plague, which broke out within less than a decade.[141] A Greenland core contains a volcanic signal corresponding to approximately 1645 BC, undoubtedly evidence of the explosive eruption of the island of Thera (Santorini) that was a catastrophe for the Minoan civilization centered on Crete. In an appropriate coincidence, that particular core was drilled at a station named Crête.[142] The climatic effects of volcanoes usually subside two or three years after the eruption ends. Volcanism would not be a major factor in long-term climate change unless there were periods of greatly increased activity.

Human-caused generation of carbon dioxide, methane, and other greenhouse gases, however, continues and increases in volume. The extent of its effect on the atmosphere and climate began to be understood in the years after the Second World War. Ironically, the rising trend of temperatures was interrupted by a period of cooling during 1940–70 in the Northern Hemisphere. Although it proved to be only a temporary pause in the overall warming trend from about 1850 to the present, it briefly reduced the level of concern about global warming. The cause of this thirty-year pause in warming is now believed to be increased levels of air pollution during and after the Second World War, including sulfur dioxide aerosols that acted to reflect solar heat back into space, very much like the volcanic aerosols just mentioned, and particulate pollution that increased the turbidity of the atmosphere. Eventually the warming produced by greenhouse gases overcame this cooling effect.

It was later noted that in the less polluted Southern Hemisphere, cooling did not occur in the years in question.

One factor thought to slow the increase of carbon dioxide in the atmosphere has been the fact that the oceans absorb carbon dioxide, providing a sink for the carbon dioxide produced by the combustion of fuels. A question existed, however, as to how rapidly this absorption worked. Radiocarbon expert Hans Suess and oceanographer Roger Revelle made studies in the 1950s that at first indicated that the oceans take up carbon dioxide in a decade or so, but they came to realize that at least 90 percent of the gas quickly evaporates from the ocean surface. The process of carbon dioxide absorption by the oceans, they concluded, does happen but takes thousands of years, leaving the unabsorbed balance in the atmosphere to increase carbon dioxide concentration. The carbon dioxide absorbed increases the acidity of the world ocean, with likely damaging effects for corals and other sea life. Revelle and Suess made an ominous conclusion: "Human beings are now carrying out a large scale geophysical experiment of a kind that could not have happened in the past nor be reproduced in the future."[143]

It was difficult to assess the problem without accurate measurement of the concentration of carbon dioxide in the atmosphere over a significant period of time. Governments granted money for such studies beginning with the International Geophysical Year of 1957–8, and Revelle encouraged a protégé, Charles David Keeling, to embark on a project to measure carbon dioxide.[144] Realizing that accurate measurements required taking samples in places far from immediate sources of anthropogenic carbon dioxide, Keeling established stations on Mauna Loa, Hawai'i, far out in the central Pacific Ocean, and in Antarctica. His careful observations first demonstrated that there are sharp seasonal variations in carbon dioxide levels with peaks reached in the Northern Hemisphere winter followed by a reduction in carbon dioxide during the northern spring and early summer each year as photosynthesis increased, using it to form plant tissue. Later it became clear that carbon dioxide concentrations rose measurably higher each year, forming an overall upward trend now termed the Keeling Curve. From 315 ppm (parts per million) in 1958, the level reached 380 in 2005, a rise of more than 20 percent in forty-seven years, and showed an unmistakable tendency to continue increasing. Since then, measurements of the Greenland and Antarctica ice cores have enabled the extension of the Keeling Curve back into the deep past, and among many other observations have established that the concentration in historic pre-industrial times before the nineteenth century was approximately 280 ppm. Further, the long-time curve of carbon dioxide concentration and the curve of average global temperature closely reflect one another. The work begun by Keeling offers graphic confirmation of the fact that human technology has become a force of worldwide geophysical change.

New methods of measuring past climates were invented during the second half of the twentieth century that enabled scientists to put recent global warming within the context of the past several hundred thousand years. Harold Urey discovered that the proportion of two oxygen isotopes to one another in the shells of tiny sea creatures called foraminifera recorded the temperatures at which they were formed. Cores from the seabeds where these shells accumulate could show changing temperatures going back 300,000 years, as Cesare Emiliani noted in 1955 and later studies confirmed. Cores have been taken from the sediments accumulated in Lake Baikal, a very ancient and stable lake with depths down to a mile below its surface, and these contain the record of environmental changes for at least 10 million years.[145] These observations supplement the ice core studies, which began in 1956 with drilling into ice sheets in northwestern Greenland as the International Geophysical Year was about to begin. The first ice cores to reach bedrock were drilled at Camp Century, Greenland in 1966, and at Byrd Station, West Antarctica in 1968.[146] They have continued to the present, with samples now from widely representative sites throughout the Earth's great ice sheets.

A third source of observations bearing on climate change has been provided by artificial satellites. The first of these were Sputnik I, launched by the Soviet Union in 1957; Explorer I, by the US in 1958; and Astérix, by France in 1965. More followed from Japan, China, the UK, and other nations. Today there are perhaps 560 operational Earth observation satellites, not counting those intended for military purposes. They provide detailed information on changes in solar irradiance, stratospheric chemistry including ozone depletion such as the "ozone hole" over Antarctica, temperature, precipitation, clouds and water vapor, wind velocity, ocean currents and surface temperatures including El Niño, sea level, vegetation and desertification, coastal configuration, volcanoes, snow cover, sea ice, glaciers and ice sheets, and human activities such as fires and the growth of cities.[147] Even this long list is incomplete. To give one example, photographs from satellites observed the breakup of the Larsen ice shelf, an area of 2,000 sq km (770 sq mi) in Western Antarctica in a matter of days in 2002.

Almost as soon as computers were developed, there were attempts to use them to analyze observations and to predict future changes both in weather and climate. In the years before the mid-1960s, however, computers were too slow and lacked adequate memory for this incredibly complex task. Around that time, efforts were made using improved computers at several centers such as the National Center for Atmospheric Research (NCAR) in Boulder, Colorado and the National Oceanic and Atmospheric Administration (NOAA) Fluid Dynamics Laboratory in Princeton, New Jersey to construct computer models that would allow credible climate prediction.[148] As might be expected, the first attempts had very limited success, and the models had to be made more sophisticated to take account of complex issues such as the interaction of the atmosphere with the oceans, the reflectivity of ice and snow, and the effects of clouds. The climate models were used to predict, among other variables, the effects an increase in greenhouse gases, such as a doubling of carbon dioxide, would have on global temperatures. By 1979 a consensus had emerged among scientists that such an increase would bring about a warmer world, perhaps an increase in the range of 2–4°C (3.6–7.2°F) by the year 2100. In the 1980s, many climate scientists concluded that this was not just a scenario, but was actually underway as observations tended to confirm the computer models. They knew that climate could change, had changed in the past, and now warned that it would change in dangerous ways.

Increasingly, the threat of global warming moved from scientific investigation to public discourse around the world, and especially in the US and Europe. Some prominent scientists, such as Stephen Schneider and James Hansen, spoke to the media and advocated programs to anticipate global warming and take action to mitigate its scale and effects, primarily through reduction in the emission of greenhouse gases such as carbon dioxide. They received widespread attention, and awareness of the likelihood of climate change spread among the general public. They also faced opposition on at least three fronts. There were a few scientists who pointed out problems with the evidence and the models used in prediction. This was to be expected, and is part of the usual process of scientific inquiry. It is positive, because it leads to open discussion, further investigation, and development of hypotheses. It is accurate to say that the weight of scientific opinion moved toward the conclusion that global warming is occurring, and that at least a major proportion of the cause is assignable to human activities.

A second source of opposition to the idea that the danger of climate change merited major interventions to alleviate it came from among the industries whose activities were evidently causing it, and therefore who might have to bear the cost of the efforts to counter it, such as oil and coal companies, other fossil fuel industries, and automobile manufacturers. They engaged in advertisements against the idea of global warming, seized on or sponsored studies that questioned it, and even created false front organizations that claimed objectivity, but in fact engaged in tendentious campaigns designed to sow doubt about the reality of global warming and the idea that human activities are an important cause of it. This was not universally true of these business organizations; some of them saw their self-interest involved in the development of environmentally friendly technology, either genuinely or at least to gain the public appearance of "greenness." A third group active in opposing programs to mitigate global warming was political, composed of right-wing organizations that fought the role of governmental intervention on principle. They feared that reduction of the emission of greenhouse gases would require regulation on national and even international levels, and used questioning of global warming and the role of humankind in causing it as a way of resisting the extension of government control.

Recognizing that scientific advances had begun to indicate that global warming was a pressing problem of worldwide scope, and that trustworthy assessment of the state of scientific knowledge was important to inform international efforts, two UN agencies – the World Meteorological Organization and the United Nations Environment Program (UNEP) – established the Intergovernmental Panel on Climate Change (IPCC) in 1988.[149] Composed of scores of scientists from many countries, the IPCC created three working groups with the following assignments: to assess existing scientific information regarding climate change; to assess environmental and economic impacts of climate change; and to formulate strategies for the management of responses to climate change. The IPCC does not directly carry out research on climate change, but it surveys published and refereed work throughout the world and issues periodic assessments. These are extensive multivolume publications that appear every few years: four so far, in 1990 (amended 1992), 1995, 2001, and 2007. Since IPCC operates by consensus, its conclusions are generally cautious and are widely regarded as authoritative, although some climate scientists and environmentalists think they may err on the conservative side. Also, the deliberations of the IPCC, while careful to maintain scientific objectivity, are not entirely insulated from pressures of the governments that are members of it.

The first report of the IPCC concluded, in part, that there is a greenhouse effect that keeps the Earth warmer than it would be without it, and that emissions from human activities are significantly adding to the concentration of greenhouse gases such as carbon dioxide, methane, chlorofluorocarbons (CFCs), and nitrous oxide. Since these gases are long-lived in the atmosphere, they will be slow to respond to reduced emissions, the report continued, recommending that reductions should be made earlier rather than later. Recognizing that predictions of future trends are uncertain, the report cautiously concluded that the global mean temperature would increase by about 1°C (1.8°F) by 2025 and by about 3°C (5.4°F) by 2100, and that sea level would rise about 0.6 m (2 ft) by 2100, mainly due to the thermal expansion of the ocean.[150]

The IPCC report became available during the preparation of the Framework Convention on Climate Change (FCCC), an international agreement that seeks to limit or reduce emissions of gases, mainly carbon dioxide and methane, with the potential to exacerbate global warming. Although the preparatory negotiations for the Climate Change Convention were a separate process, they were concluded so as to be ready for signature at the UN Conference on Environment and Development in Rio de Janeiro, 1992. The discussions took place with the positive background of the Montreal Protocol aimed at limiting the production of CFCs, compounds containing chlorine and other halogens that diminish stratospheric ozone and increase the amount of ultraviolet radiation that penetrates a weakened ozone layer. All major nations supported the Montreal agreement, and CFC production was sharply reduced as a result. Mustafa Tolba, head of UNEP, energetically moved the Montreal process along and shared credit for its success.[151]

The negotiation of the FCCC was a similar but much more difficult process. It was motivated by international concerns over the possible enhancement of the greenhouse effect by the production of carbon dioxide, methane, and other gases that tend to retain solar heat in the Earth's atmosphere, surface, and oceans. Its stated objective was

> to achieve stabilization of greenhouse gas concentrations in the atmosphere at a low enough level to prevent dangerous anthropogenic interference with the climate system. Such a level should be achieved within a time-frame sufficient to allow ecosystems to adapt naturally to climate change, to ensure that food production is not threatened and to enable economic development to proceed in a sustainable manner.[152]

Observation by atmospheric scientists had shown that a rapid increase in the concentration of these gases was occurring, and that it was associated with a rise in the average temperature of the Earth. Computer models suggest that the effects of these changes, should they continue, will vary in different regions of the Earth's surface, but could include rising temperatures, changing patterns of precipitation, an elevation of sea level, disruption of freshwater flows by removal of ice, and stresses on agricultural crops, forests, and wildlife, including coral reefs and fish because of changes in ocean temperatures and acidity. Many world leaders agree that these dangers call for an effort to reduce the level of greenhouse gas emissions, but to do this for carbon dioxide and methane, the most important heat-trapping gases, is more difficult than it was for CFCs in the Montreal process. Disagreement over the actions recommended to counter global warming figuratively raised the temperature of the meeting rooms. The US managed to keep any emission reduction goals or timetables out of the agreement. When the treaty's teeth were thus pulled, President George H.W. Bush signed it and forwarded it to the US Senate, which ratified it.

In the years after the Earth Summit, a number of international meetings were convened to carry forward the implications of the agreements reached there. Important among these were conferences to set arrangements for meeting the goals of the Framework Convention on Climate Change (FCCC), including specific reductions in emissions of greenhouse gases and timetables for achieving them. At a conference in Kyoto in 1997, representatives of the nations that had signed the treaty, including the major economic powers, reached a comprehensive agreement after difficult negotiations. According to this "Kyoto Protocol," the world's emissions of the three major greenhouse gases (carbon dioxide, methane, and nitrous oxide) were to be reduced 5.2 percent from the 1990 levels by 2008–12. Japan was to reduce by 6 percent, the United States by 7 percent, and the European Union by 8 percent.[153] Other nations accepted goals, but developing countries, including such increasingly important emitters as China, India, and Brazil, were not required to agree to any specific reductions. Three other gases (hydrofluorocarbons, sulfur hexafluoride, and perfluorocarbons) would be reduced calculated on the base year of 1995. The US successfully pushed for the inclusion of the principle of trading emission rights, that is, that nations that continue to pollute could do so by paying for that right to other nations that have reduced their pollution below the targets. As the most economically powerful nation, the US could trade for such rights and not have to reduce emissions to as great an extent. That the US initially agreed to the protocol is probably due to the direct involvement of Vice-President Al Gore in Kyoto.

Observers noted that it would take reductions several times those envisioned in the Kyoto Protocol to bring emissions down to a level that might be effective in countering global warming. The cost of measures to achieve those reductions would be high; whether it would be greater than the costs incurred by actual global warming is perhaps unlikely, but unknown. Kyoto was at best a positive step, one of many that would have to be taken as the magnitude of the effects of global warming on human health, the economy, and the natural world becomes clearer, and when most nations come to see international cooperation to initiate effective measures as being in their national interest. However, that has not yet occurred. In the case of the US, then and now the world's largest emitter of greenhouse gases (although China is not far behind and will probably soon pass the US), neither Presidents Bill Clinton nor George W. Bush (the second President Bush) sent the Kyoto Protocol to the Senate for ratification, and Bush explicitly rejected it in 2001. By 2005, enough other nations had agreed to the protocol to enter into force, and this was done formally at the United Nations Climate Change Convention in Montreal, Canada. At the same meeting, the participating nations agreed to extend the Kyoto Protocol beyond its 2012 expiration date and to conduct negotiations on deeper emissions cuts. Unfortunately there was little indication that the original target reductions would be achieved by 2012.

The IPCC issued its Fourth Assessment Report in 2007. Its important conclusions include the following.[154] There is no doubt that the Earth's climate system is warming. Concentrations of carbon dioxide, methane, and nitrous oxide in the atmosphere have increased markedly as a result of human activities since 1750 and now far exceed pre-industrial values over the past 650,000 years. Most of the increase in global average temperatures since the mid-twentieth century is very likely due to the observed increase in human-caused greenhouse gas concentrations. The probability that this is caused by natural climatic processes alone is less than 5 percent. Anthropogenic warming and sea level rise will continue for centuries due to the time lags associated with climate processes and feedbacks, even if greenhouse gas concentrations were to be stabilized, although the amount of temperature and sea level rise depends on the intensity of emissions during the next century. Both past and future anthropogenic carbon dioxide emissions will continue to contribute to warming and sea level rise for more than a millennium. World temperatures could rise by between 1.1 and 6.4°C (2–11.5°F) during the twenty-first century and sea levels will probably rise by 18–59 cm (7–23 in). There will almost certainly be more frequent warm spells, heat waves, and heavy rainfall in some areas, and probably an increase in droughts, tropical cyclones, and extreme high tides.[155] The IPCC shared the 2007 Nobel Peace Prize with former US Vice-President Al Gore, who had a prominent role in raising public awareness of the problem through a widely distributed motion picture and book.[156] The United Nations scheduled a conference in Copenhagen in 2009 to negotiate a binding treaty on climate change, which if successful would be the next major agreement after Kyoto. Meanwhile, carbon dioxide emissions are already rising faster than the IPCC 2007 report predicted, in spite of numerous conferences and the international treaties that are in place.[157]

The history of the recognition of global warming and its implications for global human society is an illustrative case of the interaction between the growth of scientific knowledge and the interests of political and economic entities. Science can tell us of the relative (not absolute) certainty of continued human-induced global warming and its probable effects. It can evaluate possible measures intended to forestall negative changes, or at least to lessen their magnitude, as well as courses of action intended to enable human societies to cope with the negative effects of the likely changes. But the decisions as to what to do will be made, as they have generally been made in recent history, by governments and corporations that have demonstrated a tendency to act in accord with what they see as their own interests in the relatively short term. The common good of humans and the Earth has unfortunately received less consideration.

Notes

1   Archibald MacLeish, *Riders on the Earth: Essays and Recollections*, Boston, MA, Houghton Mifflin, 1978, xiii–xiv.

2   Clayton R. Koppes, *JPL and the American Space Program*, New Haven, CT, Yale University Press, 1982, 79.

3   James Lovelock, *The Ages of Gaia: A Biography of Our Living Earth*, New York, W.W. Norton, 1988.

4   United Nations, *Space Activities of the United Nations and International Organizations*, doc. symbol A/AC.105/521, New York, United Nations, 1992 56–62.

5   Susanna B. Hecht, "Valuing Land Uses in Amazonia: Colonist Agriculture, Cattle, and Petty Extraction in Comparative Perspective," in Kent H. Reford and Christine Padoch, eds, *Conservation of Neotropical Forests: Working from Traditional Resource Use*, New York, Columbia University Press, 1992, 379–99, at 379.

6   Andrew Revkin, *The Burning Season: The Murder of Chico Mendes and the Fight for the Amazon Rain Forest*, Boston, MA, Houghton Mifflin, 1990.

7   Gilbert Rist, *The History of Development: From Western Origins to Global Faith*, London, Zed Books, 1997, 186–7.

8   Simon Reisman, "The Birth of a World Trading System: ITO and GATT," in Orin Kirshner, ed., *The Bretton Woods–GATT System: Retrospect and Prospect After Fifty Years*, Armonk, NY, M.E. Sharpe, 1996, 82–9.

9   Thomas Prugh, *Natural Capital and Human Economic Survival*, 2nd edn, Boca Raton, FL, Lewis Publishers, 1999, 16–17.

10 Herman E. Daly, ed., *Economics, Ecology, Ethics: Essays Toward a Steady-State Economy*, San Francisco, W.H. Freeman, 1980; Thomas Prugh, Robert Costanza, John H. Cumberland, Herman E. Daly, Robert Goodland, and Richard B. Norgaard, *Natural Capital and Human Economic Survival*, 2nd Edition, Boca Raton,FL, Lewis Publishers, 1999.

11 "Mexico vs United States. Challenge to US restrictions on tuna imports under the US Marine Mammal Protection Act," Report of the Panel, DS29/R, Geneva, GATT, 1991.

12 Joshua Karliner, *The Corporate Planet: Ecology and Politics in the Age of Globalization*, San Francisco, Sierra Club, 1997, 42–3.

13 For this apt metaphor, I am indebted to a fine analytical study by Anthony B. Anderson, Peter H. May, and Michael J. Balick, *The Subsidy from Nature: Palm Forests, Peasantry, and Development on an Amazon Frontier*, New York, Columbia University Press, 1991.

14 Eugene P. Odum, *Ecology and Our Endangered Life-Support Systems*, Sunderland, MA, Sinauer Associates, 1993, 131–2.

15 Prugh, *Natural Capital and Human Economic Survival*, 16–17.

16 C. Fred Bergtsen, "The Rationale for a Rosy View," *The Economist* 328, September 1993 (7828), 57.

17 Donella H. Meadows, Dennis L. Meadows, Jørgen Randers, and William W. Behrens III, *The Limits to Growth*, New York, Universe Books, 1972.

18 Mauricio Schoijet, "Limits to Growth and the Rise of Catastrophism," *Environmental History* 4, 4, October 1999, 515–30, gives an analysis of various critics of the report.

19 Donella H. Meadows, Dennis L. Meadows, and Jørgen Randers, *Beyond the Limits: Confronting Global Collapse and Envisioning a Sustainable Future*, Post Mills, VT, Chelsea Green Publishing, 1992, xv.

20 World Commission on Environment and Development, *Our Common Future*, Oxford, Oxford University Press, 1987, 43.

21 Robert Costanza, John Cumberland, Herman Daly, Robert Goodland, and Richard Norgaard, *An Introduction to Ecological Economics*, Boca Raton, FL, St. Lucie Press, 1997, 8–13.

22 Herman E. Daly, *Beyond Growth: The Economics of Sustainable Development*, Boston, MA, Beacon Press, 1996.

23   Brian McDonald, *The World Trading System: The Uruguay Round and Beyond*, Houndmills, UK, Macmillan, 1998, 259.

24   Edward B. Barbier, Joanne C. Burgess, and Carl Folke, *Paradise Lost?: The Ecological Economics of Biodiversity*, London, Earthscan, 1994, 229.

25   Edward O. Wilson, *The Diversity of Life*, Cambridge, MA, Belknap Press of Harvard University, 1992, 278–80.

26   Ibid., 32. See also Paul and Anne Ehrlich, *Extinction: The Causes and Consequences of the Disappearance of Species*, New York, Random House, 1981.

27   Kathryn Phillips, *Tracking the Vanishing Frogs: An Ecological Mystery*, New York, St. Martin's Press, 1994.

28   Michael J. Eden, *Ecology and Land Management in Amazonia*, London, Belhaven Press, 1990, 94.

29   Judith Gradwohl and Russell Greenberg, *Saving the Tropical Forests*, London, Earthscan, 1988, 36.

30   Edward O. Wilson, "Threats to Biodiversity," in Jonathan Piel *et al.*, *Managing Planet Earth: Readings from Scientific American Magazine,* New York, W.H. Freeman, 1990, 56.

31   Edward A. Parson, Peter M. Haas, and Marc A. Levy, "A Summary of the Major Documents Signed at the Earth Summit and the Global Forum," *Environment* 34, 8, October 1992, 14.

32   John Rodman, "Paradigm Change in Political Science: An Ecological Perspective," *American Behavioral Scientist* 24, 1, 1980, 49–78, at 68.

33   Wilson, "Threats to Biodiversity," 81.

34   Eden, *Ecology and Land Management in Amazonia*, 39.

35   John C. Kricher, *A Neotropical Companion: An Introduction to the Animals, Plants, and Ecosystems of the New World Tropics*, Princeton, NJ, Princeton University Press, 1989, 122–3.

36   Ronald A. Foresta, *Amazon Conservation in the Age of Development: The Limits of Providence*, Gainesville, University of Florida Press, 1991, 32–55.

37   Eneas Salati, Marc J. Dourojeanni, Fernando C. Novaes, Adélia Engrácia de Oliveira, Richard W. Perritt, Herbert Otto Roger Schubart, and Julio Carrizosa Umana, "Amazonia," in B.L. Turner II *et al.*, eds, *The Earth as Transformed by Human Action*, Cambridge, Cambridge University Press, 1990, 479–94, see 479.

38   Ibid., 484.

39   Kricher, *Neotropical Companion*, 73.

40   Carl F. Jordan, ed., *An Amazonian Rain Forest: The Structure and Function of a Nutrient Stressed Ecosystem and the Impact of Slash-and-Burn Agriculture*, Paris, UNESCO, 1989, 1; Anthony B. Anderson, ed., *Alternatives to Deforestation: Steps Toward Sustainable Use of the Amazon Rain Forest*, New York, Columbia University Press, 1990.

41   Anna C. Roosevelt, ed., *Amazonian Indians from Prehistory to the Present*, Tucson, University of Arizona Press, 1994.

42   John Hemming, *Amazon Frontier: The Defeat of the Brazilian Indians*, Cambridge, MA, Harvard University Press, 1987.

43   Michael Goulding, Nigel J.H. Smith, and Dennis J. Mahar, *Floods of Fortune: Ecology and Economy Along the Amazon*, New York, Columbia University Press, 1996, 27–9.

44   Roberta M. Delson and John Dickenson, "Conservation Tendencies in Colonial and Imperial Brazil: An Alternative Perspective on Human Relations to the Land," *Environmental Review* 8, 3, Fall 1984, 270–83.

45   Barbara Weinstein, *The Amazon Rubber Boom, 1850–1920*, Stanford, CA, Stanford University Press, 1983, 8.

46   Michael Edward Stanfield, *Red Rubber, Bleeding Trees: Violence, Slavery, and Empire in Northwest Amazonia, 1850–1933*, Albuquerque, University of New Mexico Press, 1998.

47   Christine Padoch, "The Economic Importance and Marketing of Forest and Fallow Products in the Iquitos Region," in William M. Denevan and Christine Padoch, eds, *Swidden-Fallow Agroforestry in the Peruvian Amazon*, New York, New York Botanical Garden, 1988, 74–89, at 75.

48   Donald Kennedy and Marjorie Lucks, "Rubber, Blight, and Mosquitoes: Biogeography Meets the Global Economy," *Environmental History* 4, 1999, 369–83.

49   Roger D. Stone, *Dreams of Amazonia*, New York, Viking, 1985, 77–81.

50   Eduardo Bedoya and Loren Klein, "Forty Years of Political Ecology in the Peruvian Upper Forest: The Case of Upper Huallaga," in Leslie E. Sponsel, Thomas N. Headland, and Robert C. Bailey, eds, *Tropical Deforestation: The Human Dimension*, New York, Columbia University Press, 1996, 165–86.

51   Salati *et al.*, "Amazonia," 488.

52 Goulding *et al.*, *Floods of Fortune*, 48.
53 Napoleon A. Chagnon, *Yanomamö: The Last Days of Eden*, San Diego, CA, Harcourt Brace Jovanovich, 1992, 248–55.
54 Marianne Shmink and Charles H. Wood, *Contested Frontiers in Amazonia*, New York, Columbia University Press, 1992, 253–343.
55 Paul Sabin, "Searching for Middle Ground: Native Communities and Oil Extraction in the Northern and Central Ecuadorian Amazon, 1967–1993," *Environmental History* 3, 2, April 1998, 144–68; Joe Kane, *Savages*, Vancouver, Douglas and McIntyre, 1995.
56 Bertha K. Becker, "Undoing Myths: The Amazon – An Urbanized Forest," in M. Clusener-Godt and I. Sachs, eds, *Brazilian Perspectives on Sustainable Development of the Amazon Region*, Paris, UNESCO, 1995, 54–90, at 72–3.
57 John R. McNeill, "Agriculture, Forests, and Ecological History: Brazil, 1500–1984," *Environmental Review* 10, 2, Summer 1986, 122–33, at 128.
58 Betty J. Meggers, *Amazonia: Man and Culture in a Counterfeit Paradise*, revised edn, Washington, Smithsonian Institution Press, 1996, vii.
59 Amazon Cooperation Treaty, *Proposal of Criteria and Indicators for Sustainability of the Amazon Forest*, Protem Secretariat, Lima, Peru, 1995.
60 Nigel J.H. Smith, Emanuel Adilson S. Serrão, Paulo T. Alvim, and Italo C. Falesi, *Amazonia: Resiliency and Dynamism of the Land and Its People*, Tokyo, United Nations University Press, 1995, 208–9.
61 Richard O. Bierregaard, Jr., and Virginia H. Dale, "Islands in an Ever-Changing Sea: The Ecological and Socioeconomic Dynamics of Amazonian Rainforest Fragments," in John Schelhas and Russell Greenberg, eds, *Forest Patches in Tropical Landscapes*, Washington, Island Press, 1996, 187–204.
62 Salati *et al.*, "Amazonia," 491.
63 Nigel J.H. Smith, *Man, Fishes, and the Amazon*, New York, Columbia University Press, 1981, 132; Michael Goulding, *The Fishes and the Forest: Explorations in Amazonian Natural History*, Berkeley and Los Angeles, University of California Press, 1980.
64 J.H.C. Gash, ed., *Amazonian Deforestation and Climate*, New York, John Wiley, 1996.
65 Harald Sioli, "The Effects of Deforestation in Amazonia," in John Hemming, ed., *Change in the Amazon basin*, Vol. 1, *Man's Impact on Forests and Rivers*, Manchester, Manchester University Press, 1985, 58–65, at 63.
66 Anthony B. Anderson, Peter H. May, and Michael J. Balick, *The Subsidy from Nature: Palm Forests, Peasantry, and Development on an Amazon Frontier*, New York, Columbia University Press, 1991.
67 Erin Moore Daly, "New Orleans, Invisible City," *Nature and Culture* 1, 2, Autumn 2006, 133–48, at 135.
68 Ari Kelman, *A River and Its City: The Nature of Landscape in New Orleans*, Berkeley and Los Angeles, University of California Press, 2006.
69 Ari Kelman, "City of Nature: New Orleans' Blessing; New Orleans' Curse," Slate Podcast, August 31, 2005, www.slate.com/id/2125346/.
70 Craig E. Colten, *An Unnatural Metropolis: Wresting New Orleans from Nature*, Baton Rouge, LA, Louisiana State University Press, 2005, 26–8.
71 Craig E. Colten, "Paradise Lost?" in Donald E. Davis, Craig E. Colton, Megan Kate Nelson, Barbara L. Allen, and Mikko Saikku, eds, *Southern United States: An Environmental History*, Santa Barbara, CA, ABC-CLIO, 2006, 183–221, at 206.
72 Mikko Saikku, *This Delta, This Land: An Environmental History of the Yazoo–Mississippi Floodplain*, Athens, GA, University of Georgia Press, 2005, 163–4.
73 Colten, *Unnatural Metropolis*, 142.
74 Ibid., 31.
75 John V. Dennis, *The Great Cypress Swamps*, Baton Rouge, Louisiana State University Press, 1988, 2.
76 Craig E. Colten, "Cypress in New Orleans: Revisiting the Observations of Le Page du Pratz," *Louisiana History* 64, Fall 2003, 463–77.
77 William H. Conner, Thomas W. Doyle, and Ken W. Krauss, eds, *Ecology of Tidal Freshwater Forested Wetlands of the Southeastern United States*, Dordrecht, Netherlands, Springer, 2007, 448–50.
78 Mikko Saikku, "Down by the Riverside: The Disappearing Bottomland Hardwood Forest of Southeastern North America," *Environment and History* 2, 1, February 1996, 77–95, at 88.
79 Colten, *Unnatural Metropolis*, 73–5.
80 Kelman, *River and Its City*, 154–5.

81 Ivor van Heerden and Mike Bryan, *The Storm: What Went Wrong and Why During Hurricane Katrina: The Inside Story from One Louisiana Scientist*, New York, Penguin Books, 2007, 90.

82 Federal Emergency Management Association (FEMA) Release R6-04-093, July 23, 2004.

83 Stephen A. Curtis, ed., *Hurricane Katrina Damage Assessment: Louisiana, Alabama, and Mississippi Ports and Coasts*, Reston, VA, American Society of Civil Engineers, 2007, iii, 3, 98–101.

84 Kelman, *River and Its City*, xviii.

85 Todd Shallat, "In the Wake of Hurricane Betsy," in Craig E. Colten, ed., *Transforming New Orleans and Its Environs: Centuries of Change*, Pittsburgh, PA, University of Pittsburgh Press, 2000, 121–37.

86 Sara Shipley Hiles, "The Environment," in Jenni Bergal, Sara Shipley Hiles, Frank Koughan, John McQuaid, Jim Morris, Katy Reckdahl, and Curtis Wilkie, eds, *City Adrift: New Orleans Before and After Katrina*, Baton Rouge, Louisiana State University Press, 2007, 16.

87 National Research Council, *Drawing Louisiana's New Map: Addressing Land Loss in Coastal Louisiana*, Washington, DC, National Academies Press, 2006.

88 Bill Streever, *Saving Louisiana? The Battle for the Coastal Wetlands*, Jackson, University Press of Mississippi, 2001.

89 Robin Clarke and Imogen Bertin, *The African Elephant*, UNEP/GEMS Environment Library 3, Nairobi, UNEP, 1989.

90 Anthony Fagin, ed., *UNEP Biennial Report, 1996–1997*, London, December 1997, 1.

91 Mark Allan Gray, "The United Nations Environment Programme: An Assessment," *Environmental Law* 20, 1990 291–319, at 294.

92 Fagin, *UNEP Biennial Report, 1996–1997*, 49.

93 John McCormick, *Reclaiming Paradise: The Global Environmental Movement*, Bloomington, Indiana University Press, 1989, 97. Other NGOs convened a parallel Environmental Forum that discussed the same issues as the conference, while observing and criticizing it.

94 P. Birnie, "The UN and the Environment," in Roberts and Kingsbury, eds, *United Nations, Divided World: The UN's Roles in International Relations*, Oxford, Clarendon Press, 1993, 338.

95 Mostafa Kamal Tolba, ed., *Evolving Environmental Perceptions: From Stockholm to Nairobi*, London, Butterworths, 1988, 97–8; Philip Shabecoff, *A New Name for Peace: International Environmentalism, Sustainable Development, and Democracy*, Hanover, NH, University Press of New England, 1996, 39–40.

96 Stockholm Declaration, Stockholm 1972, Principle 1.

97 *UNEP Biennial Report, 1996–1997*, 64.

98 Said Mahmoudi, "The United Nations Environment Programme (UNEP): An Assessment," *Asian Yearbook of International Law* 5, 1995, 175–98, at 197.

99 "UNEP: The First Twenty Years," *Environmental Policy and Law* 23, 1, February 1993, 14–15.

100 Mahmoudi, "UNEP: An Assessment," 191.

101 Carole Annette Petsonk, "The Role of the United Nations Environment Programme (UNEP) in the Development of International Environmental Law," *American University Journal of International Law and Policy* 5, 1990, 351–91, esp. 355n.

102 Donald Kaniaru, Manjit Iqbal, Elizabeth Mrema, and Siraj Chowdhury, "UNEP's Programme of Assistance on National Legislation and Institutions," in Sun Lin and Lal Kurukulasuriya, eds, *UNEP's New Way Forward: Environmental Law and Sustainable Development*, Nairobi, UNEP, 1995, 153–70.

103 Petsonk, "The Role of UNEP in the Development of International Environmental Law," 354.

104 Mostafa K. Tolba and Iwona Rummel-Bulska, *Global Environmental Diplomacy: Negotiating Environmental Agreements for the World, 1973–1992*, Cambridge, MA, MIT Press, 1998, 55–88.

105 Ibid., 371.

106 UNEP has an international press service, and puts out *UNEP News* and the quarterly *Our Planet*.

107 Officially called the Intergovernmental Conference of Experts on the Scientific Basis for the Rational Use and Conservation of the Resources of the Biosphere, held in Paris, September 1968. UNESCO's General Conference approved the MAB Programme in 1970.

108 UNESCO, *Backgrounder: The MAB Programme*, Paris, UNESCO, 1982, 3.

109 Peter I. Hajnal, *Guide to UNESCO*, London, Oceana Publications, 1983, 119–20.

110 Michel Batisse, "Developing and Focusing the Biosphere Reserve Concept," *Nature and Resources* 22, 3, 1986, 2–10, see 4.

111 J. Ronald Engel, in "Keeping Faith with Life," the text of a dialogue in Steven C. Rockefeller and John C. Elder, eds, *Spirit and Nature*, Boston, MA, Beacon Press, 1992, 188.

112 David Lovatt Smith, *Amboseli: Nothing Short of a Miracle*, Nairobi, East African Publishing House, 1986.

113 Craig L. Shafer, *Nature Reserves: Island Theory and Conservation Practice*, Washington, DC, Smithsonian Institution Press, 1990.

114 M.I. Dyer and M.M. Holland, "The Biosphere–Reserve Concept: Needs for a Network Design," *Bio-Science* 41, 5, 1991, 319–25, see 319.

115 Adam Rogers, *The Earth Summit: A Planetary Reckoning*, Los Angeles, CA, Global View Press, 1993, 269.

116 Ranee K.L. Panjabi, *The Earth Summit at Rio: Politics, Economics, and the Environment*, Boston, MA, Northeastern University Press, 1997.

117 Rogers, *Earth Summit*, 193.

118 Stephen Luper-Foy, "International Justice and the Environment," in David E. Cooper and Joy A. Palmer, eds, *Just Environments: Intergenerational, International and Interspecies Issues*, London, Routledge, 1995, 91–107, esp. 95.

119 Ross McCluney, "Sustainable Values," in Noel J. Brown and Pierre Quiblier, eds, *Ethics and Agenda 21: Moral Implications of a Global Consensus*, New York, United Nations Publications, 1994, 13–27, esp. 16–17.

120 The principles to which the US objected were Principles 3, 7, 12, and 23.

121 United Nations, *Agenda 21: The United Nations Programme of Action from Rio*, New York, United Nations Publications, 1993; see also Dale Boyd, "UNEP after Rio," *Our Planet* 4, 4, 1992, 8–11.

122 Mark F. Imber, "The United Nations' Role in Sustainable Development," in Caroline Thomas, ed., *Rio: Unravelling the Consequences*, Ilford, UK, Frank Cass, 1994, 122–36, esp. 128–9.

123 Edward A. Parson, Peter M. Haas, and Marc A. Levy, "A Summary of the Major Documents Signed at the Earth Summit and the Global Forum," *Environment* 34, 8, October 1992, 14.

124 Rogers, *Earth Summit*, 294.

125 Mahathir Mohamad, "Statement to the UN Conference on Environment and Development," *Environmental Policy and Law* 22, 4, 1992, 232.

126 Pratap Chatterjee and Matthias Finger, *The Earth Brokers: Power, Politics and World Development*, London, Routledge, 1994, 167.

127 John Gordon, "Letting the Genie Out: Local Government and UNCED," in Thomas, ed., *Rio: Unravelling the Consequences*, 137–55, at 153.

128 Karliner, *Corporate Planet*, 10–11.

129 Matthew 10. 16.

130 NICL is a joint program funded by the US Geological Survey and the National Science Foundation. It is located in the Denver Federal Center, west of Denver, Colorado. Scientific aspects of NICL'S work are coordinated by a Science Management Office in the Climate Change Research Center at the University of New Hampshire.

131 Richard B. Alley, *The Two-Mile Time Machine: Ice Cores, Abrupt Climate Change, and Our Future*, Princeton, NJ, Princeton University Press, 2000.

132 Richard B. Alley, ed., *Abrupt Climate Change: Inevitable Surprises*, Washington, DC, National Academy Press, 2002, 24–36. See also Alley, *Two-Mile Time Machine*, 111–15.

133 Ben Orlove, Ellen Wiegandt, and Brian H. Luckman, eds, *Darkening Peaks: Glacier Retreat, Science, and Society*, Berkeley and Los Angeles, University of California Press, 2008.

134 Spencer R. Weart, *The Discovery of Global Warming: Revised and Expanded Edition*, Cambridge, MA, Harvard University Press, 2008. This section is based, in large part, on Weart's ground-breaking work.

135 Richard W. Battarbee and Heather A. Binney, eds, *Natural Climate Variability and Global Warming: A Holocene Perspective*, Chichester, Wiley-Blackwell, 2008.

136 William F. Ruddiman, *Plows, Plagues and Petroleum: How Humans Took Control of Climate*, Princeton, NJ, Princeton University Press, 2005.

137 Svante August Arrhenius, "On the Influence of Carbonic Acid in the Air Upon the Temperature of the Ground," *Philosophical Magazine* 41, 1896, 237–76.

138 James Rodger Fleming, *Historical Perspectives on Climate Change*, New York, Oxford University Press, 1998, 55–64.

139 Milutin Milankovitch, *Canon of Insolation and the Ice-age Problem*, Jerusalem, Israel Program for Scientific Translations, 1969.

140 L.B. Larsen *et al.*, "New Ice Core Evidence for a Volcanic Cause of the A.D. 536 Dust Veil," *Geophyical Research Letters* 35, 2008, L04708, doi: 10.1029/2007GL032450.

141 The Plague of Justinian, believed to be an early occurrence of the Bubonic Plague, reached Constantinople in 642.

142 Willi Dansgaard, *Frozen Annals: Greenland Ice Cap Research*, Aage V. Jensens Fonde, Odder, DK, Narayana Press, 2004, 81.

143 Roger Revelle and Hans E. Suess, "Carbon Dioxide Exchange between Atmosphere and Ocean and the Question of an Increase of Atmospheric CO_2 During the Past Decades," *Tellus* 9, 1957, 18–27.

144 Spencer R. Weart, *The Discovery of Global Warming*, Cambridge, MA, Harvard University Press, 2003, 34–8.

145 Kenji Kashiwaya, ed., *Long Continental Records from Lake Baikal*, Tokyo, Springer, 2003.

146 Dansgaard, *Frozen Annals*, 54; Alley, *Two-Mile Time Machine*, 18–19.

147 R.J. Gurney, J.L. Foster, and C.L. Parkinson, eds, *Atlas of Satellite Observations Related to Global Change*, Cambridge, Cambridge University Press, 1993.

148 Weart, *Discovery of Global Warming: Revised*, 90–113.

149 Bert Bolin, *A History of the Science and Politics of Climate Change: The Role of the Intergovernmental Panel on Climate Change*, Cambridge, Cambridge University Press, 2007.

150 J.T. Houghton, G.J. Jenkins and J.J. Ephraums, eds, *Scientific Assessment of Climate Change: Report of Working Group I*, Cambridge, Cambridge University Press, 1990; W.J. McG Tegart, G.W. Sheldon, and D.C. Griffiths, eds, *Impacts Assessment of Climate Change: Report of Working Group II*, Canberra, Australian Government Publishing Service, 1990; *The IPCC Response Strategies: Report of Working Group III*, Covelo, CA, Island Press, 1990.

151 Rummel-Bulska, *Global Environmental Diplomacy*, 55–88.

152 Framework Convention on Climate Change, Article 2.

153 Elizabeth J. Rowbotham, "Legal Obligations and Uncertainties in the Climate Change Convention," in Tim O'Riordan and Jill Jager, eds, *Politics of Climate Change: A European Perspective*, London, Routledge, 1996, 32–50.

154 Rajendra K. Pachauri and Andy Reisinger, eds, *Climate Change 2007: Synthesis Report*, Geneva, World Meteorological Organization, 2008.

155 See also Weart, *Impacts of global warming*, http://www.aip.org/history/climate/impacts.htm#impacts.

156 Al Gore, *An Inconvenient Truth: The Planetary Emergency of Global Warming and What We Can Do About It*, Emmaus, PA, Rodale, 2006.

157 Robert Kunzig, "A Sunshade for Planet Earth," *Scientific American* 299, 5, November 2008, 46.

10 A general conclusion

Looking back over our journey through the history of humankind's changing role in the community of life, and our glimpses of particular places and periods of time, we may ask what this historical experience offers in understanding what is happening today. What has really been going on, ecologically speaking, during human history? Humans have related in multiple ways to the Earth's systems; some of these ways promise a sustainable balance with them, while others are destructive. Experience could teach us which are which.

Can the processes we see happening now continue indefinitely? No, since activities that are immensely destructive of the biophysical environment now dominate human efforts. The results, mentioned many times in the preceding pages, form an ominous litany: polluted water and air, acidic precipitation, diminution of the ozone layer, global warming, the spread of radioactive materials, deforestation, loss of biodiversity, extinction of species, soil erosion, overpopulation. As we have seen, similar destructive phases, if not as intense, have occurred in the past in more limited areas of the Earth, and they have ended with degraded ecosystems that were able to support neither the continued growth of human numbers nor the level of culture and economic prosperity that then existed. The examples are numerous: the Mesopotamian experience with soil salinity, the southern lowland Mayan experience with deforestation, the experience of the Romans with depletion of forests, soils, and wildlife in the Mediterranean basin, and others.

Have humans achieved sustainable lifestyles in particular times and places? If so, it would appear that human exploitation of ecosystems can be kept within limits, and that with appropriate attitudes and actions both human societies and the entire community of life may be spared destruction. It seems the answer is yes: the indigenous agriculture of the Hopi and the Balinese, the ancient Egyptians' beneficial interaction with the Nile, and the promise of the Inca economy, albeit truncated in each case, give reason for positive evaluations. Even though these are older, agriculture-based economies, the success of such peoples over various periods is worth study for possible applications. It is harder to find examples in the modern industrial world because change is rapid and time has been short; the returns are not yet in, so to speak. But the northernmost European countries, with a near stable population, a relatively clean environment, and a high level of public and governmental environmental concern within the context of reliable democracy, may serve as potential models.

Some of the trends visible in recent history resist the dominant pattern of destruction noted above. One of these is increasing knowledge of the workings of natural systems, and the advances of the science of ecology. This is not simply growth in information, but hard-won understanding. If our society is to be in a sustainable dynamic relationship with that which supports it, every decision we make must be arrived at in respect to the ecosystem.

We must understand our collective actions in terms of what science tells us about the operation of the natural systems. But science is not a dogma; it is a search for understanding that always continues. This age, with its characteristic skepticism, tends only slowly to accept what science demonstrates, but it will not accept anything that does not seem to have a scientific basis.

A second contemporary trend with some promise is that of appropriate technology. The dominant trend of technology in the modern age is undoubtedly toward destruction of landscape, vegetation, biodiversity, and often human health as well. Biotechnology presents a bewildering variety of potential dangers.[1] We should not trust those who design technology to find a solution for every environmental problem. But applications are being found that may enable us to work with natural processes rather than against them.[2] In a village in India, I saw biogas generators adapted to local conditions that enabled people to get power and natural fertilizer from effluents that would otherwise have polluted sources of water. Inexpensive insulated stoves offer Africans the ability to use fuel much more efficiently and to use less wood from depleted forests. Recycling is of questionable value if it uses too much energy, or produces pollution, but recycling could be designed to fit into the natural cycles of the ecosystem.

A third historic trend that may be emerging is one that demonstrates consideration for the community of life in thought and practice. The Chipko movement in India was one example of this, briefly described in Chapter 7. Villagers, mainly women, in the Himalayan region of Uttar Pradesh opposed the deforestation of their district by hugging trees, putting their bodies in the way of loggers wielding axes and chain saws. They wanted to save the forest because it was their source of fuel – there women are the wood gatherers – and because experience had taught them that when steep hillsides are clearcut, the villages below them often are devastated by floods carrying mud and boulders. But it was also because they honored the trees, singing songs such as:

> What do the forests bear?
> Soil, water and pure air.
> Soil, water and pure air
> Sustain the Earth and all she bears.[3]

Chipko had some successes, but those who demonstrate concern for the relationship of humans to nature have often suffered for it. Wangari Maathai, who began the Green Belt Movement in Kenya to advocate the planting and care of trees by women and children, was beaten and imprisoned. Since then she received the Nobel Peace Prize. Chico Mendes, who organized the *seringueiros*, rubber tappers in the Amazon rainforest, to defend the forest and their livelihood against illegal clearing, was murdered, doubtless by those whose financial interests in forest removal he threatened.[4]

There is some historical evidence of a search by human beings for a positive role in the ecosystem. Human thought about the future has generated a series of utopian pictures indicating various possible modes of interaction with the environment. This is a useful enterprise, since humans must be able to imagine possible futures before they can choose among them. A mindful role, one that does not destroy the community or seriously degrade it, is possible. The ecosystem has an integrity that we must respect, at the risk of disaster: not the integrity of ecosystems as something outside ourselves that needs to be preserved, but the integrity that we share with the community of life. Biodiversity must continue if ecosystems are to continue. We are part of the community of life, but have not acted as if we were. It is difficult, perhaps impossible, to justify destroying other species and ecosystems so that we can have luxuries. When we choose to act as if we were not part of the community, we run the risk of destroying the very supports on which we depend. We need an effective environmental ethics.

It is too easy to find historical evidence of humans who act as if they had resigned from the community of life. That evidence is plentiful in the preceding chapters. Governments often seek short-term advantage in the balance of trade instead of exercising trusteeship of their natural resources. Officials are liable to corruption. Entrenched corporations resist practices aimed at conservation, reducing pollution, or protecting their workers. The terrible Bhopal incident, in which isocyanate released in an explosion at the Union Carbide pesticide plant killed thousands of people and injured hundreds of thousands, is one example of the latter.[5] Urban residents of the industrialized nations leave a huge "ecological footprint" across the world. Villagers who live on the edge of nature reserves are often protectors of them, but sometimes they are forced into poaching by economic circumstances.

The rapid growth of the human population, particularly in the nineteenth and twentieth centuries, seems certain to continue to an unsupportable height. Will the food supply be adequate to sustain it? Ecology shows that when one species dominates an ecosystem, it is in the process of collapse, since overpopulation leads to a crash. A totally objective observer from another planet might conclude that that is what is happening to the human species. Being in the ecosystem, with its immense variety of life, has made humans what we are and made human culture and even existence possible. We are now destroying that milieu. Has ecology, as once was said of economics, turned out to be a "dismal science"? Should we be optimistic or pessimistic about the human future?

What possibilities exist for a sustainable balance between human technology and a flourishing environment with healthy ecosystems? Given the unpredictability inherent in large systems such as human culture and the natural environment, any predictions must be tentative. But in order to reach even a provisional answer to the question, it is necessary to do what this book attempts to do, namely, to examine the past human experience of technological growth and environmental change. On one hand, the course of human thought and action in the past seems to provide little hope that the interests of our species and life everywhere on Earth will prevail over narrow, short-term considerations. On the other hand, the growth in scientific knowledge, the existence of a subtle technology, the availability of a body of ethical considerations, and the certainty of threats to human survival unless controls are placed on destructive activities and population increase, have produced a situation unique in human history. There is a reasonable opportunity for an effective response to that situation, but it will require the most creative efforts of which humankind is capable. Usually our frames of reference are too small. Economists think only of economic factors, artists only of the artistic realm, and so forth. But all human activities are deeply linked to the ecosystem, and take place within it. Every area of human endeavor needs to be guided by the consideration of sustainable balance with the ecosystem. Humans need to consider not what benefits only our own species, but what benefits life as a whole system, since we are part of the whole, and our welfare and fulfillment as individuals and communities ultimately depends on the whole. Our community, in the deepest sense, is the community of life.

Notes

1   Vandana Shiva, *Stolen Harvest: The Hijacking of the Global Food Supply*, Cambridge, MA, South End Press, 2000.

2   Some of these are suggested by Paul Hawken, Amory Lovins, and L. Hunter Lovins in *Natural Capitalism: Creating the Next Industrial Revolution*, Boston, Little, Brown & Co., 1999.
3   Vandana Shiva, *Staying Alive: Women, Ecology and Development*, London, Zed Books, 1989, 77.
4   Alex Shoumatoff, *The World is Burning: Murder in the Rain Forest*, New York, Avon Books, 1991.
5   Paul Shrivastava, *Bhopal: Anatomy of a Crisis*, Cambridge, MA, Ballinger Publishing, 1987.

Suggestions for further reading on world environmental history

Interest in world environmental history has increased exponentially in the last decade or two, and the number of books and articles available now has improved in quality as well as quantity since the first edition of this book appeared in 2001. In this essay, I recommend a selected number of works that will supplement this book and provide the reader with opportunities for further exploration. An overview of writing in the entire field of environmental history, including a chapter on global environmental history, may be found in J. Donald Hughes, *What is Environmental History?* (2006).[1]

I begin with recent books covering the sweep of world environmental history both in time and geographical space. Joachim Radkau wrote *Natur und Macht*, which appeared in 2008 in English as *Nature and Power*.[2] The title reflects an ambitious program, and the author goes far toward fulfilling it. It is collective human power, often political, to which he refers. He asks which forms of power, and on what scale, have had the most important impacts on nature, and which can produce desired effects. Radkau contemplates history from primeval symbioses of humans and nature to today's global economy. Looking at modes of human interaction with nature, he discerns five eras within this span. They are the times of subsistence hunting and agriculture, of great civilizations dependent on water and wood, of colonialism, of the Industrial Revolution, and of globalization. These are historical configurations, not strict chronological units.

An analysis of the subject that places it in the context of culture, is *Global Environmental History* by the geographer Ian G. Simmons (2008).[3] It portrays the successive ways in which human societies have related to the natural environment ("gatherer–hunters," agriculture, the Industrial Revolution, and the post-industrial era). Where Radkau's approach is socio-political, Simmons's is cultural–intellectual but with a strong scientific element. Simmons has taken on this theme before, with *Changing the Face of the Earth* (1989) and *Environmental History: A Concise Introduction* (1993).[4]

Sing C. Chew, a sociologist, has written a competent environmental history of 5,000 years, from the appearance of the first cities to the present, *World Ecological Degradation* (2001).[5] His thesis is that urbanized societies have exploited and depleted the environment everywhere and throughout history, the most powerful engines of destruction being accumulation, urbanization, and population growth. Among the processes of ecological degradation, Chew treats one in detail: deforestation, an excellent choice as an example, since it occurred from the discovery of fire to the present, and can be documented. An original element in his analysis is the idea that "dark ages" are the result of expanding cultures exhausting the resources available to them, an idea that he develops further in *The Recurring Dark Ages* (2007).[6]

Undoubtedly the environmental history books most widely read by the public today are Jared Diamond's *Guns, Germs, and Steel: The Fates of Human Societies* (1997)[7] and *Collapse: How Societies Choose to Fail or Succeed* (2005),[8] which treat the influence of geography and biology on history, and human cultural responses, in a thematic approach with case studies. In *Guns, Germs, and Steel*, Diamond asks why technologically advanced civilizations appeared among some societies and not others. Rejecting the idea that certain peoples might be more intelligent and inventive than others, he finds the answer in differences of geography and environment. Among these are the availability of domesticable plants and animals, and the orientation of arable continental lands. Reviewers have called this line of argument environmental determinism, but *Collapse* can be seen as a defense against that criticism. Examining the historical collapse of a number of societies, Diamond groups the causes in five categories: climatic change, hostile neighbors, trade partners, environmental problems, and a society's response to environmental problems. It is in the last category that a society may "choose" to fail or succeed, and if it can choose, then environment does not totally determine the outcome. Diamond provides illustrations of cases where two societies existed at the same time in much the same place, but one failed and the other succeeded because of differences in their environmental relationships.

A very useful contribution by two Scandinavian historians to the literature on world environmental history is *Jorden en Ö: En Global Miljöhistoria (Earth an Island: A Global Environmental History)* by Sverker Sörlin and Anders Öckerman (1998), an outline of global environmental history primarily focused on the modern world.[9]

Somewhat older, but still popular, is Clive Ponting's *Green History of the World* (1991)[10] a survey of environmental issues through history, which begins with the problem of the destruction of the ecosystems of Easter Island as a parable for environmental history, and proceeds topically. Although his style is journalistic and his documentation inadequate, Ponting handles most of the salient themes. The book has appeared in a new edition that is little changed from the first.[11]

Next I turn to recent studies that encompass the globe but concentrate on a particular time period, and I begin with a synoptic world environmental history of the last century by John R. McNeill: *Something New Under the Sun: An Environmental History of the Twentieth-Century World* (2000).[12] McNeill traces the environmental and related social changes that characterize the period, and maintains that the twentieth century was different in kind, not only in degree, from any previous one, in that "the human race, without intending anything of the sort, has undertaken a gigantic uncontrolled experiment on the earth."[13] He explains that contemporary cultural behavior is adapted to abundant resources, fossil fuel energy, and rapid economic growth, behavior that has increased the inevitability of change in those conditions, but will not easily be altered when circumstances demand it. This book is a classic of modern environmental history.

A study of a previous historical period is John F. Richards, *The Unending Frontier: The Environmental History of the Early Modern World* (2003),[14] covering the fifteenth through eighteenth centuries. The argument of the book is that the salient patterns of the world were the expansion of Europeans across much of the globe and progress in human organization in Europe, India, and East Asia. He considers geographical settings, biological factors, the indigenous peoples whom he portrays neither as helpless victims nor as ecological saints, and the adaptations of the Europeans and their exported domestic animals, plants, and pathogens. The last section, entitled "The World Hunt," gives an overview of the way in which Europeans ranged the world in search of organic resources and, treating them as inexhaustible, reduced the abundance and diversity of wildlife to a waning remnant. This substantial volume can stand beside John McNeill's *Something New Under the Sun* as a complementary work. The two together almost cover the modern world; obviously what is now needed is an environmental history of the nineteenth century to bridge the gap between them.

A new perspective is offered by Robert B. Marks in *The Origins of the Modern World* (2002).[15] Spanning the early modern and modern worlds from 1400 to 1850, Marks places China, instead of Europe, in the center of world history. From this viewpoint, the "Rise of the West" was not inevitable or the result of Europe's inherent superiority, but "the story of how some states and peoples benefited from historically contingent events and geography to be able, at a certain point in time (a historical conjuncture), to dominate others and to accumulate wealth and power."[16]

A number of collections of articles on world environmental history have appeared. Among these is Donald Worster's *The Ends of the Earth* (1988).[17] Still timely, this choice volume includes articles on population, the Industrial Revolution, India, Africa, the Soviet Union, and three on the US in addition to Worster's useful introduction and widely cited appendix, "Doing Environmental History."[18] My collection, *The Face of the Earth: Environment and World History* (2000),[19] contains essays on biodiversity, the Pacific, Australia, Russia, India, and eco-racism in the US. *Encountering the Past in Nature*, edited by Timo Myllyntaus and Mikko Saikku (1999),[20] has discussions of environmental history as a discipline, and articles on forests in Karelia, Thailand, and the US South.

Another category consists of studies and collections that are global in scope, but deal with special topics. These include books on world forest history, such as the major recent monograph by Michael Williams, *Deforesting the Earth: From Prehistory to Global Crisis* (2003),[21] an authoritative masterwork that relates the historical course of human impact on forests across the world's continents and islands. There are good collections on world forests, such as *Global Deforestation and the Nineteenth-Century World Economy* (1983), edited by Richard P. Tucker and John F. Richards;[22] and *Tropical Deforestation: The Human Dimension*(1996), edited by Leslie E. Sponsel, Thomas N. Headland, and Robert C. Bailey.[23] On the history of fire, Stephen J. Pyne has given environmental historians a series of excellent books on fire in selected parts of the world, "Cycle of Fire," and an overview, *World Fire: The Culture of Fire on Earth* (1995).[24] The latter book is a global history of human involvement with the element of fire in all its forms from its origin in geological epochs to the high-tech fire powering the information revolution and the world market economy. On soils, a subject too often neglected, there is *Soils and Societies: Perspectives from Environmental History* (2006), edited by J.R. McNeill and Verena Winiwarter.[25] On climate, a fine study of the history of scientific ideas about climate change is Spencer R. Weart, *The Discovery of Global Warming* (2008).[26] Richard Grove and John Chappell edited a volume investigating the worldwide effects on human history of oscillating oceanic temperatures called *El Niño* (and its cooler counterpart *La Niña*).[27]

Among a number of works on the environmental impacts of imperialism, there is Alfred Crosby's noted *Ecological Imperialism* (1986; 2nd edn 2004).[28] Another landmark book is Richard Grove's *Green Imperialism* (1995),[29] which traces the origin of modern ecological thought and environmental history to scientific civil servants in the French, British, and Dutch empires in the seventeenth through nineteenth centuries. A book that traces the same theme through 1895–1945, emphasizing the fledgling science of ecology, is Peder Anker's *Imperial Ecology* (2001).[30] Richard Drayton's *Nature's Government* (2000)[31] sees science as an instrument of imperialism and racism in the British Empire to 1903. He emphasizes the role of botanical gardens, particularly the Royal Botanic Gardens at Kew. A related theme in India is examined by Deepak Kumar, *Science and the Raj, 1857–1905* (1995).[32] The role of Scotland in the environmental story of the British Empire is explored by John MacKenzie in *Empires of Nature and the Nature of Empires: Imperialism, Scotland and the Environment* (1997).[33] A good collection of articles on imperialism and the environment is *Ecology and Empire: Environmental History of Settler Societies* (1997), edited by Tom Griffiths and Libby Robin.[34]

The environmental effects of what many have called the American Empire extend far beyond areas administered by the US. Richard Tucker takes on this subject in *Insatiable Appetite: The United States and the Ecological Degradation of the Tropical World* (2000),[35] concerned with the 1890s to the 1960s. Tucker portrays ways in which American business and government impacted the warmer regions of the globe, describing the unsustainability of much of the development that occurred, and the damage that resulted. Thomas Dunlap's *Nature and the English Diaspora* (1999)[36] is an environmental history of Britain and the "Neo-Britains:" Canada, the US, Australia, and New Zealand.

For the environmental history of globalization, there are two important collections: *Rethinking Environmental History: World-System History and Global Environmental Change*, edited by Alf Hornborg, J.R. McNeill, and Joan Martinez-Alier (2007),[37] and *The Globalization of Environmental Crisis*, edited by Jan Oosthoek and Barry K. Gills (2008).[38]

Histories of the environmental movement around the world include John Young's *Sustaining the Earth* (1990)[39] and Ramachandra Guha, whose *Environmentalism: A Global History* (2000)[40] offers an analysis of environmental objectives and activism in many corners of the world, spanning history from nationalistic ruralism to social ecology, from the Roman poet Virgil to the Nobel Prize-winner Wangari Maathai. John McCormick's *Reclaiming Paradise: The Global Environmental Movement* (1989)[41] emphasizes international aspects of the environmental movement from the founding of the United Nations in 1945 through the Brundtland Commission report of 1987, including the creation of the United Nations Environment Programme. The idea that technological and financial fixes do not address the deeper changes called for by activist environmental movements is the theme of Carolyn Merchant's *Radical Ecology: The Search for a Livable World* (1992),[42] which discusses deep ecology, social ecology, Green politics, ecofeminism, Earth First!, Chipko, and indigenous rainforest action groups.

Searches for articles dealing with aspects of world environmental history may well begin with two journals: *Environmental History*, the journal of the American Society for Environmental History and the Forest History Society, and *Environment and History*, published by the White Horse Press in England. A number of other journals often publish articles in the field. The new journal, *Nature and Culture* (Berghahn), already has published several useful essays. Some have devoted special issues to the topic; *Pacific Historical Review*, for example, did so in August 1985 and again in February 2001. *Globalizations* (Routledge) had a special issue on "The Globalization of Environmental Crisis" in December 2005, and *Leidschrift: Historisch Tijdschrift* (Leiden University) issued one on "Culture and Nature: History of the Human Environment" in April 2006. For further reference to articles in the field of world environmental history, the Environmental History Resources website is a treasure trove (www.eh-resources.org).

Finally, an essential reference work covering worldwide topics is the *Encyclopedia of World Environmental History* in three volumes, edited by Shepard Krech III, J.R. McNeill, and Carolyn Merchant (2004).[43]

Notes

1   J. Donald Hughes, *What is Environmental History?* Cambridge, UK, Polity Press, 2006.
2   Joachim Radkau, *Nature and Power: A Global History of the Environment*, New York, Cambridge University Press, 2008.
3   I.G. Simmons, *Global Environmental History: 10,000 BC to AD 2000*, Edinburgh, Edinburgh University Press, 2008.
4   I.G. Simmons, *Changing the Face of the Earth: Culture, Environment, History*, Oxford, Blackwell, 1989; *Environmental History: A Concise Introduction*, Oxford, Blackwell, 1993.
5   Sing C. Chew, *World Ecological Degradation: Accumulation, Urbanization, and Deforestation, 3000 B.C.–A.D. 2000*, Walnut Creek, CA, Rowman & Littlefield, 2001.
6   Sing C. Chew, *The Recurring Dark Ages: Ecological Stress, Climate Changes, and System Transformation*, Lanham, MD, Altamira Press, 2007.
7   Jared Diamond, *Guns, Germs, and Steel: The Fates of Human Societies*, New York, W.W. Norton, 1997.
8   Jared Diamond, *Collapse: How Societies Choose to Fail or Succeed*, New York, Viking, 2005.
9   Sverker Sörlin and Anders Öckerman, *Jorden en Ö: En Global Miljöhistoria (Earth an Island: A Global Environmental History)*, Stockholm, Natur och Kultur, 1998.
10 Clive Ponting, *A Green History of the World*, New York, St. Martin's Press, 1991.
11 Clive Ponting, *A Green History of the World: The Environment and the Collapse of Great Civilizations*, New York, Penguin, 2007.
12 John R. McNeill, *Something New Under the Sun: An Environmental History of the Twentieth-Century World*, New York, W.W. Norton, 2000.
13 Ibid., p. 4.
14 John F. Richards, *The Unending Frontier: The Environmental History of the Early Modern* World, Berkeley and Los Angeles, University of California Press, 2003.
15 Robert B. Marks, *The Origins of the Modern World: A Global and Ecological Narrative*, Lanham, MD, Rowman & Littlefield, 2002.
16 Ibid., p. 151.
17 Donald Worster, ed., *The Ends of the Earth: Perspectives on Modern Environmental History*, Cambridge, Cambridge University Press, 1988.
18 Ibid., pp. 289–308.
19 J. Donald Hughes, ed. *The Face of the Earth: Environment and World History*, Armonk, NY, M.E. Sharpe, 2000.
20 Timo Myllyntaus and Mikko Saikku, eds, *Encountering the Past in Nature*, Athens, OH, Ohio University Press, 1999.
21 Michael Williams, *Deforesting the Earth: From Prehistory to Global Crisis*, Chicago, University of Chicago Press, 2003.
22 Richard P. Tucker and John F. Richards, eds, *Global Deforestation and the Nineteenth-Century World Economy*, Durham, NC, Duke University, 1983.
23 Leslie E. Sponsel, Thomas N. Headland, and Robert C. Bailey, eds *Tropical Deforestation: The Human Dimension*, New York, Columbia University Press, 1996.
24 Stephen J. Pyne, *World Fire: The Culture of Fire on Earth*, New York, Holt, 1995. Pyne also has a number of regional studies on the subject of fire.
25 J.R. McNeill and Verena Winiwarter, eds, *Soils and Societies: Perspectives from Environmental History*, Isle of Harris, UK, White Horse Press, 2006.
26 Spencer R. Weart, *The Discovery of Global Warming*, Cambridge, MA, Harvard University Press, 2008.
27 Richard H. Grove and John Chappell, eds, *El Niño: History and Crisis*, Cambridge, UK, White Horse Press, 2000.
28 Alfred W. Crosby, *Ecological Imperialism: The Biological Expansion of Europe, 900–1900*, Cambridge, Cambridge University Press, 2004 (1st edn, 1986).
29 Richard H. Grove, *Green Imperialism: Colonial Expansion, Tropical Island Edens and the Origins of Environmentalism, 1600–1860*, Cambridge, Cambridge University Press, 1995.
30 Peder Anker, *Imperial Ecology: Environmental Order in the British Empire, 1895–1945*, Cambridge, MA, Harvard University Press, 2001.

31   Richard Drayton, *Nature's Government: Science, Imperial Britain, and the "Improvement" of the World*, New Haven, CT, Yale University Press, 2000.

32   Deepak Kumar, *Science and the Raj, 1857–1905*, Delhi, Oxford University Press, 1995.

33   John M. MacKenzie, *Empires of Nature and the Nature of Empires: Imperialism, Scotland and the Environment*, East Linton, Tuckwell Press, 1997.

34   Tom Griffiths and Libby Robin, eds, *Ecology and Empire: Environmental History of Settler Societies*, Edinburgh, Keele University Press, 1997.

35   Richard P. Tucker, *Insatiable Appetite: The United States and the Ecological Degradation of the Tropical World*, Berkeley and Los Angeles, University of California Press, 2000.

36   Thomas Dunlap, *Nature and the English Diaspora: Environment and History in the United States, Canada, Australia, and New Zealand*, Cambridge, Cambridge University Press, 1999.

37   Alf Hornborg, J.R. McNeill, and Joan Martinez-Alier, eds, *Rethinking Environmental History: World-System History and Global Environmental Change*, Lanham, MD, Altamira Press, 2007.

38   Jan Oosthoek and Barry K. Gills, eds, *The Globalization of Environmental Crisis*, London, Routledge, 2008.

39   John Young, *Sustaining the Earth: The Story of the Environmental Movement*, Cambridge, MA, Harvard University Press, 1990.

40   Ramachandra Guha, *Environmentalism: A Global History*, New York, Longman, 2000.

41   John McCormick, *Reclaiming Paradise: The Global Environmental Movement*, Bloomington, Indiana University Press, 1989.

42   Carolyn Merchant, *Radical Ecology: The Search for a Livable World*, New York, Routledge, 1992.

43   Shepard Krech III, J.R. McNeill, and Carolyn Merchant, eds, *Encyclopedia of World Environmental History*, 3 vol, New York, Routledge, 2004.

Index